An Introduction to Satellite Image Interpretation

An Introduction to
Satellite Image Interpretation

Eric D. Conway and
The Maryland Space Grant Consortium

The Johns Hopkins University Press
Baltimore and London

Printed in the United States of America on acid-free paper
9 8 7 6 5 4 3

The Johns Hopkins University Press
2715 North Charles Street
Baltimore, Maryland 21218-4363
www.press.jhu.edu

ISBN 0-8018-5576-4
ISBN 0-8018-5577-2 (pbk.)

Library of Congress Cataloging-in-Publication Data will be found at the end of this book.

A catalog record for this book is available from the British Library.

Table of Contents

List of Figures and Tables

Figures

Tables

Preface

On April 1, 1960, TIROS 1 was launched by the United States as a test satellite designed to monitor the Earth's atmosphere from space. In the decades that followed, there evolved a global fleet of environmental satellites that use their vantage point in space to study the interactions between all aspects of the Earth's environment. Today, environmental satellites are used to monitor developing storms and track their movement, measure ocean surface temperature, monitor global temperature change, measure changes in atmospheric composition, study vegetation characteristics, monitor volcano emissions, measure the loss of rainforests, measure ocean productivity, monitor land use practices, provide the data used in computer-based weather prediction models, and perform hundreds of other observational tasks related to the study of the planet Earth.

In all of these applications, satellite imagery provides environmental information on a spatial and temporal scale that is not possible through traditional methods. By allowing us repeated views of large portions of the Earth, satellites have given us a new perspective in which we are able to see our planet as a global system. This system, as we have discovered, is composed of many different parts, which interact in a way that allows life to develop as it has for billions of years. With this new understanding, we can begin to study the oceans, the atmosphere, and land surfaces as parts of a whole system, each influencing the other. Once we better understand the intricacies of these interactions, we can make informed decisions and predictions about how our actions will affect the global environment.

Traditionally, the majority of satellite data has only been available to research scientists and private organizations. Recently, however, satellite imagery has become readily available to educators, pilots, and the general public. With advances in technology and reductions in equipment costs, many schools, television stations, research organizations, and other interested groups across the country have installed direct readout stations that allow a user to directly access a satellite in order to obtain imagery. The recent explosion of the Internet as a method for data exchange has also made satellite imagery available to almost anybody with a personal computer and a modem. In addition, most research organizations that specialize in satellite remote sensing now distribute datasets in CD-ROM format, often at very low cost to the user.

To gather meaningful information from a satellite image, one must learn how to interpret satellite data. Through careful study and practice, one can easily learn the patterns to look for in order to identify various geographic, oceanic, and atmospheric features found in a satellite image. The purpose of this book is to provide you with the information necessary to interpret the pictures of the Earth-atmosphere-ocean system that come from satellites. More specifically, using this manual you should be able to:

—understand the basic phenomena that can be observed in satellite imagery, and

—recognize the basic patterns that help locate and identify these features in a satellite image.

This book will focus on the interpretation of imagery collected from polar orbiting and geostationary weather satellites operated by the U.S. National Oceanic and Atmospheric Administration (NOAA). The concepts, however, will be applicable to a wide variety of situations, including interpretation of imagery from satellites operated by other countries and imagery from environmental satellites that are not primarily weather satellites. The manual has been written in such a manner that a beginner with relatively little scientific background can utilize the information and learn basic image interpretation. There is also plenty of information that more advanced users will find valuable.

The book is written in a sequential format; therefore, each chapter builds upon information presented in previous chapters. The first three chapters cover the basics of using satellite imagery, including the foundations of remote sensing, the satellites, and the basics of image interpretation. Chapter 4 discusses geographical applications of satellite imagery, including the location of surface features and snow identification. Chapters 5–13 deal with various aspects of atmospheric science, including winds, jet streams, storm systems, thunderstorms, tropical storms, and air quality. Chapter 14 discusses various applications of weather satellites to oceanography, including methods of locating currents, sea ice, icebergs, and areas of upwelling.

An indexed glossary is included at the end of the text to help locate key terms and acronyms. Each of the central concepts is also illustrated with high-quality satellite images from NOAA satellites and simple line drawings designed to help you visualize the concept being discussed. As you read through this manual, it would be very helpful for you to keep an atlas nearby. As you will soon see, a working knowledge of geography is essential to satellite studies. Most of the satellite images have not been annotated to highlight the features of interest; instead, geographical references in the text will help you locate the various examples discussed. It would also be helpful for you to have many other satellite images available. Although learning how to interpret a satellite image is a relatively simple process, it requires a great deal of practice. Image interpretation depends on one's ability to recognize patterns in the data that indicate various features. Frequent observations and practice will make you more proficient in identifying these.

Beginners and experts alike will agree that seeing a satellite image can be an exciting and moving experience. Beyond the aesthetic aspects of an image, however, lies a wealth of meaningful information that can further our understanding of the Earth. Often, in just one image, hundreds of features are present that are related to the geography, atmosphere, and oceans of the Earth. This book is intended to provide you with the tools needed to discover this information and use satellite imagery to better understand the world in which we live.

Acknowledgments

I would like to gratefully acknowledge the role of H. Michael Mogil, formerly of the National Oceanic and Atmospheric Administration's National Environmental Satellite Data and Information Service (NOAA/NESDIS), in the completion of this project. His support during all stages of this book's creation—including multiple roles as mentor, adviser, editor, content reviewer, and much more—was critical to its success. I would also like to thank his wife, Barbara Levine, for the support she gave us as we produced this manual. Without their aid this project would not have been possible. I would also like to thank the following people for their various roles: Anne Anikis and Dr. Richard Henry, of the Maryland Space Grant Consortium (MSGC), for providing the resources necessary to complete this project; Gary Ellrod, Jim LaDue, and Dick Pritchard, of NOAA/NESDIS, for supplying GOES imagery and technical review; Ralph Meiggs and Will Gould, at the Satellite Data Services Division (SDSD) of NOAA, for supplying the HRPT imagery used in this document; and Phil Golden, of the Interactive Processing Branch (IPB) of NOAA, for high-quality photographic processing. Additional thanks go to all others who contributed to the production of this document at NOAA/NESDIS, including Nancy Everson, Ed Fischer, Larason Lambert, Paige Bridges, John Shadid, John Schmidt, Vernon DeVorak, Frank Smigelski, Kelly Taggert, Eileen Maturi, and Rod Scofield. Thanks to Bob Popham, the architect and facilitator of the Maryland Space Grant Consortium "Earth Observations from Space" course that gave me the introduction to aerospace science education I needed to apply for the internship at NOAA/NESDIS that resulted in the creation of this book; his strong personal letter of recommendation to NOAA/NESDIS undoubtedly helped balance the scales of the internship selection process in my favor; and special appreciation to Ana Swamy, director of the Center for Math and Science Excellence at Morgan State University, for her enthusiasm and support of the project. Thanks also to Sharylyn Young, at the Joint Ice Center, for information on sea ice forecasting, and to Charles Davis, of Dallas Remote Imaging, for technical information on operational weather satellites. Thanks to Beth Jones and the Lockheed Martin Graduate Fellows Program for continued support throughout this project. Thanks to all reviewers of this document, and all others who contributed to its publication and were inadvertently omitted from this list. Most importantly, I would like to thank my wife, Crissy, for her continuing love, support, and patience through many long days and nights while I was writing this book. I never could have done it without her.

E.D.C.

Introduction to Satellite Image Interpretation

CHAPTER 1

Foundations of Remote Sensing

Introduction to remote sensing

Remote sensing is a term that is used to describe the study of something without making actual contact with the object of study. It involves making measurements of the physical properties of an object from a remote distance. Satellite technology is an example of remote sensing, since satellite sensors are designed to study energy reflected and energy emitted from the Earth. Using data transmitted from a satellite in orbit around the Earth, individuals located at receiving stations on the surface of the Earth can measure properties of the Earth without having to actually go to the area of interest and make measurements.

Electromagnetic radiation

Electromagnetic radiation is the basis for all remote sensing of the Earth. **Radiation** is energy that is emitted in wave form by all substances that are not at **absolute zero** (–273° C or –459° F). A wave of radiated energy is not a material object. Although it has no mass, a wave is able to transmit energy from one place to another. A wave of electromagnetic radiation can be thought of as a pattern of disturbance of the electromagnetic field. As a wave of radiation passes through this field the energy level fluctuates up and down in a regular pattern. The pattern of a repeating wave and the terms that describe it are illustrated in figure 1.1.

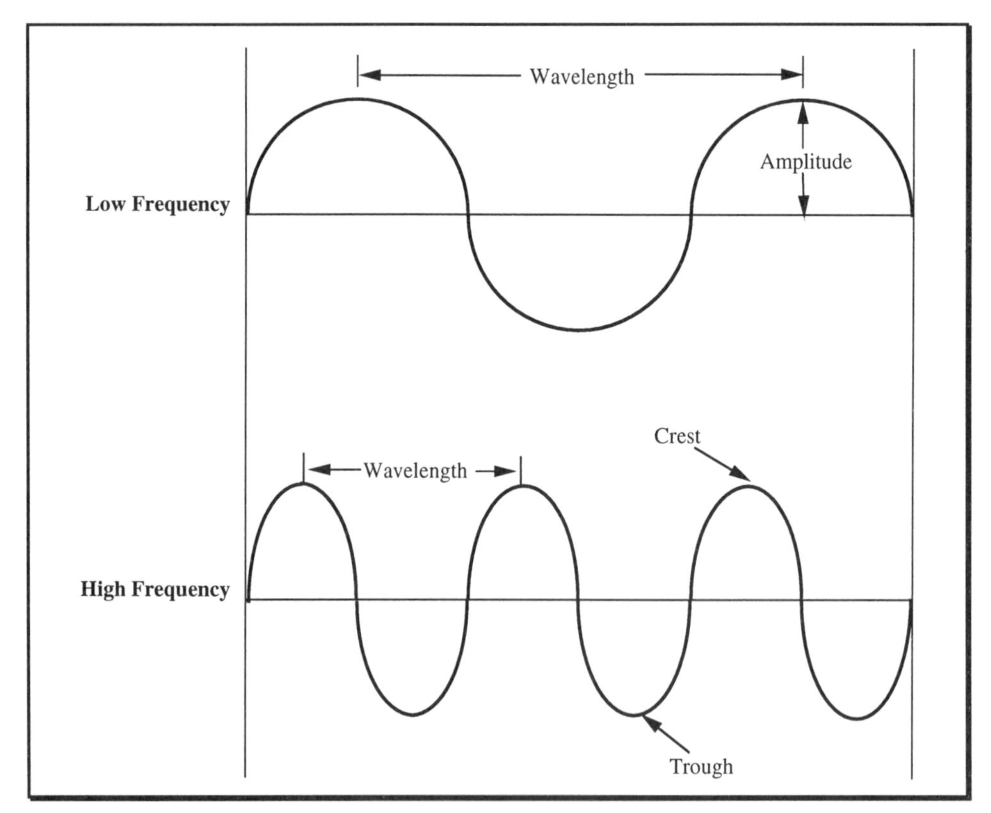

Figure 1.1. Characteristics of an electromagnetic wave

An electromagnetic wave can be characterized by its amplitude, its wavelength, and its frequency. A **crest** (or ridge) is the maximum upward displacement of a wave; a **trough** is the maximum downward displacement. **Amplitude** measures the magnitude of the wave and refers to the amount of displacement that occurs in it. The **wavelength** of a wave is measured as the distance between two successive troughs or crests. The

frequency of a wave is determined by the number of waves that pass a certain point in a given period of time. All types of electromagnetic energy travel at the same speed (the speed of light). In the case of shorter wavelengths, more waves are able to pass a point within a given period of time. Thus, shorter wavelengths produce high-frequency waves and longer wavelengths produce low-frequency waves.

The **electromagnetic spectrum** is a continuum of all the types of electromagnetic radiation (figure 1.2). On the spectrum, each type of energy is ordered according to wavelength. **Gamma rays** and **x-rays** are found at the end of the spectrum with the shortest wavelengths. X-rays may be familiar to you if you have ever had an x-ray examination at the hospital or dental office. At the longer-wavelength end of the spectrum there are **radio** waves; without these there would be no television and radio broadcasts. Our eyes detect a small portion of the spectrum called **visible light**, while we feel **infrared** radiation as heat. We use **microwave** radiation to cook food, and **ultraviolet** radiation from the sun is what causes sunburn, and possibly skin cancer.

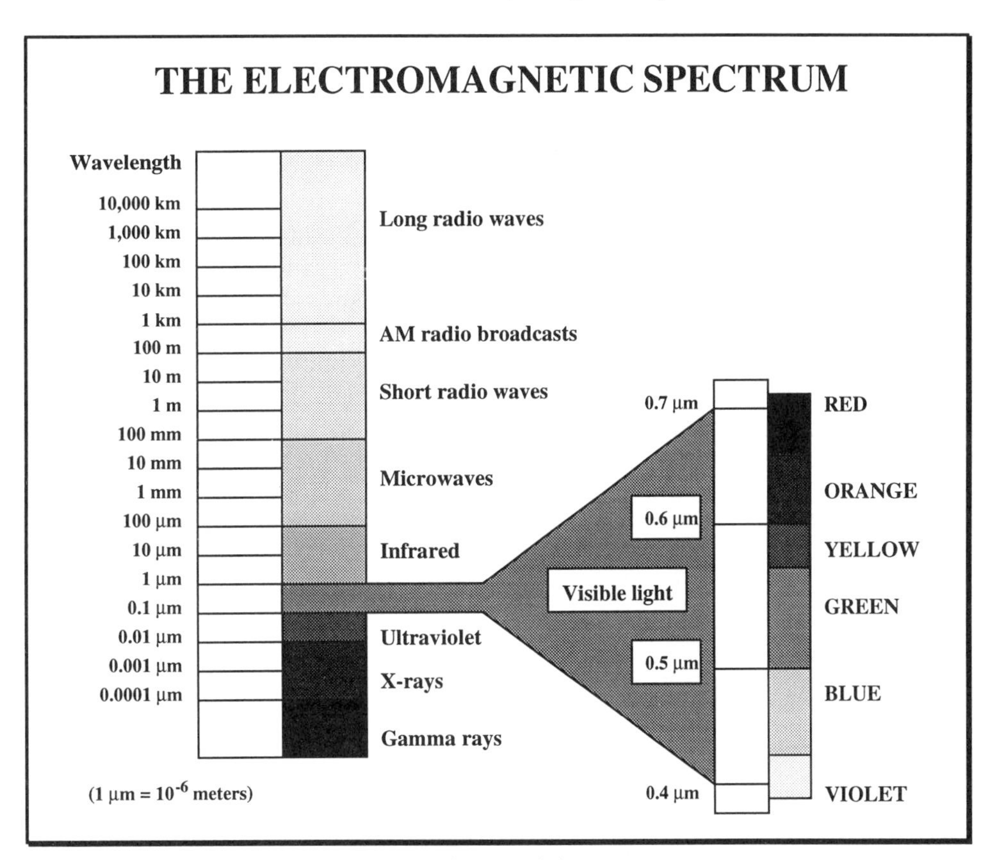

Figure 1.2

Emission of electromagnetic radiation

Radiation is **emitted** (given off) by all objects that are not at absolute zero. Objects do not necessarily emit radiation in only one wavelength; instead, most things radiate energy in specific ranges of wavelengths, often referred to as the object's **spectrum**. The **temperature** of an object determines the characteristics of its radiated energy spectrum. An object with a very high surface temperature will emit very high energy radiation at shorter wavelengths, while a cooler object will emit a lower energy spectrum at longer wavelengths. This relationship between wavelength of maximum emission and surface temperature can be expressed by the equation in figure 1.3. Therefore, the sun, with a surface temperature of approximately 6000 K, has its maximum emission at a wavelength of 0.48 **microns**, which falls in the range of visible light. On the other hand, the much cooler Earth, with a surface temperature of about 300 K, has its maximum emission at 9.4 microns. This falls within the range of infrared wavelengths.

$$\textbf{wavelength of maximum emission (microns)} = \frac{\textbf{2832}}{\textbf{temperature (K)}}$$

[1 micron (μm) = 10^{-6} meters]

Figure 1.3. Relationship between wavelength and surface temperature

Figure 1.4 is a graph that compares the emission spectra of the sun and the Earth. The three curves on the graph represent the spectra of the Earth, the sun, and the sun's energy at the Earth. The Earth's emission is primarily in the long-wavelength portion of infrared spectrum (as predicted by the equation in figure 1.3). This portion of the spectrum is known as **thermal infrared** because it is dependent on the temperature of the surface that is emitting the radiation. Satellite sensors that detect infrared radiation in these wavelengths can be used to study the thermal properties of the Earth and the atmosphere.

In figure 1.4, notice that the intensity of the sun's actual emission is much greater at all wavelengths than that of the Earth's emission. As the sun's radiation travels through space, however, its intensity dissipates. By the time it has reached the Earth, its intensity has decreased. However, the majority of the sun's emission remains in the visible portion of the spectrum. The sun is also radiating strongly in the short-wavelength portion of the infrared spectrum. This radiation, known as **near infrared**, involves reflected solar energy along with thermal radiation. Satellite sensors that detect visible and near-infrared radiation can measure the amount of solar energy reflected off, as well as the energy radiated by, the earth and the clouds.

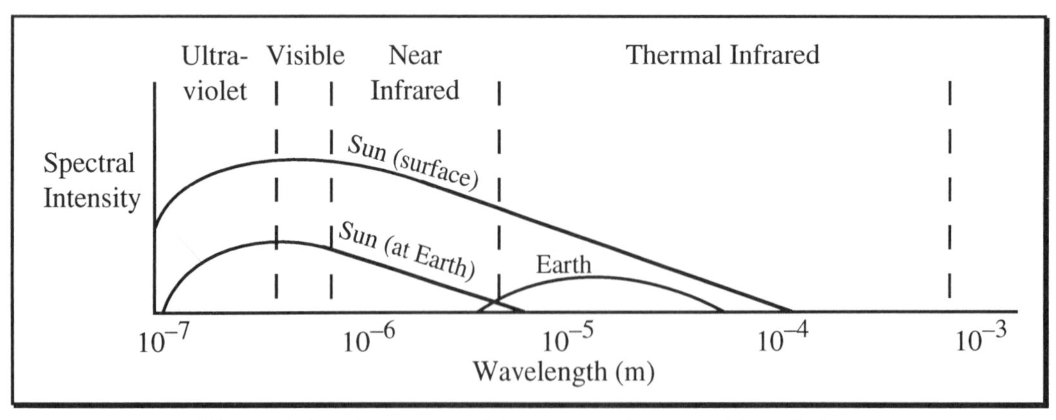

Figure 1.4. The spectra of the sun's and the Earth's radiation

Radiation that strikes the Earth

The Earth is considered an **opaque** surface; in other words, it does not allow any light to pass through. When radiation strikes an opaque surface, such as solid rock, it is either absorbed by the rock or reflected away from it (lefthand side of figure 1.5). The **albedo** of a surface expresses the fraction of visible radiation that is reflected away from the surface. Objects such as newly fallen snow or thick clouds have a very high albedo.

When seen from space, these objects are very bright because they reflect a large amount of incoming solar radiation. Forests and dark soils have lower albedos (that is, they reflect less visible radiation); therefore, they appear dark. Table 1.1 lists the albedos of many common features on the Earth's surface and in the atmosphere. A totally black object, with an albedo of 0, would appear as a "black hole," and you would see none of its features

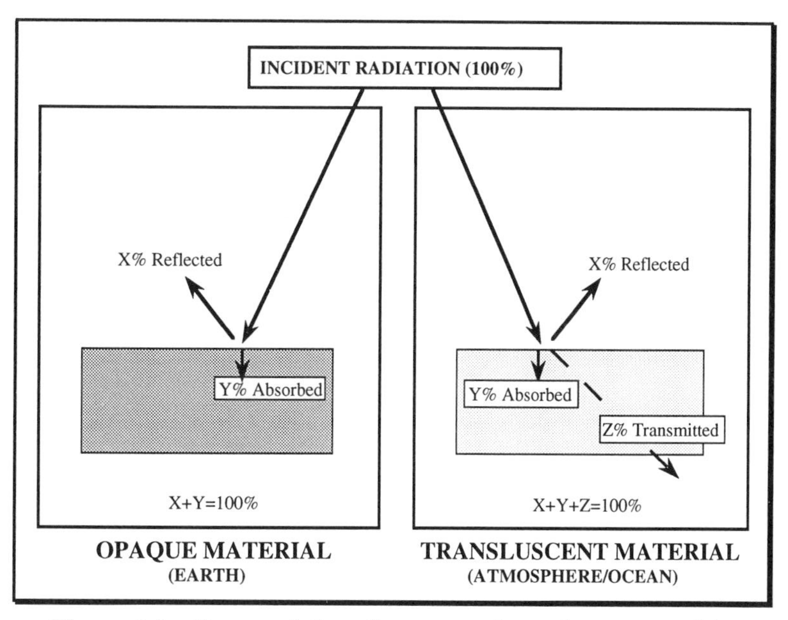

Figure 1.5. Characteristics of opaque and translucent materials

Feature	Albedo
Clouds	
Cumulonimbus, large and tall	92
Cumulonimbus, small, tops at 6 km	86
Cirrostratus, thick, with lower clouds	74
Cumulus with stratocumulus	69
Stratocumulus	68
Stratus, thick (0.5 km), over ocean	64
Stratocumulus masses, with cloud sheet, over ocean	60
Stratus, thin, over ocean	42
Cirrus, alone, over land	36
Cirrostratus, alone, over land	32
Cumulus, fair weather	29
Water features	
Sunglint on Gulf of Mexico	17
Lake, Great Salt lake, Utah	9
Ocean, Gulf of Mexico	9
Ocean, Pacific	7
Surface features—bare areas/soils	
Snow, freshly fallen	75–90
Snow, 3 to 7 days old	40–70
White Sands, New Mexico	60
Sand dune, dry	35–45
Soil, dry light sand	25–45
Soil, dry clay or gray	20–35
Sand dune, wet	20–30
Concrete, dry	17–20
Soil, moist gray	10–20
Soil, dark	5–15
Road, blacktop	5–10
Vegetative zones	
Desert	25–30
Savanna, dry season	25–30
Crops	15–25
Savanna, wet season	15–20
Tundra	15–20
Chaparral	15–20
Meadows, green	10–20
Forest, deciduous	10–20
Forest, coniferous	5–15

Table 1.1. Approximate albedos for various features in visible satellite imagery (albedo is expressed as the percentage of light reflected by the surface)

except its profile. Such an object, which would absorb radiation at all wavelengths, is called a **black body**. Since no object exists that absorbs all light that reaches it, black bodies do not occur in the natural world. However, scientists use the concept of black bodies to study radiation theory.

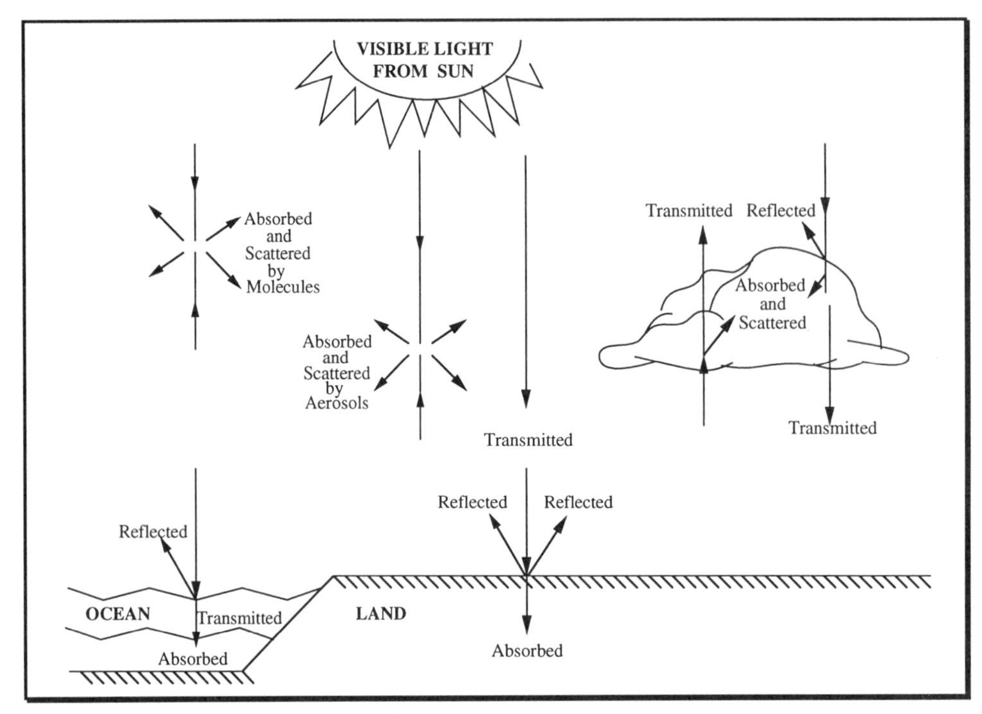

Figure 1.6. Visible light in the atmosphere

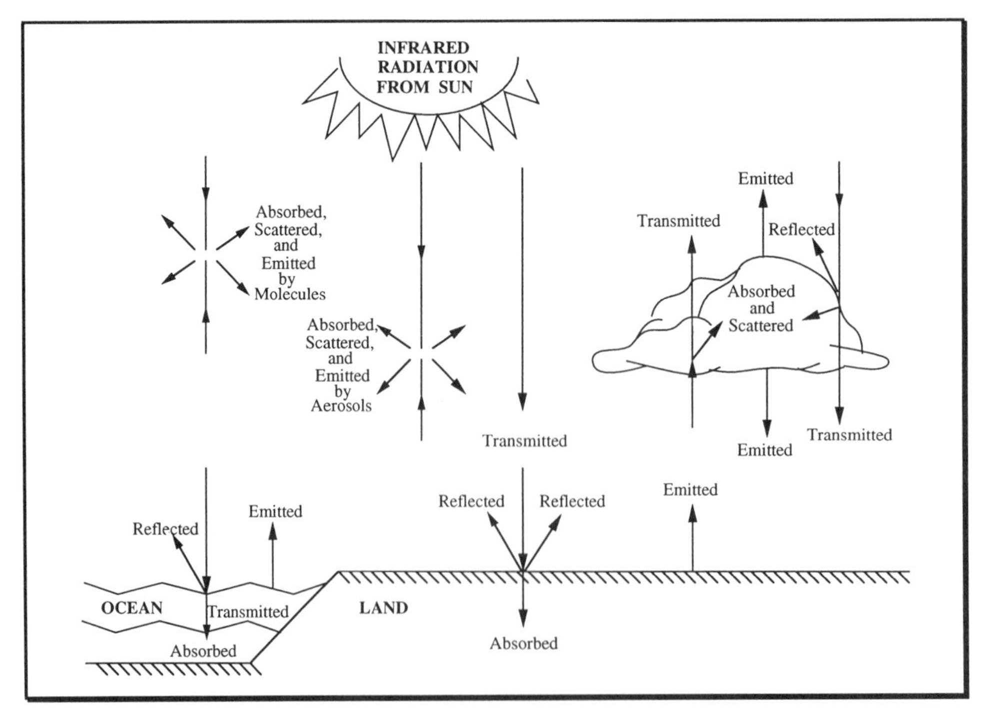

Figure 1.7. Infrared energy in the atmosphere

Radiation in the atmosphere

The atmosphere is a **translucent** medium. A translucent medium **transmits** (allows to pass through) a portion of the radiation that strikes it, while absorbing or reflecting the rest (righthand side of figure 1.5). As radiation enters and passes through a translucent medium, such as the atmosphere or the ocean, a variety of things can occur. Visible radiation can be absorbed, transmitted, or reflected by the molecules, aerosols, ice crystals, and water droplets that make up the atmosphere. These particles can also **scatter** visible light into its component colors, including red, orange, yellow, green, blue, and violet (when visible light is scattered by particles in the atmosphere, the blue wavelengths spread out the most, giving the sky a blue color). Visible radiation can also be absorbed or reflected by the Earth's various surfaces. These processes are illustrated in figure 1.6. Infrared radiation can be absorbed, transmitted, reflected, or scattered as it passes through the atmosphere. In addition, the various surfaces of the Earth and clouds can absorb infrared radiation and re-emit it as heat into the atmosphere or space (figure 1.7).

Transmissivity of the atmosphere

Some types of electromagnetic radiation easily pass through the atmosphere, while others are prevented from passing through. The ability of the atmosphere to allow radiation to pass through is referred to as its **transmissivity**. The transmissivity of radiation through an average, cloud-free atmosphere varies with the wavelength of the radiation. Gases in the atmosphere absorb radiation in specific wavelength bands and allow radiation with other wavelengths to pass through. Figure 1.8 is a graph that illustrates the varying levels of transmission for different wavelengths of radiation. The wavelength (in meters) is shown across the top of the graph, and the percentage of radiation transmitted by the atmosphere is shown on the vertical axis of the graph.

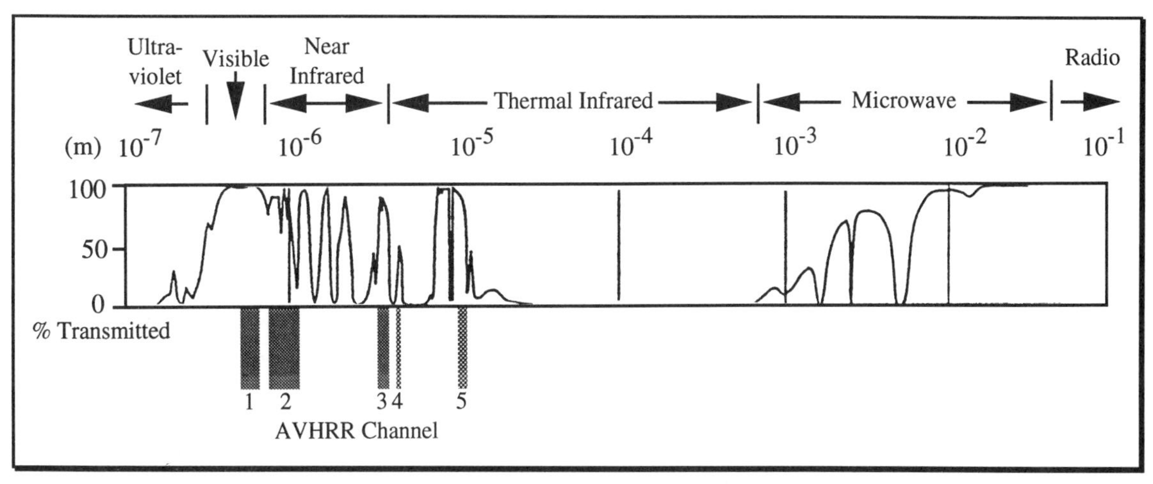

Figure 1.8. Transmissivity of the atmosphere and the spectral range of the five main sensors on the Advanced Very High-Resolution Radiometer (AVHRR)

The atmosphere is opaque to short-wavelength, high-energy radiation such as ultraviolet, gamma, and x-rays. This is because the atmosphere absorbs energy at these particular wavelengths. A common example of this phenomenon is the absorption of high-energy radiation by atmospheric ozone. Ozone is a gas in the atmosphere that absorbs nearly 100% of all radiation with a wavelength less than about 0.3 µm (3×10^{-7} m). If there were no ozone layer in the atmosphere, much of this radiation would reach the Earth's surface and make it impossible for most living organisms to survive.

The atmosphere also absorbs energy in specific wavelength bands, especially in the infrared and microwave portions of the electromagnetic spectrum. The areas of the electromagnetic spectrum that are absorbed by atmospheric gases are collectively known as **absorption bands**. In figure 1.8, absorption bands are shown by a low transmission value that is associated with a specific range of wavelengths.

Along with absorption bands, there are areas on the electromagnetic spectrum where the atmosphere is transparent to specific wavelengths. These wavelength bands are known as **atmospheric windows**, since they allow the radiation to easily pass through the atmosphere. Atmospheric windows are shown in figure 1.8 where the transmission values are the highest. One such window exists within the visible light portion of the spectrum, which is at the peak of the sun's energy output. On a cloud-free day, most of the visible light energy from the sun is able to pass through the atmosphere without being absorbed by atmospheric gases. There is also a series of atmospheric windows in the infrared portion of the spectrum. These windows allow infrared radiation at specific wavelengths to penetrate the atmosphere. The infrared windows are important to remote sensing since they occur at the same wavelength as the peak of the Earth's radiation output. These IR windows allow thermal energy emitted from Earth to penetrate the atmosphere and reach into space, where a satellite's sensors can detect it.

Sensing radiation in the atmosphere

The sensors on meteorological satellites, known as **radiometers**, are designed to take advantage of the atmospheric windows. These instruments measure radiation brightness in specific, narrow wavelength bands known as **channels**. The bottom portion of figure 1.8 shows the specific wavelength ranges of the five channels of the **Advanced Very High-Resolution Radiometer (AVHRR)**, which is the main sensor on board the United States polar orbiting weather satellites (see chapter 2 for a discussion of these satellites and their instrumentation).

Data from **infrared (IR)** sensors reveal specific thermal characteristics of the Earth's surface, the ocean surface, and the cloud tops. Like most IR sensors, the AVHRR measures infrared radiation emitted from the Earth at wavelengths that are able to penetrate the atmosphere. This allows surface temperature measurements to be made from space. If the atmosphere were opaque at these wavelengths it would not be possible to study these characteristics from space.

Although most satellite radiometers are designed to make use of atmospheric windows, there is an exception. The IR sensors at 6.7 and 7.3 microns detect energy that is unable to penetrate the atmosphere. Water vapor in the upper atmosphere absorbs energy at these wavelengths; therefore, the atmosphere is opaque to this radiation. By studying the radiation in these bands, important data regarding water vapor content of the upper atmosphere can be inferred without interference from radiation emitted from the Earth's surface.

The **visible (VIS)** channel on the AVHRR and other satellites takes advantage of the atmospheric window in the visible light portion of the spectrum. Visible light sensors measure the amount of solar radiation that is reflected away from Earth by clouds and the Earth's surface. Since each feature on Earth and in the atmosphere has a different albedo (see table 1.1), these sensors can detect different land and water features as well as distinguish between various cloud types.

CHAPTER 2

The Satellites

Introduction to meteorological satellites

In the United States, the **National Oceanic and Atmospheric Administration (NOAA)** operates the meteorological satellite program. Data from NOAA satellites can be used to study the interactions between the Earth's atmosphere, oceans, and landmasses. In addition, many other countries, including Japan, China, Russia, and India, operate environmental satellites that supply similar data. In Europe, a consortium of countries called the **European Space Agency (ESA)** operates meteorological satellites. The imagery from these satellites is readily available to anybody with the proper equipment. This chapter describes the weather satellites that currently provide data and discusses how you can obtain imagery from these satellites. The chapter divides weather satellites into two groups—**polar orbiting** and **geostationary** satellites—on the basis of their orbital characteristics. It should be noted that in the future, changes in the status of existing satellites and the introduction of new satellites may lead to differences in imaging capabilities; however, the general concepts discussed in this chapter, and in later chapters, will remain essentially the same.

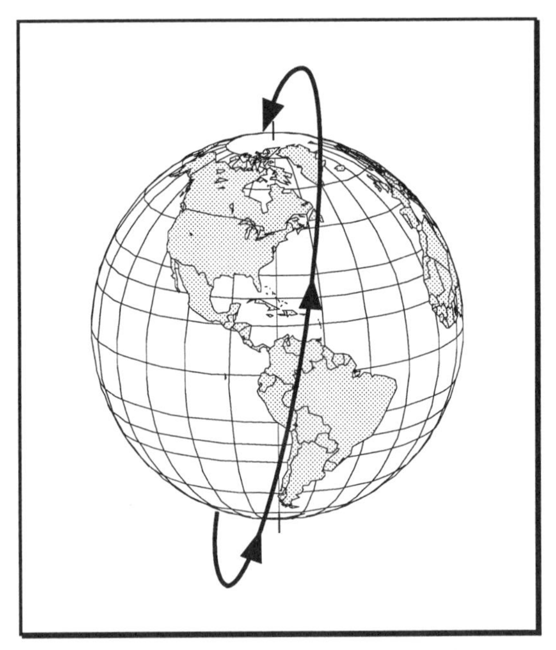

Figure 2.1. Schematic of a polar orbit

Polar orbiting satellites

The orbits of the polar orbiting satellites are nearly from pole to pole, with an inclination of 98° (figure 2.1). This means that these satellites pass within 8° of the poles during each orbit. Polar orbiting satellites circle the Earth in a **sun-synchronous** orbit: the orbital plane of a polar orbiting satellite remains stationary with respect to the sun. As the satellite moves through its orbit, the Earth rotates below it. The result is that the satellite scans a different strip of the Earth during each orbit (figure 2.2). From a fixed point on Earth, a polar orbiting satellite will always cross the equator at approximately the same local time relative to the sun. Each orbit has a period of approximately 102 minutes. Therefore, in one day, the satellite makes roughly 14 orbits (1440 minutes per day ÷ 102 minutes per orbit).

The very first U.S. polar orbiting satellite was launched on April 1, 1960. This satellite, known as **Television and Infrared Observational Satellite (TIROS 1)**,

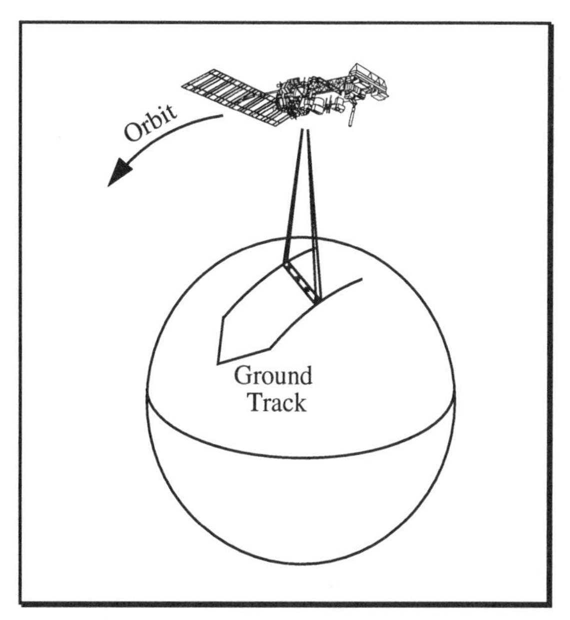

Figure 2.2. Scanning from a polar orbiting satellite
(vertical scale exaggerated)

was a meteorological satellite that eventually led to the establishment of the U.S. operational weather satellite system. Between 1960 and 1965, ten TIROS satellites were launched for the purpose of research and development. Between 1966 and 1969, the **TIROS Operational System (TOS)** became the first operational weather satellite system. The improved TIROS-type satellites (including **ITOS 1** and NOAA 2–5) followed between 1970 and 1978. The advanced TIROS N (figure 2.3) became the improved prototype for the modern NOAA satellites in use today. The satellites in this series are called **NOAA satellites** (after the organization responsible for their daily operation and maintenance) and are numbered after they are placed in orbit. In 1995, NOAA 9–12 and NOAA 14 were operational; NOAA 13 stopped functioning in 1993. Table 2.1 is a summary of the history of U.S. polar orbiting meteorological satellites.

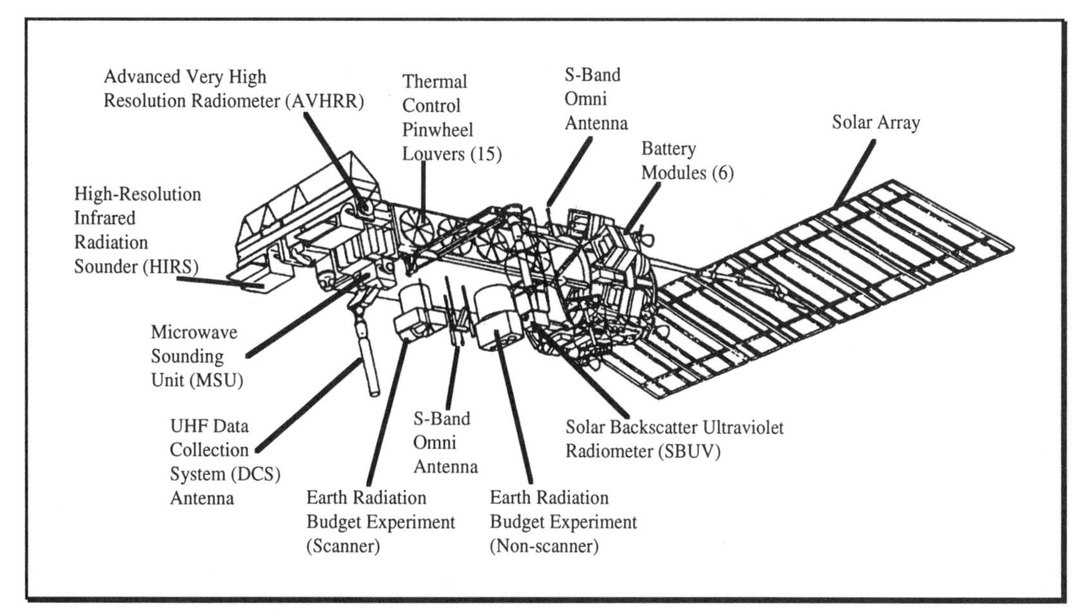

Figure 2.3. Advanced TIROS N weather satellite

MILESTONES IN THE HISTORY OF THE U.S. POLAR ORBITING METEOROLOGICAL SATELLITES		
DATE	**EVENT**	**SIGNIFICANCE**
April 1960	TIROS 1 launched	First meteorological satellite
1960–1965	TIROS 1–10	Research and development
December 1963	TIROS 8 launched	First satellite transmitting Automatic Picture Transmission (APT)
1966–1969	ESSA 1–9 (Environmental Science Services Administration)	Established TIROS Operational System (TOS), the world's first operational weather satellite system
1970–1978	Improved TIROS Operational System (ITOS) satellites, ITOS 1 and NOAA 2–5	Very High-Resolution Radiometer (VHRR), atmospheric soundings
October 1978	TIROS N launched	Prototype for today's polar orbiting satellites
1978–1981	TIROS N and NOAA A–D operational system	Advanced Very High-Resolution Radiometer (AVHRR), TIROS Operational Vertical Sounder (TOVS), Data Collection System (DCS), Solar Environmental Monitor (SEM), High-Resolution Picture Transmission (HRPT)
1983–present	Advanced TIROS N system (NOAA 9–14)	Search and Rescue (SAR), Earth Radiation Budget Experiment (ERBE), Solar Backscatter Ultraviolet (SBUV) Radiometer

Table 2.1

Currently the United States maintains at least two operational satellites in polar orbits. Each satellite will pass within radio range of a ground station once a day traveling from north to south, known as a **descending node**, and once a day traveling from south to north, known as the **ascending node**. The descending and ascending passes are approximately 12 hours apart for one satellite. The two polar orbiting satellites are in such orbits that one satellite (NOAA 10 or NOAA 12) will make a descending pass in the morning, while the second satellite (NOAA 9, NOAA 11, or NOAA 14) will make an ascending pass in the afternoon, approximately 6 hours later. The two satellites will then cross approximately the same point going in the opposite direction 12 hours later. This allows a ground station to receive images of the local area at least every 6 hours.

Each of the currently operational polar orbiting satellites has an array of different sensors, each with a specific purpose. Table 2.2 lists the main sensors used on these satellites, and their functions. The primary set of meteorological sensors on board the NOAA polar orbiting satellites is known as the Advanced Very High-Resolution Radiometer.

PRIMARY SENSORS ON THE TIROS N POLAR ORBITING SATELLITE	
SENSOR	**FUNCTION**
1. Advanced Very High-Resolution Radiometer (AVHRR)	Provides VIS and IR imagery for weather, ocean, and land studies
2. TIROS Operational Vertical Sounder (TOVS)	Measures temperature profile of atmosphere; water vapor, carbon dioxide, ozone, and oxygen content of atmosphere
3. ARGOS Data Collection System (DCS)	Monitors position of transmitters located on ships, ocean buoys, weather balloons, and animals for tracking purposes
4. Solar Backscatter Ultraviolet (SBUV) Radiometer	Measures vertical distribution and total content of atmospheric ozone
5. Search and Rescue (SAR)	Locates radio signals from Emergency Locator Transmitters (ELT) in emergency situations to aid in rescue operations
6. Earth Radiation Budget Experiment (ERBE)	Measures all incoming and outgoing solar radiation in order to determine Earth's heat budget
7. Space Environment Monitor (SEM)	Detects radiation at various energy levels in space

Table 2.2

This five-channel scanning radiometer with 1.1 km resolution is sensitive in the visible, near infrared, and thermal infrared portions of the spectrum. By using both VIS and IR sensors, the satellite is able to collect data on a 24-hour basis. As table 2.3 illustrates, each of the five channels is sensitive to a specific wavelength band within the electromagnetic spectrum and is used for a different purpose.

The Direct Readout System

Data from the NOAA polar orbiting satellite sensors are transmitted continually and can be received by any properly equipped **ground station** within radio range. This type of service is known as **direct readout**. To receive data from the AVHRR there are two categories of direct readout services: **High-Resolution Picture Transmission (HRPT)** and **Automatic Picture Transmission (APT)**.

The APT signal from U.S. polar orbiting satellites is broadcast at 137.50 or 137.62 **megahertz (MHz)** in analog transmission at 120 lines per minute. The ground stations required to receive APT use little more than an FM antenna and a radio receiver tuned to the proper frequency. The radio signal is converted to a digital image, which can be displayed on a personal computer. The relatively simple and inexpensive equipment needed to capture this imagery makes APT the most commonly used form of direct readout.

SUMMARY OF CHANNELS ON THE ADVANCED VERY HIGH-RESOLUTION RADIOMETER		
CHANNEL	**SPECTRAL BANDWIDTH**	**TYPES OF DATA PROVIDED**
1 (Visible)	0.58–0.68 μm	Daytime cloud cover, snow cover, ice studies, mapping, pollution
2 (Near infrared)	0.73–1.10 μm	Daytime cloud cover, surface water, vegetation/agricultural assessment
3 (Thermal infrared)	3.55–3.93 μm	Black body temperatures, nighttime cloud cover, sea surface temperatures, forest fire and volcano monitoring
4 (Thermal infrared)	10.50–11.50 μm (NOAA 6, 8, 10) or 10.30–11.30 μm (NOAA 7, 9, 11, 12)	Daytime/nighttime cloud cover, land and sea temperature patterns
5 (Thermal infrared)	11.5–12.5 μm (NOAA 7, 9, 11, and 12 only) *	Water vapor correction when paired with channel 4, daytime/nighttime cloud cover, land and sea temperature patterns

* On NOAA 6, 8, and 10, channel 5 is a repeat of channel 4.

Table 2.3

APT transmits data from two channels of the AVHRR at a reduced resolution of 4 km. The reduced resolution allows a larger area of view than HRPT data, although smaller features may not show up. During the day, both a VIS and an IR channel are used, while nighttime transmissions usually consist of two different IR channels. The images are received simultaneously, resulting in a side-by-side double image. An example of this can be seen in the APT image shown in figure 2.4. The left half of this picture is a VIS image of the east coast of North America, while the right half is an IR image of the same scene. Comparisons between the two different channels can help an investigator to infer various characteristics of the clouds, oceans, and land surfaces that could not be determined with data from only one channel.

The HRPT signal from U.S. polar orbiting satellites is broadcast at 1698.0 MHz. The receiving stations for HRPT are more complex and expensive than those designed for receiving APT; therefore HRPT imagery can be more difficult to obtain. However, advances in receiver technology and innovations in building advanced receiving stations are bringing HRPT close to an affordable level for use by many. Additionally, organizations equipped with HRPT direct readout systems often make this imagery available to the public through various electronic bulletin boards.

An HRPT transmission consists of data from all five channels of the AVHRR at a full 1.1 km resolution. While the resolution of an HRPT image is very high, the area covered in each image is smaller. Figure 2.5 is an example of an HRPT image from the VIS

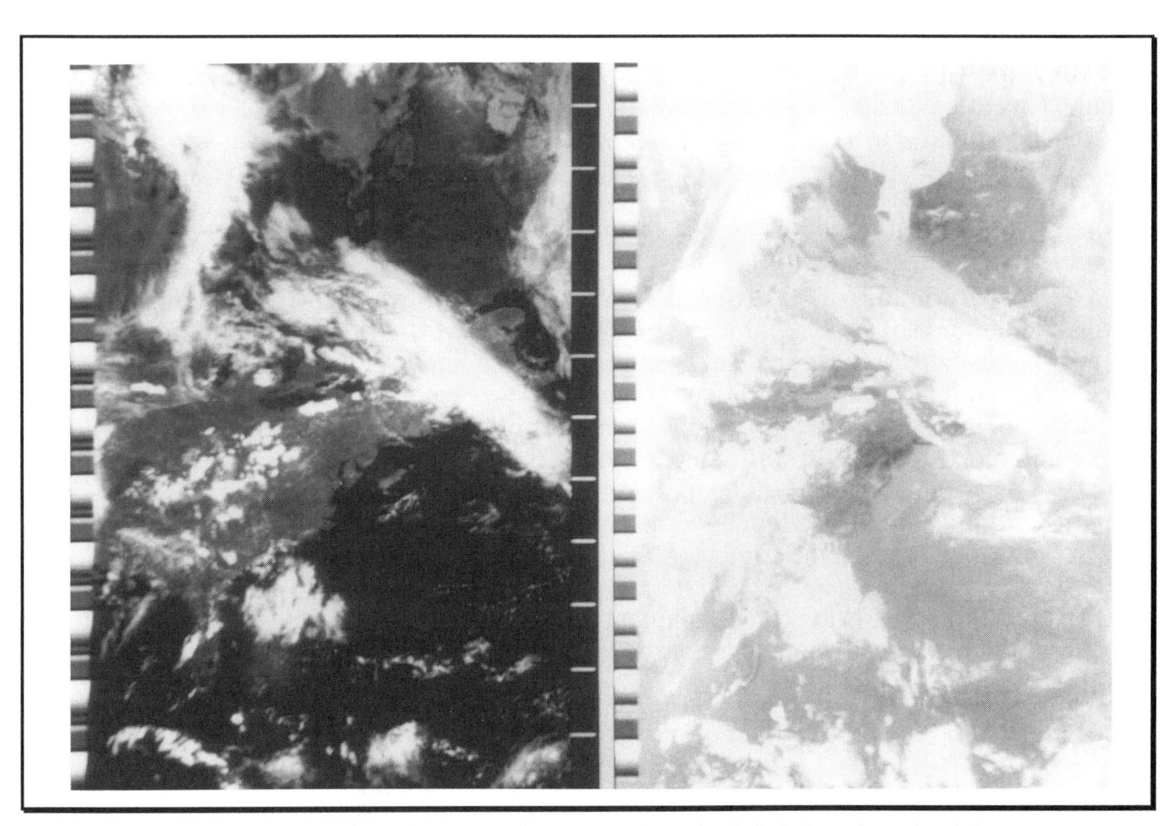

Figure 2.4. An APT image (visible channel on the left, infrared on the right)

Figure 2.5. An HRPT image (HRPT VIS, June 25, 1987)

channel of the AVHRR. Compare this image to the VIS APT image in figure 2.4. Note that while the APT image covers a larger area of the Earth, the HRPT image offers more detail. This often makes HRPT imagery more useful when studying smaller-scale features such as individual storms or surface characteristics of the land and oceans.

Geostationary satellites

The **Geostationary Operational Environmental Satellites (GOES)** orbit the Earth at an altitude of about 35,000 km (21,000 miles) above the equator. At this height, the angular velocity of the spacecraft is equal to the angular velocity of the Earth (each travels 360°, or one complete orbit, in 24 hours). As a result, each satellite remains over the same point on the Earth throughout its entire orbit. This type of orbit, known as a **geosynchronous orbit**, allows frequent monitoring of the same portion of the Earth. Since the Earth does not move in relation to the satellite, a series of images can be constructed into a time-lapse motion loop. A loop plays back several images, at speeds that can be controlled through the computer, which allows one to see cloud motion.

The GOES mission is to provide frequent repetitive observations necessary to detect, track, and predict severe weather systems. The GOES satellites are able to observe the full

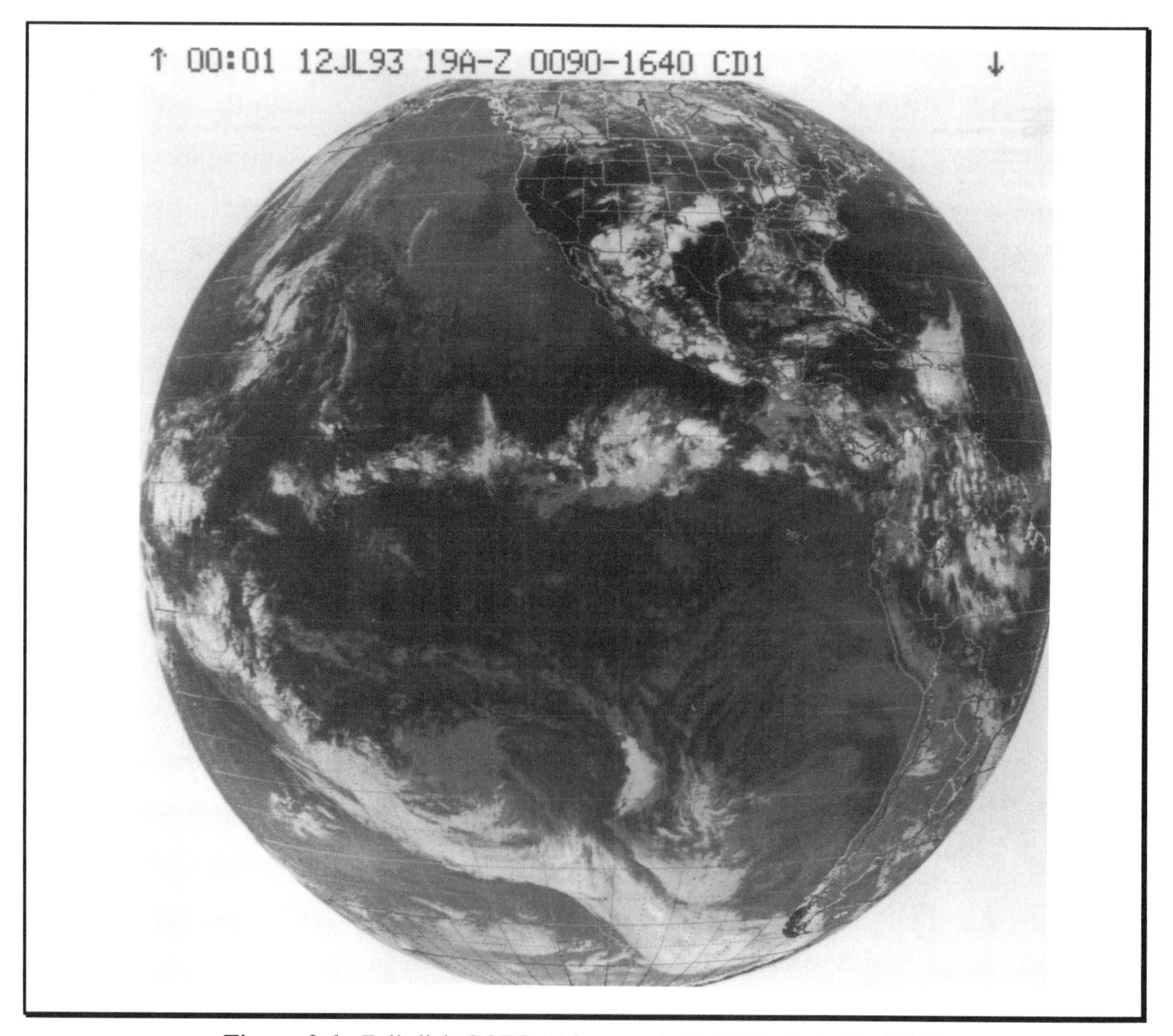

Figure 2.6. Full-disk GOES IR image (GOES IR, July 12, 1993)

Earth disk in both VIS and IR regions of the spectrum. Figure 2.6 is an example of a full-disk VIS GOES image. The GOES satellites can observe and measure cloud cover, snow and ice cover, surface temperatures, and vertical distribution of atmospheric temperature and wind. They also measure solar x-rays and other forms of radiation in space and act as data relay platforms by distributing information from ground-based environmental sensors (e.g., rainfall and river data from hydrological data collection platforms) to various locations on Earth.

In order to monitor North America at frequent intervals, two GOES satellites are typically used to provide coverage of the entire hemisphere. When two-satellite coverage is fully operational, one satellite occupies the **GOES East** position, located over the equator at 75° W, and one occupies the **GOES West** position over the equator at 135° W. These two satellites have an overlapping area of coverage and provide imagery for the entire North and South American continents, as well as the Pacific and Atlantic Oceans.

GOES history

The first combined geostationary weather and communication satellite launched by the United States was the **Applications Technology Satellite (ATS 1)** in late 1966 (table 2.4). Following the ATS series, **SMS 1 (Synchronous Meteorological Satellite)** was launched as a prototype for a U.S. operational satellite program. The United States' first operational geostationary satellites were part of a series that began with GOES 1, launched in 1974, and ended with GOES 7, which was scheduled to run out of fuel in early 1995. The early GOES meteorological satellites established the two-satellite coverage of the Western Hemisphere. This dual coverage continued until GOES 6 ran out of fuel in 1989.

In response to the GOES 6 failure, the United States moved GOES 7 to 112° W to monitor the central and western portions of the country. To reestablish two-satellite coverage, it then leased **METEOSAT 3** from the European Space Agency. This satellite was moved to 75° W to monitor the eastern portion of North and South America and the Atlantic Ocean until a new U.S. geostationary satellite could become operational.

The GOES I–M series of geostationary satellites represents the future generation of U.S. geostationary environmental satellites. This series of improved GOES satellites is scheduled to extend into the first decade of the twenty-first century. The United States officially entered this new phase of satellite meteorology with the 1994 launch of GOES I, which was renamed GOES 8 once it was placed in orbit. Following extensive testing and calibration, GOES 8 was placed in orbit over 75° W longitude in early 1995 to fill the GOES East position. The next GOES satellite, GOES J, was launched on May 23, 1995. Renamed GOES 9, this satellite was moved to replace GOES 7 at 135° W to fill the GOES West position. These two satellites once again established overlapping coverage of much of the Western Hemisphere.

Early GOES satellites: GOES 1–7

A diagram of the GOES 7 satellite showing some of its main components can be seen in figure 2.7. The main imaging sensor on the early GOES satellites was the **Visible and Infrared Spin Scan Radiometer (VISSR)**. The VISSR, which first appeared on GOES 4, is a **multi-spectral imaging tool**. This means that it can collect and transmit VIS data and up to three bands of IR data for every scan line. The VISSR sensor scans the Earth in horizontal lines, starting at the north and working south. This type of GOES

MILESTONES IN THE HISTORY OF THE U.S. GEOSTATIONARY METEOROLOGICAL SATELLITE PROGRAM		
DATE	**EVENT**	**SIGNIFICANCE**
December 1966	Applications Technology Satellite (ATS 1) launched	First geostationary meteorological satellite, Spin Scan Cloud Cover camera (SSCC), WEFAX transmissions, first full-disk Earth images
1967–1974	ATS 3 and ATS 6	Research and development, communication experiments, Geosynchronous Very High-Resolution Radiometer (GVHRR)
1974–1975	Synchronous Meteorological Satellites (SMS 1 and 2)	Prototypes for future GOES series, Visible and Infrared Spin Scan Radiometer (VISSR)
1975–1978	Geostationary Operational Environmental Satellites (GOES 1–3)	Developed overlapping satellite coverage of the Western Hemisphere using two satellites over the equator; GOES East at 75° W and GOES West at 135° W
1980–1987	GOES 4–7	VISSR Atmospheric Sounder (VAS)
1989	GOES 6 failure	Return to one U.S. GOES satellite coverage of North America
1992–1995	METEOSAT 3 leased from European Space Agency (ESA)	Moved to 75° W in early 1993 to reestablish two-satellite coverage of Western Hemisphere
April 13, 1994	GOES 8 launched	First in GOES I–M series, three-axis stabilization, simultaneous sounding and imaging. Moved to fill GOES East position in early 1995
May 23, 1995	GOES 9 launched	Placed in GOES West position to reestablish U.S. dual GOES coverage of North America and Western Hemisphere

Table 2.4

satellite spins at 100 rpm. With each spin the sensors are lowered slightly and a new strip of the Earth is observed (figure 2.8). A full-Earth disk image consists of 1821 sweeps at 100 rpm; therefore, it takes 18.2 minutes to obtain a full-Earth disk image. After each full-Earth scan, the sensors are reset and smaller portions of the Earth may be imaged to provide data about smaller-scale features of interest. The next full-Earth scan begins 30 minutes after the previous full scan. Therefore, along with images of smaller sections of the Earth, the ground station receives one full-Earth disk every half hour. Under some special conditions, full-disk imagery may be suspended to allow for more frequent imaging of severe thunderstorm or hurricane conditions.

The VIS sensor on the VISSR consists of eight detectors arranged vertically. This means that with each sweep of the sensor there are 8 lines of data. Since a full-disk image consists of 1821 sweeps, this results in 14,568 lines of data. GOES VIS imagery has a resolution of up to 1 km (0.6 miles). The IR sensors on the GOES 1–7 spacecraft use only one detector for each scan, so IR imagery has a lower resolution (7 km, 4.2 miles) than VIS imagery.

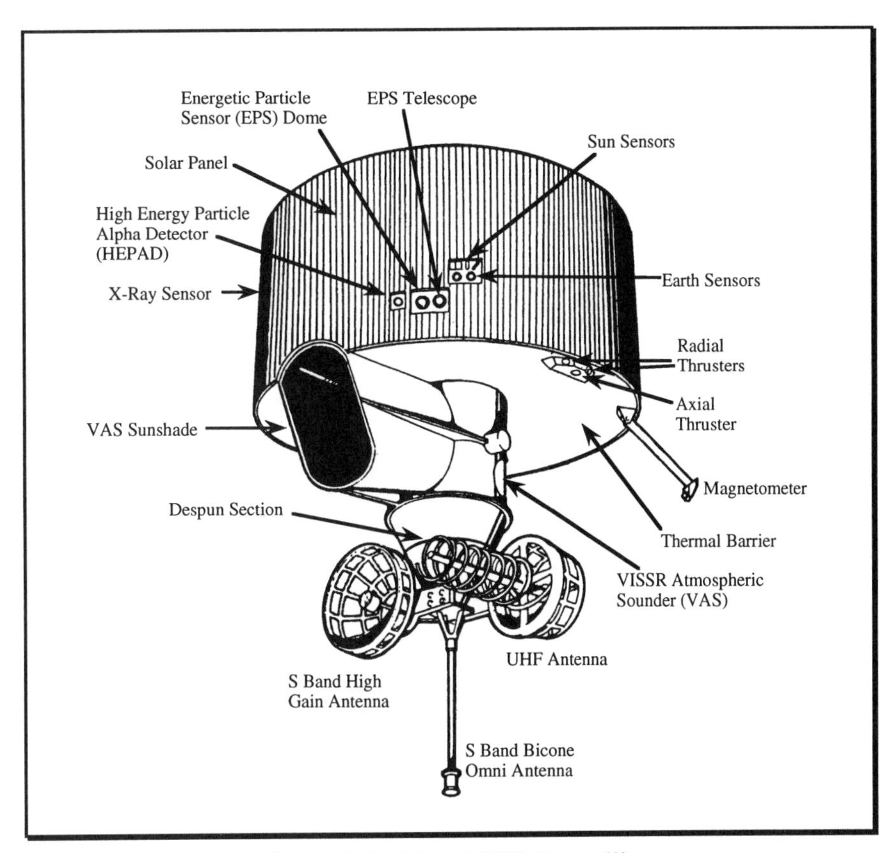

Figure 2.7. The GOES 7 satellite

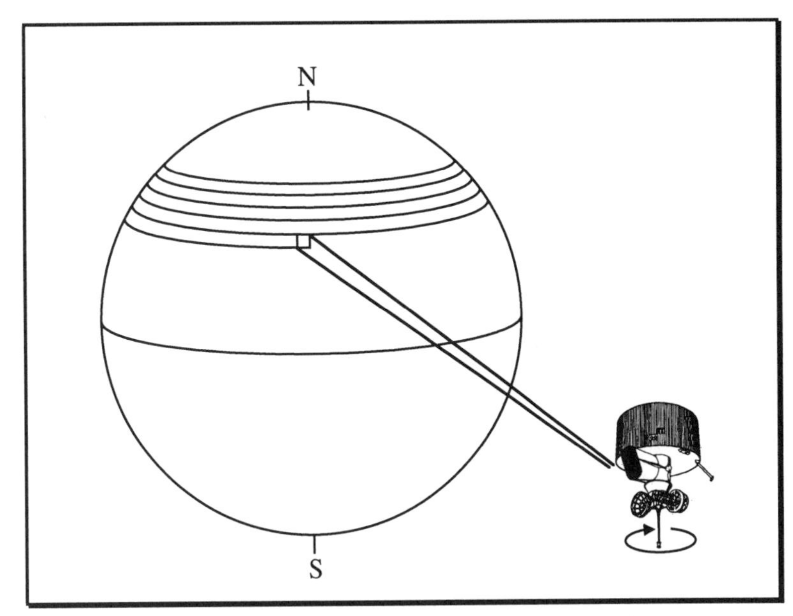

Figure 2.8. Scanning the Earth from GOES 7

Future GOES satellites: GOES 8 and GOES J–M

The new series of satellites is much different from the previous GOES series. These new geostationary satellites utilize a **three-axis stabilized** system instead of the spin-scan system used previously. The three-axis stabilization of the craft allows for frequent scanning up to a 1-minute interval. Time-lapse imagery this frequent is especially helpful in research programs for severe thunderstorms and hurricanes. A diagram of the first satellite in this series can be seen in figure 2.9.

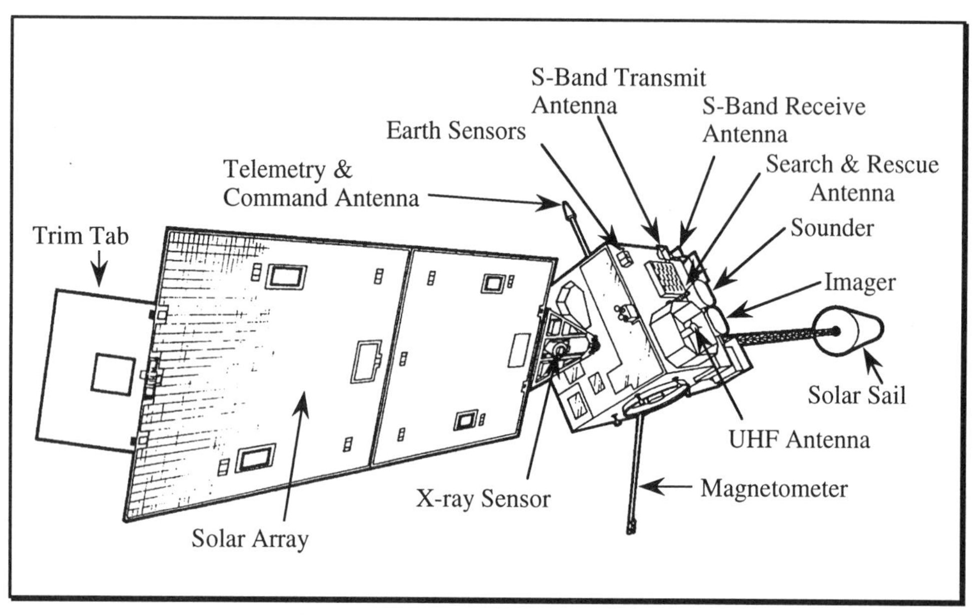

Figure 2.9. The GOES 8 satellite

The new GOES satellites are also equipped with a new array of sensors. On previous GOES satellites, the imager and the sounder were part of the same instrument (the VAS), which led to scheduling conflicts between the two sensors. The new GOES spacecraft utilizes two separate instruments for the imager and the sounder. Designing these sensors as two different components allows for more frequent and flexible imaging.

Like the imagers in the previous GOES satellites, the GOES I–M imager is a multi-channel instrument that is capable of measuring both reflected visible light and emitted infrared radiation. There are five channels, including VIS, near IR, and thermal IR channels, on the GOES I–M series imager. In some of these channels, the new imager offers a larger spectral response. In addition, both the imager and sounder operate with reduced noise levels and a higher spatial, temporal, and thermal resolution than the previous GOES series. Table 2.5 summarizes the major functions and improvements for each of the channels on the new GOES imager.

To image the Earth (from north to south), the scanner sweeps across the Earth from east to west, then moves down before sweeping in a west-to-east direction. The imager on the new GOES satellites can produce full-Earth images, sectored images, and local area scans. These images are then broadcast to ground stations on the Earth, where they are processed and redistributed to users through traditional broadcasting methods, including WEFAX. The many improvements on the GOES I–M imaging system will lead to better imaging capabilities and new satellite-derived products that will be extremely useful in understanding the Earth's atmosphere, oceans, and land surfaces.

CHANNEL	SPECTRAL BANDWIDTH	SPATIAL RESOLUTION	MAJOR FUNCTIONS AND IMPROVEMENTS
1	0.55–0.75 µm (visible)	0.5 x 1 km	Improved cloud edge detection; cloud top feature detection; severe storm identification; pollution and haze detection; cloud height measurements
2	3.80–4.00 µm (near infrared)	2 x 4 km	Improved fog detection at night; detection of water clouds vs. snow and ice clouds; detection of hot spots (fire, volcano); nighttime sea surface temperatures (SST); location of hurricane eye through thin cirrus
3	6.50–7.00 µm (water vapor)	2 x 8 km	Improved study of large-scale weather patterns; better wind speed determinations in middle troposphere
4	10.20–11.20 µm (thermal infrared)	2 x 4 km	Improved cloud edge and cloud top detection; heavy precipitation estimates; severe storm identification; wave cloud detection
5	11.50–12.50 µm (thermal infrared)	2 x 4 km	When used in combination with channel 4, will help determine low-level moisture; improved SST; volcanic ash detection

SUMMARY OF CHANNELS ON THE GOES I–M SERIES OF U.S. GEOSTATIONARY METEOROLOGICAL SATELLITES

Table 2.5

GOES data broadcasting

Data from GOES satellites is broadcast by two methods that are available to anybody within receiving range. Primarily, the VISSR sensor and spacecraft data are sent directly to ground stations operated by NOAA. This data is not processed and it is not considered a public service, although it is available to anybody with the right equipment. The data is processed at the command center and sent back to the satellite for retransmission at a slower data rate.

The second method of data distribution is through the **Weather Facsimile (WEFAX)** system, which allows private users to access imagery directly from GOES satellites. WEFAX is the retransmission of reformatted low-resolution imagery in a form that is available to most users. The U.S. GOES WEFAX service broadcasts processed environmental information at a frequency of 1691.0 MHz. Standard meteorological charts and processed GOES imagery are transmitted along with imagery from NOAA polar orbiting satellites. WEFAX data is transmitted in a nearly continuous broadcast according to a specific schedule. Users wishing to receive this data can obtain a schedule of transmissions to determine which images can be captured at a certain time.

Developers of WEFAX took into consideration that many people were already receiving APT. Thus, WEFAX operates in much the same way. Most APT receiving stations can easily be modified to receive WEFAX by the addition of an antenna dish and a down-converter to change the 1691.0 MHz signal to a 137.62 MHz signal. Once captured, the signal is converted into a digital image and displayed on a personal computer.

Satellites operated by other countries

Japan, Russia, India, China, and the European Space Agency all operate environmental satellites that supply imagery over all parts of the Earth's surface. Most of these satellites transmit APT and/or WEFAX, making data available for direct readout. If the satellites are out of radio range, imagery from these satellites can often be obtained using **telecommunication** or **CD-ROM** technology.

Russian polar orbiters

Russian polar orbiters include the **OKEAN** (meaning "ocean") satellites, which transmit on a frequency of 137.3 MHz, and the **METEOR** satellites, which transmit at 137.3 MHz, 137.4 MHz, and 137.85 MHz. The METEOR series is the only series that is regularly heard from in the United States. There are two different types of METEOR satellites: the METEOR 2 series, which transmits VIS imagery, and the METEOR 3 series, which transmits VIS and IR imagery. METEOR 2-17 through 2-21 and METEOR 3-1 through 3-6 are all capable of transmitting data; however, they are not all operational at the same time. Instead, the Russians frequently turn these satellites on and off. Each satellite is capable of transmitting data at several frequencies; these are often changed without notice. Additionally, the orbits of the Russian satellites are frequently shifted. The combined effect of all these factors can make tracking Russian satellites challenging. Furthermore, imagery from the METEOR satellites can be difficult to interpret, since it differs from imagery from U.S. polar orbiters. The IR sensors on the Russian satellites are designed to sense moisture, water vapor, and clouds; hence, there is very little ground imaging. This makes it difficult to locate land features that can help you determine where the image was taken. Russian IR imagery also uses a gray scale different from that used by U.S. satellites; cold temperatures appear black and warm temperatures appear white.

Other non-U.S. polar orbiting satellites

Cosmos. A Russian oceanographic satellite series similar to the OKEAN series. Cosmos 1602 is the only known satellite in this series that is transmitting APT. Since 1990, this spacecraft has rarely been heard from over the United States.

Feng Yun. A Chinese polar orbiter that transmits APT and HRPT with both VIS and IR imagery. These satellites are plagued by instrumentation problems and are known for infrequent transmissions.

Non-U.S. geostationary satellites

Geosynchronous Meteorological Satellite (GMS). A Japanese geostationary satellite that provides VIS and IR imagery of the western Pacific, eastern Asia, and Australia. GMS provides WEFAX service. Figure 2.10 is a GMS IR image showing most of eastern Asia, Australia, and the western Pacific.

GOMS. A Russian version of the geostationary satellite, similar to the U.S. GOES series. The first IR images from GOMS were received in February 1995. Scheduled to be fully operational by late 1995, it will provide coverage, including WEFAX, over central Asia and the Indian Ocean.

INSAT. An Indian geostationary weather satellite that doubles as a communication satellite. INSAT provides WEFAX service.

METEOSAT. Operated by the European Space Agency, this series of geostationary satellites provides coverage of Europe and Africa. METEOSAT 5 is located at 0° and METEOSAT 3 was temporarily over 75° W to provide coverage of the eastern United States until GOES 8 became fully operational in 1995. The METEOSAT series also provides WEFAX service.

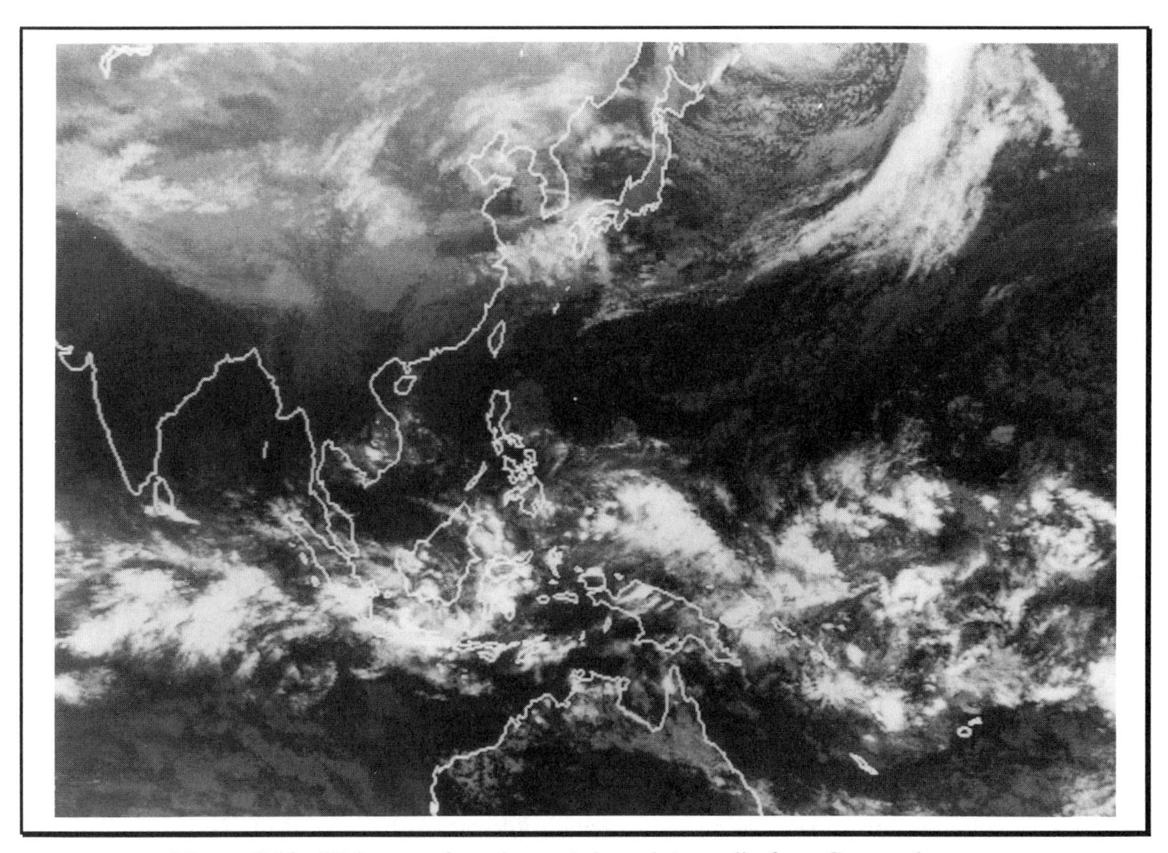

Figure 2.10. IR image of southeast Asia and Australia from Geosynchronous Meteorological Satellite (GMS IR, December 10, 1994)

Methods of obtaining satellite imagery

There are many different ways, including direct readout, telecommunications, CD-ROM, and television broadcasts, to acquire imagery from environmental satellites. Each of these methods has its own unique set of advantages, usually determined by availability, quality, and selection of imagery. Constraints can include budgets, availability of technology, degree of technical expertise, and time.

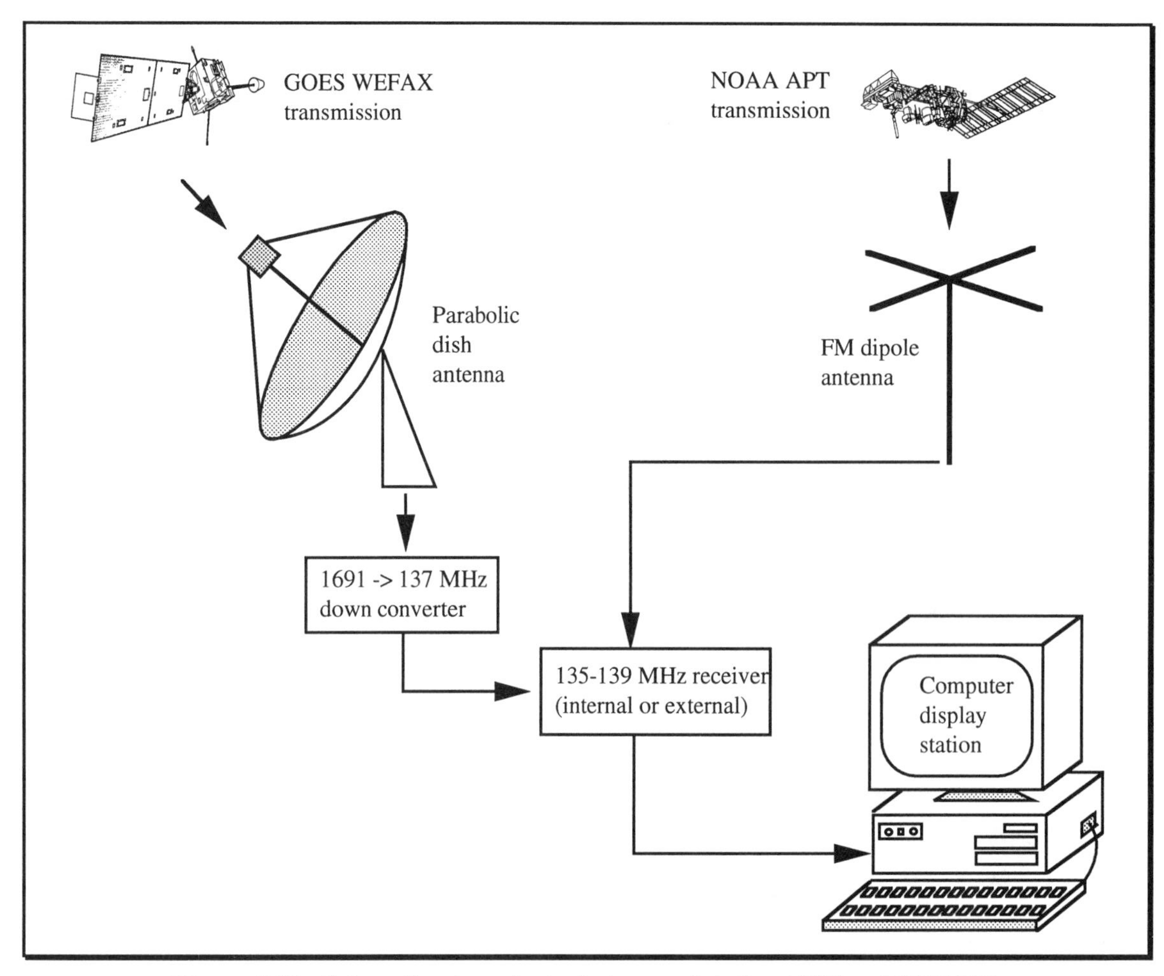

Figure 2.11. Schematic view of a typical ground station (APT and WEFAX)

Ground stations

Satellite direct readout is one of the most exciting ways to acquire satellite imagery. A direct readout **ground station** consists of an antenna, cable, and a radio receiver designed to capture real-time data directly from a satellite via APT, HRPT, or WEFAX transmission (figure 2.11). The complexity of a ground station is determined by the type of data it is designed to capture, with APT being relatively simple and WEFAX and HRPT more complex. Whatever type of data is desired, a direct readout ground station provides large amounts of real-time data at regular and predictable intervals. Once captured, the live images can be enhanced, animated, and archived using image-processing software. In addition, tracking satellites on a daily basis and capturing live imagery can be a stimulating and rewarding experience. Some enterprising amateurs have even built their own direct readout stations; however, operating a ground station does requires some technical aptitude, and obtaining the equipment necessary for a ground station can be difficult or expensive. Recently, several commercial vendors began offering comprehensive packages that include all of the equipment and software required to install and operate a ground station at any location. Most of these systems are capable of capturing APT and WEFAX, are very reliable and easy to operate, and can be purchased within a modest budget.

Telecommunications

Computer networking is an inexpensive method for gathering satellite data that is becoming more popular as the number of people using telecommunication technology increases. The best example of this growing technology is the recent explosive growth of the **Internet**. The Internet is a global network of scientists, researchers, educators, and others who exchange data and information through computers linked together by phone lines. Anybody with access to a computer, a phone line, and a modem can access hundreds of locations on the Internet (and various other independent bulletin board services) which supply satellite images, image processing software, weather forecasts, weather maps, and other information relevant to satellite research (see Appendix B). Weather satellite images are often posted on these sites nearly in real-time and are updated regularly. In addition, imagery is often available from satellites that cannot be accessed through a local ground station. These files can be transferred (or **downloaded**) to a personal computer and enhanced using image-processing software.

Though the promise of computer networking as a means of data collection is great, telecommunications remains a growing technology, and there can be problems when too many people attempt to go on-line simultaneously. Also, satellite images tend to be very large files, and it can take a long time to transfer an image to another computer. This can use up valuable computer time and lead to expensive line charges if a long distance dial-up service is used. Advancements in communication technology will undoubtedly minimize these problems in the future and allow networking to become the preferred method of data acquisition for most people.

CD-ROM

Many scientific organizations are also supplying satellite data on CD-ROM disks. A CD-ROM (Read Only Memory) is a storage device that is able to hold large amounts of software and data that can be read by a personal computer. These disks are capable of containing hundreds of stored satellite images, along with software that allows a user to manipulate and enhance the data. A CD-ROM drive is a relatively inexpensive piece of equipment to add to a personal computer, and many disks are available free or at low cost. Some disks can be expensive, however, and the selection of imagery on a disk is often limited.

Television broadcasts

If none of the above methods is within grasp, television broadcasts can be a source of satellite imagery. Local and national broadcasts often use satellite imagery during daily weather forecasts. Many of the cloud patterns discussed in this manual are commonly seen in these broadcasts. Since most people have access to a television, this method of obtaining imagery is readily available at very little cost. In addition, weather broadcasts often include radar summaries, surface maps, and other pertinent information along with the satellite imagery. However, the satellite imagery in television broadcasts is often of reduced quality and cannot be enhanced, manipulated, or digitally stored.

CHAPTER 3

The Basics of Image Interpretation

Introduction to image interpretation

All U.S. weather satellites take images of the Earth in selected spectral bands that are in both the visible and the infrared portions of the electromagnetic spectrum. In addition, many satellites provide a type of infrared imagery known as **water vapor (WV) imagery**. All three types of imagery are important for different reasons, and, in some cases, all three are needed to accurately interpret atmospheric conditions. The overall purpose of this chapter is to describe the basics of satellite imagery and provide the background necessary for accurate image interpretation. The chapter discusses the features common to all three types of weather satellite imagery and the characteristics that make each one unique.

Image composition

Although it may look like an actual picture of the Earth, a weather satellite image is composed of thousands of points known as **pixels** (short for "picture elements"). This is analogous to a picture in a newspaper. If you examine a newspaper photograph with a magnifying lens, you will see that it is simply a collection of different-sized dots. When you view the image, your eyes blend these dots together to form a picture. In a satellite image, each pixel has its own tone (or color), and when viewed together, the pixels form an image of the Earth-atmosphere system.

Most weather satellite images are collected in a **gray tone display**. In this format, each pixel is assigned a tone that represents a level of energy (called the **brightness value**) sensed by the satellite. The tone is either white, black, or an array of intermediate gray shades (known as a **gray scale**). Typically, there are 256 possible brightness values or shades of gray in a satellite image. Different features on the Earth or in the atmosphere have different brightness values; therefore the relative brightness aids in the identification of features in a satellite image.

In all types of imagery the degree of **contrast**, or gray tone difference, between an object and its background is important. The greater the contrast, the easier it is to identify features in satellite imagery. When contrast is poor, enhancement techniques can be used to make accurate interpretation easier. To enhance an image, all the pixels in a specific range of brightness values are highlighted to locate the features of interest. For example, flood forecasters use IR imagery to look for specific cloud top temperatures that indicate heavy precipitation. By highlighting all the pixels with the corresponding brightness values, one can locate the areas within a storm where heavy precipitation is most likely.

Finally, satellite images are often described in terms of their **resolution**. Resolution refers to the size of the smallest feature that can be seen in an image. Since one pixel is the smallest element in an image, the area represented by one pixel is equal to the image resolution. In a satellite image with a resolution of 1 km (about 0.6 miles), the smallest features clearly seen will be 1 km by 1 km. Each pixel represents the average brightness over an area this size. Image resolution is determined by the satellite sensor, the type of transmission used, and the type of display hardware used to view the imagery. It should also be noted that satellite resolution will be best at the point on the Earth's surface directly below the satellite, known as the **satellite subpoint**. As you move further away from the satellite subpoint, the satellite's angle of view changes and the resolution decreases. In most imagery, the satellite subpoint is near the center of the image; therefore, resolution usually decreases toward the edges of the image.

Characteristics of visible imagery

VIS imagery indicates the amount of solar radiation reflected from the Earth. A VIS image is an approximation of the Earth's albedo, that is, the percentage of incoming sunlight reflected by a surface. In satellite VIS imagery, light tones represent areas of high reflectivity and darker tones represent areas of low reflectivity. Features on the surface of the Earth or in the atmosphere vary in their reflectivity and can therefore be discerned on a VIS image (refer to table 1.1 for a list of the albedos of various features of the Earth and the atmosphere). In figure 3.1, a VIS image, the large, thick clouds appear white since they have a high albedo. Thinner clouds appear light to medium gray. The ocean, with a very low albedo, appears nearly black. The land, characterized by albedos that depend on the nature of the surface, appears as various shades of gray.

Figure 3.1. A GOES VIS image

The angle of illumination by the sun also affects brightness in VIS imagery. Midday VIS images will be brightest, while images with low sun angles will be less bright. Often, GOES images taken early or late in the day will include the sunrise-sunset line known as the **terminator**. This is the actual dividing line between day and night on the Earth. Figure 3.2 is a VIS GOES image showing the terminator over the Pacific Ocean. In this image, darkness is to the east of the terminator, indicating that this is an evening image. In a morning image, darkness will appear to the west of the terminator.

The angle of the sun with respect to the satellite sensors also creates shadows that are quite helpful in analyzing cloud types in VIS satellite imagery. In images taken with a lower sun angle, tall clouds or clouds with sharp edges can cast shadows on lower cloud layers or the ground. Clouds with a bumpy top can create a great deal of shadow and will appear lumpy in VIS imagery. Clouds that have flat tops do not exhibit shadows and appear smooth in a VIS image. In figure 3.3, a tall cloud has cast a shadow on a lower

Figure 3.2. GOES image with terminator (GOES VIS, July 25, 1992)

cloud layer at point *A*. At point *B*, a shadow is being cast on the cloud top by the overshooting top of a thunderstorm, giving this cloud top a lumpy appearance. Figure 3.4 is an HRPT VIS image in which many shadows are visible. At points *A* and *B*, higher layers of clouds are casting shadows on the lower cloud layers. At point *C*, a layer of clouds is casting a shadow on the ocean surface. By measuring the width of these shadows and by knowing the sun angle at this time of day, the cloud heights can be estimated.

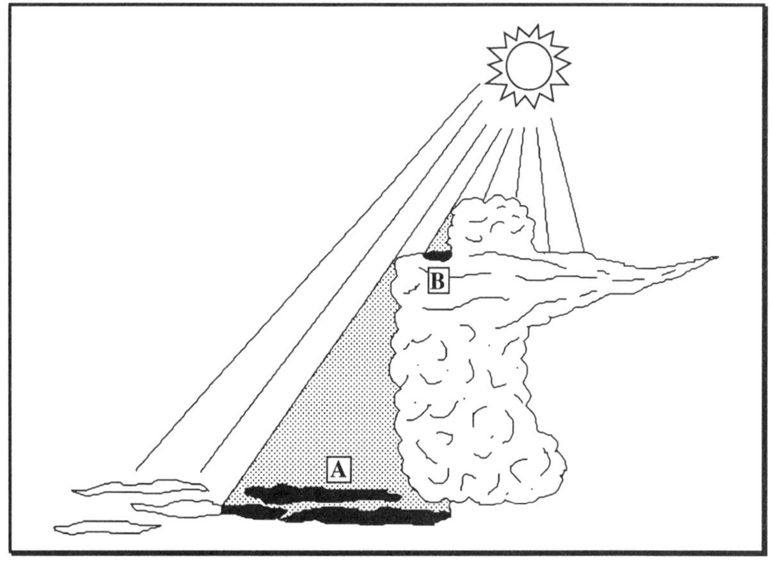

Figure 3.3. Shadows (not to scale)

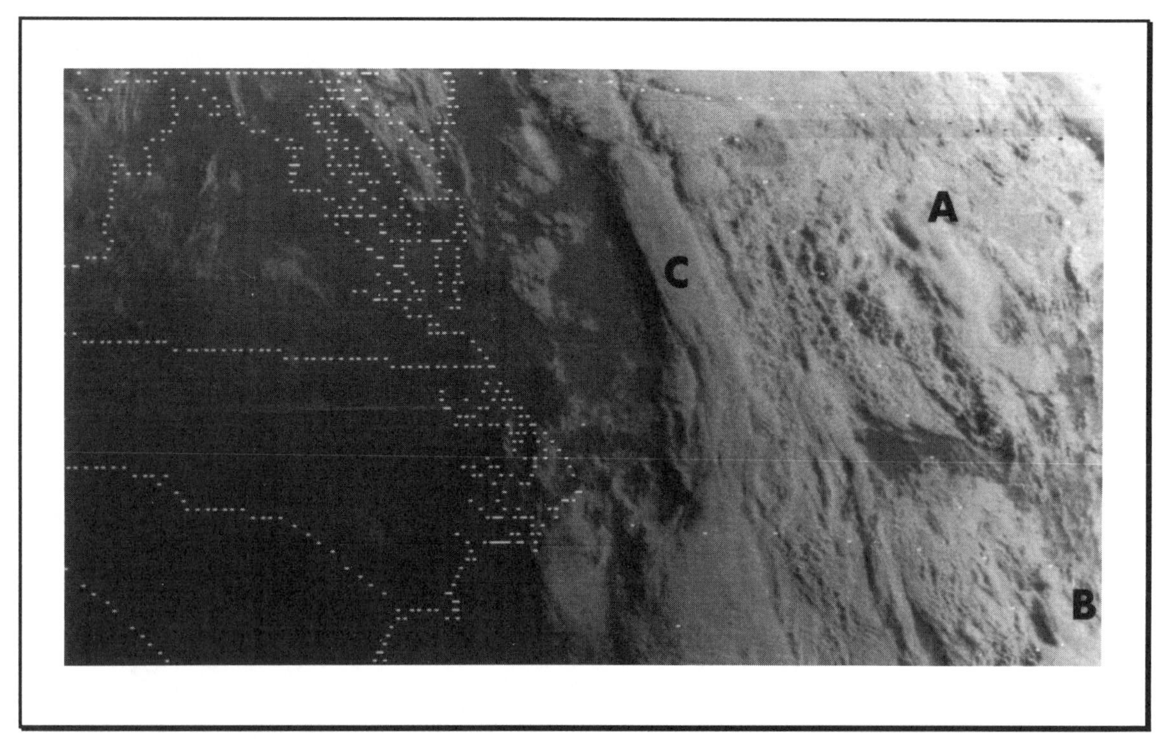

Figure 3.4. Shadows in VIS imagery (HRPT VIS, May 22, 1993)

Often in a VIS image a large bright area will be seen on water features such as oceans, lakes, and large rivers. This is known as **sunglint**, and it is simply the reflection of the sun off the water surface. Sunglint can range from a small bright spot to a large dull area with diffuse boundaries, depending on the roughness or smoothness of the water surface.

Figure 3.5 illustrates why sunglint appears in VIS satellite images. If the water surface is smooth, the reflectance is similar to that of a mirror and the sun's rays will be reflected at point *B* into the satellite's sensor. *B* will appear very bright in imagery. Outside this area, at points *A*, sunlight reflected from the sea's surface completely misses the satellite sensor, as shown by *C*. At *A*, the satellite sees only reflections from outer space (coming from *D*). On the image, *A* appears as darkness.

At first, sunglint can seem like a nuisance; however, it can offer some interesting clues about the weather. Since the size and brightness of a sunglint depend on the roughness of the water surface, sunglint is an indicator of surface-level winds. The higher the waves (and generally the stronger the wind), the larger and more diffuse the sunglint area. Very little sunglint will appear in areas where the seas are rough. When calm waters exist, a very bright sunglint will appear in a localized area. In this case, the sunglint is often the brightest feature on a VIS image. In figure 3.6, a GOES VIS image, a very bright sunglint is located over the Pacific Ocean west of the Hawaiian Islands, indicating very calm seas in this region. The terminator is also visible in this image. This image was taken in the evening (local time in Hawaii), as indicated by the darkness to the east of the terminator. Sunglint can also appear on inland bodies of water. Figure 3.7 shows a very bright sunglint on southern Lake Michigan. Cool, sinking air over the lake has resulted in very light winds over the lake. Therefore, the surface of the lake is very calm, resulting in the bright sunglint in this image.

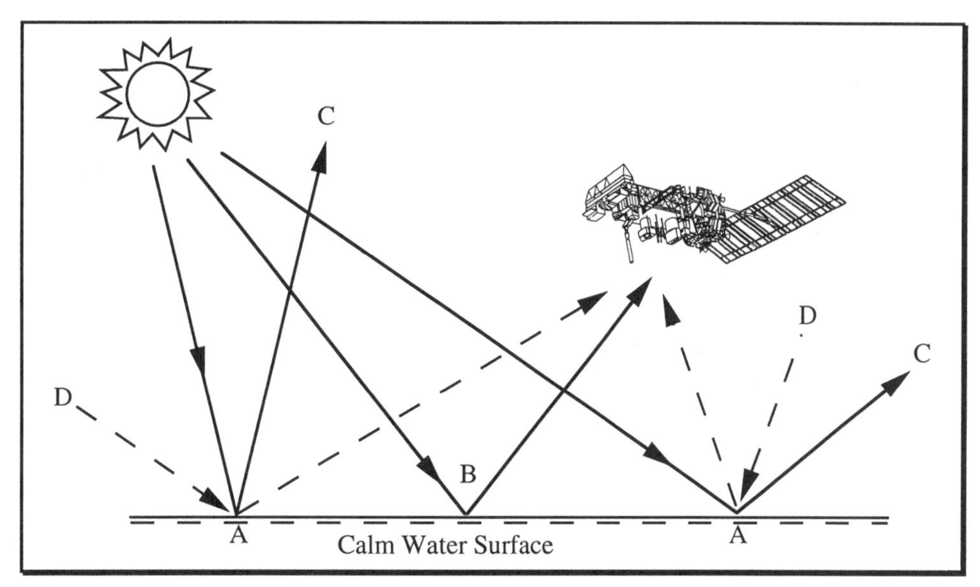

Figure 3.5. Creation of sunglint (not drawn to scale)

Figure 3.6. Sunglint on the Pacific Ocean (GOES VIS, May 22, 1993)

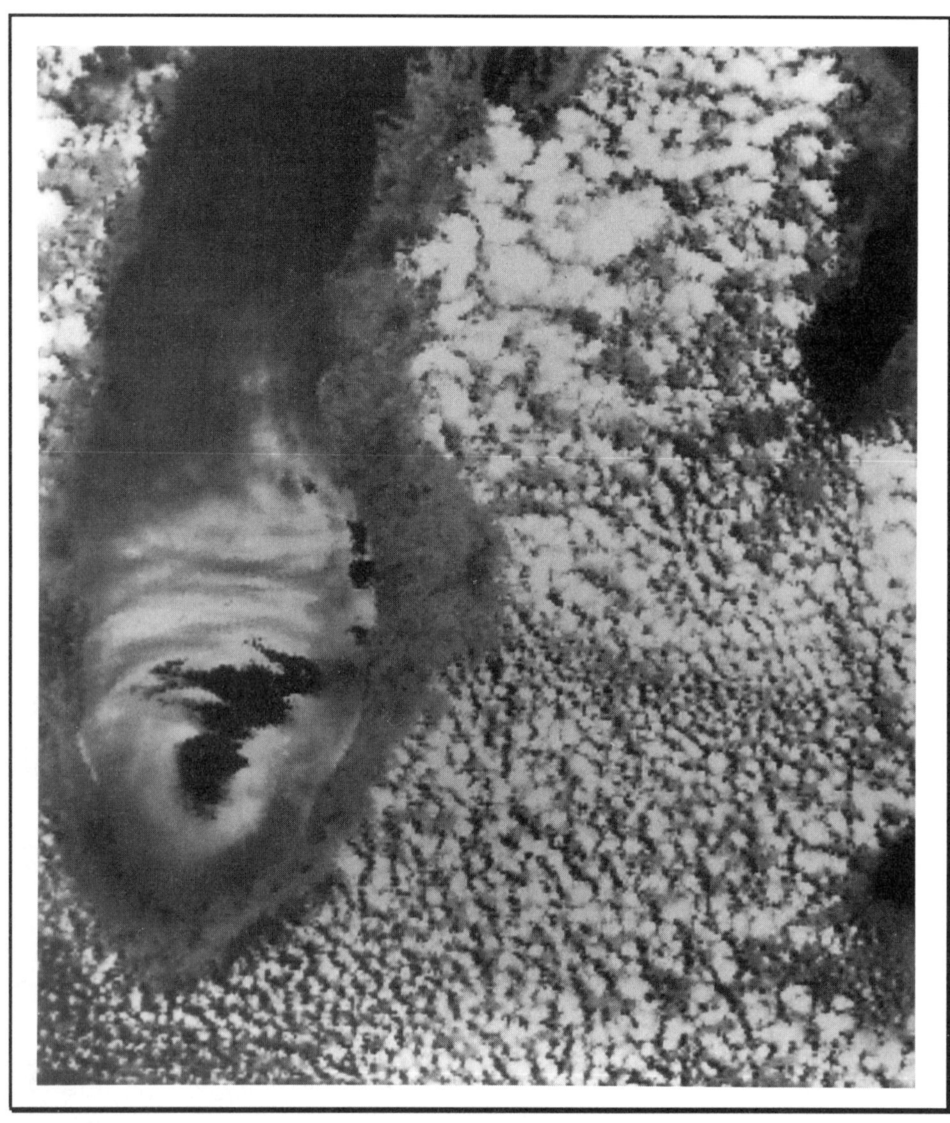

Figure 3.7. Sunglint on Lake Michigan (HRPT VIS, May 28, 1992)

The location of a sunglint depends on the location of the satellite subpoint (the point directly below the spacecraft) and the **solar subpoint** (the point where the sun's most direct rays strike the Earth). A sunglint will always appear between these two points if conditions are favorable. With geostationary satellites, the satellite subpoint remains approximately in the same position year round, while the solar subpoint continually changes. This means that sunglint in GOES imagery will move from east to west each day and migrate between 11.75° N and 11.75° S with the seasons. The shape of a sunglint on GOES imagery will also change, being circular when the subpoints are very close, and oval when they are farther apart.

Sunglint in imagery from the NOAA polar orbiters is shaped differently because the sun-satellite geometry is changing in a different manner; however, it also migrates with the seasonal and daily cycles. In NOAA polar orbiter imagery, sunglint often appears as an elongated region that is oriented approximately north-south, parallel to the path of the satellite. The sunglint appears to the east of the satellite subpoint track on local morning passes and to the west of the subpoint track during local afternoon passes.

Characteristics of infrared imagery

The IR sensors on board the polar orbiting and geostationary satellites measure the amount of infrared energy emitted by the Earth and the atmosphere. Because the amount of energy emitted depends on the temperature of the surface, IR imagery is essentially a picture of the surface and cloud top temperatures portrayed in black, white, or gray shades. This information can be used to observe thermal properties of the Earth and the atmosphere. In conventional IR imagery, colder areas appear as white or light gray tones and warm areas appear black or dark gray.

On most display systems, the gray scale of an IR image is composed of 256 gray shades ranging from white (coolest temperatures) to black (warmest temperatures). The data correlates temperature with gray shade in a simple linear relationship, shown in figure 3.8. Figure 3.9 is an example of a typical IR image. In this image, the highest (and therefore coldest) cloud tops appear white. Lower clouds appear as lighter shades of gray, and warmer land and water surfaces appear as darker shades of gray.

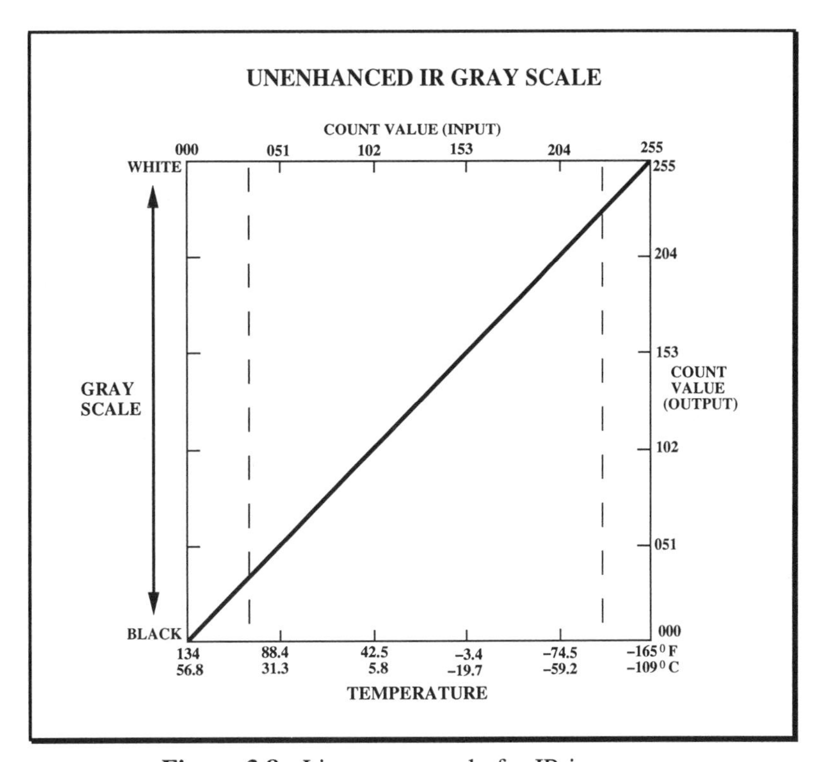

Figure 3.8. Linear gray scale for IR imagery

In IR imagery, land-water contrast is usually fairly sharp. However, during a 24-hour cycle, land heats up and cools off much more rapidly than nearby water. In the early mornings, the land and the water are usually closer in temperature. IR images from this time of day can have poor land-water contrast. As the land heats up, however, it appears darker in IR imagery and the contrast with water features usually improves. In figure 3.9, taken at night when the temperatures of the land and water surfaces are similar, the land-water contrast is poor. Figure 3.10 is an image that was taken in the mid-morning on the same day. The land-water contrast has improved as the land heated up during daylight hours. Sometimes, the land will cool so rapidly at night that the water will appear darker than the land. Seasonal temperature changes will also affect the land-water contrast in an IR image. During warm months, the land will usually appear darker (warmer) than water features; however, in winter months the water features may appear darker than the land

Figure 3.9. Nighttime IR image (GOES IR, 0931 UTC, June 2, 1993)

Figure 3.10. Mid-morning IR image (GOES IR, 1631 UTC, June 2, 1993)

features. At other times throughout the year, land and water temperature may be very similar, resulting in poor contrast.

Features in IR imagery stand out best when the temperature difference between the feature and its background is at a maximum. When two features in an IR satellite image have similar temperatures, it can be very difficult to distinguish between them. For example, low clouds (stratus and fog) are often difficult to distinguish from the surface of the Earth in IR imagery because their temperatures are very close to those of the Earth's surface. When this happens, the contrast between the clouds and the land is very poor and the clouds may not be detected. Snow can also be difficult to identify in IR images because of its similarity in temperature to the clouds and the surrounding land. One way to overcome this limitation of IR imagery is to compare both IR and VIS images of an area taken at approximately the same time. A cloud layer or snow line that is difficult to locate

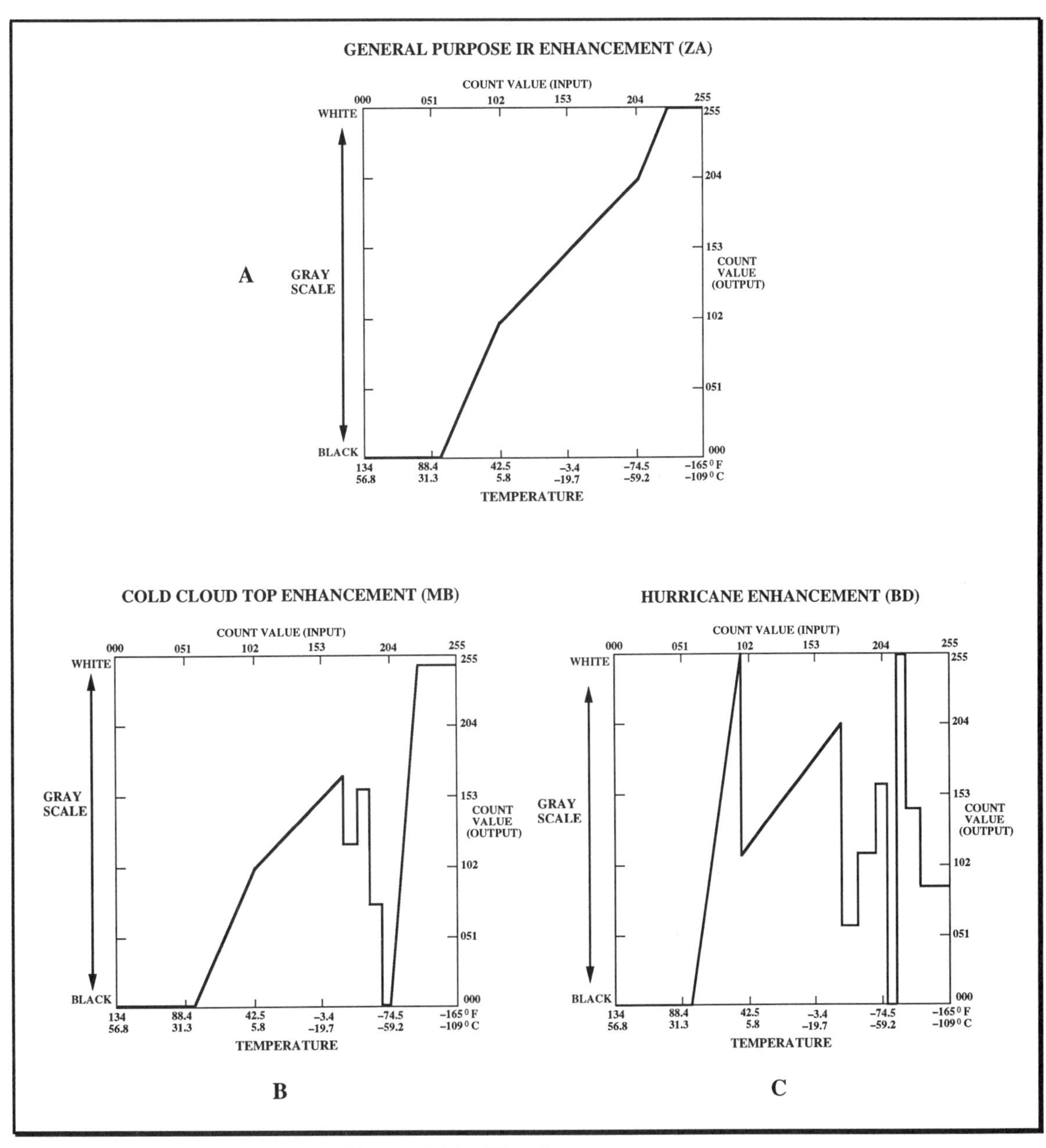

Figure 3.11. IR enhancement curves

in an IR image can often be easily detected in a VIS image of the same area. Alternatively, contrast in IR imagery can be improved by the use of enhancement techniques for increasing contrast between certain features.

Enhancement of IR imagery

In the standard temperature–gray scale relationship shown in figure 3.8, many of the 256 gray shades are wasted in the very warm and very cold temperature ranges, since the temperatures of meteorological significance rarely exceed 40° C (104° F) or fall below –80° C (–112° F). A solution to this is to make all temperatures above 40° C black and all temperatures below –80° C white. In this manner, the 256 shades of gray are used over a smaller temperature range. This improves the contrast in the temperature range that is more significant for weather studies, allowing smaller differences in temperature to be detected in the image. This **enhancement**, illustrated by the graph in figure 3.11A, is called the **ZA enhancement curve** and is used as a general-purpose infrared enhancement technique.

Figure 3.12 is an unenhanced IR image of a thunderstorm on the gulf coast of Texas. Figure 3.13 is an IR image of the same storm, enhanced with the ZA enhancement curve. In the unenhanced image, it is difficult to distinguish the colder cloud tops associated with the main portion of the storm from the warmer clouds. In the enhanced image, the contrast between the coldest cloud tops and the warmer clouds is improved, especially along the western edge of the storm. This aids the interpreter in identifying the exact location of the main part of the thunderstorm.

Other enhancement curves can be used to highlight areas of particular interest. When thunderstorms are studied, IR images can be enhanced to show where the strongest rainfall is occurring. In the **MB enhancement curve**, the gray scale is altered to make the very cold, high, overshooting cumulonimbus or thunderstorm cloud tops stand out (figure 3.11B). The effect is a contoured pattern in the cloud top that highlights areas of intense and/or potentially severe weather. Figure 3.14, an IR image of the same storms shown in figures 3.12 and 3.13, has been enhanced using the MB enhancement curve. The tallest convection and often the heaviest rainfall is found where the cloud tops exhibit the highly contoured black and white pattern.

Hurricanes can be studied in a similar manner using the **BD enhancement curve** (figure 3.11C). This curve highlights certain temperatures in the **eye** and **eye wall** of the storm system known to be related to the intensity of the hurricane. Figure 3.15 is an IR image of Hurricane Andrew that has been enhanced using the MB enhancement curve. Notice the lack of contrast in the clouds immediately surrounding the eye of the hurricane. In figure 3.16, the image has been enhanced with the BD enhancement curve to show the small temperature differences that occur between the eye of the hurricane and its surrounding eye wall.

Water vapor imagery

As the Earth and the atmosphere emit energy, specific wavelengths are absorbed by the atmosphere, especially by clouds and suspended water vapor. At other wavelengths, the energy is not absorbed and is transmitted through the atmosphere. Most IR sensors on meteorological satellites take advantage of the infrared bands that are transmitted through the atmosphere. This allows accurate measurements of the temperatures of the Earth and cloud tops to be made. Some satellite sensors, however, study radiation at wavelengths

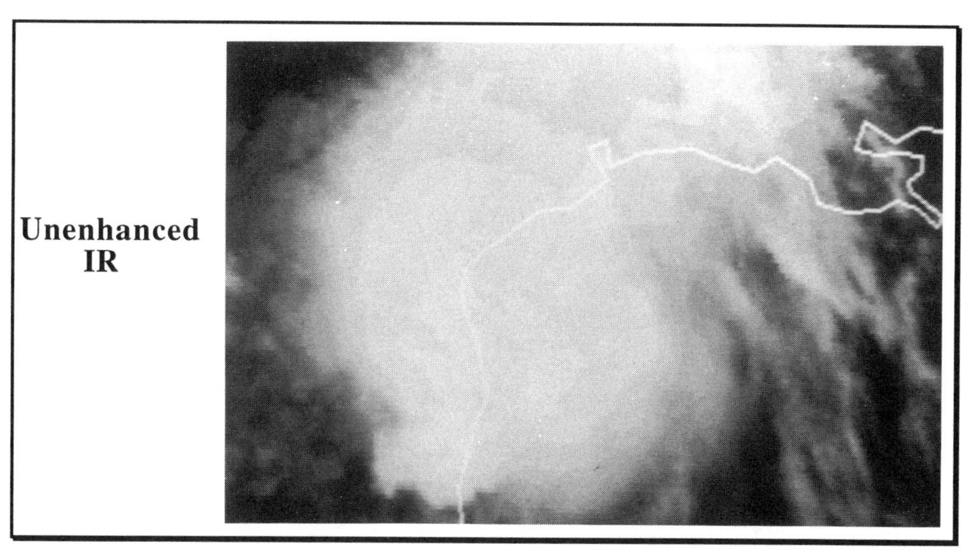

Figure 3.12. IR image of a Texas thunderstorm (GOES IR, October 8, 1994)

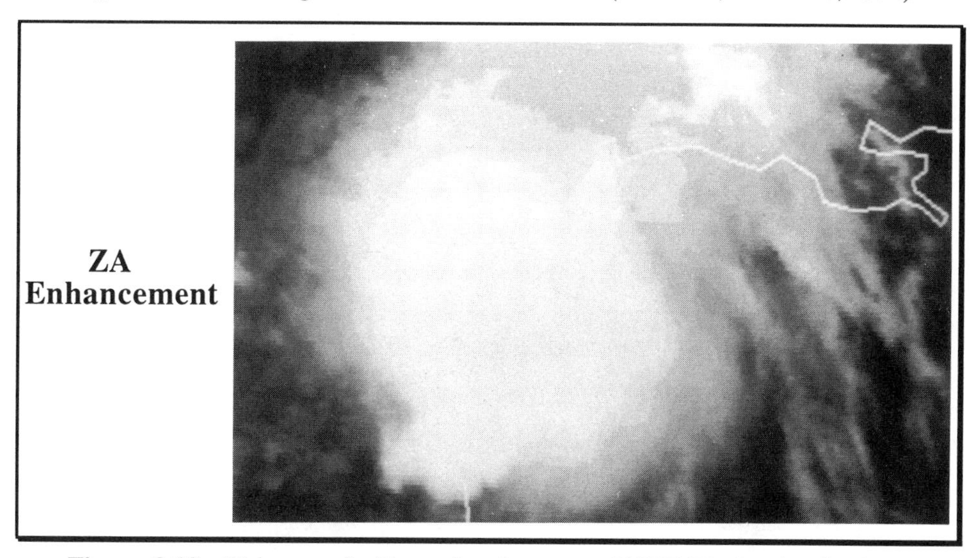

Figure 3.13. IR image of a Texas thunderstorm (GOES IR, October 8, 1994)

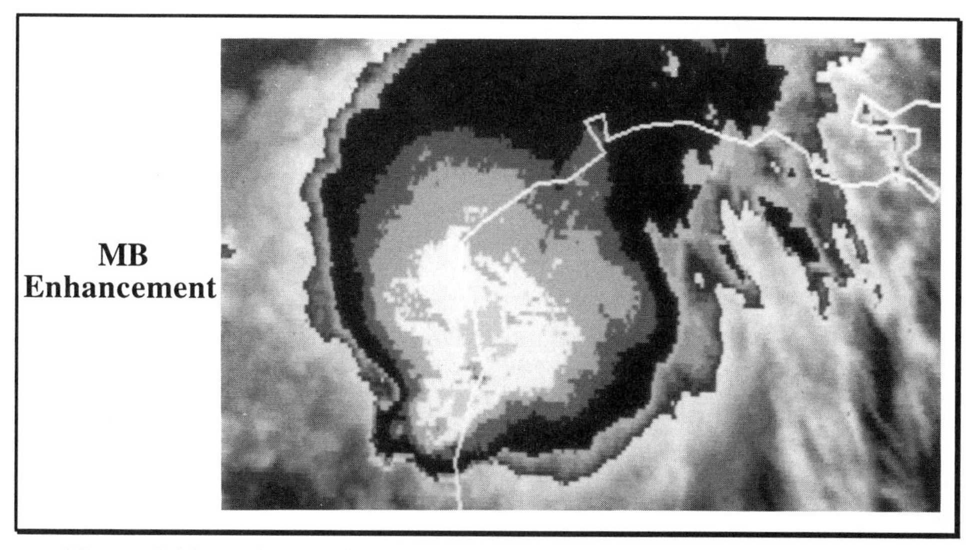

Figure 3.14. IR image of a Texas thunderstorm (GOES IR, October 8, 1994)

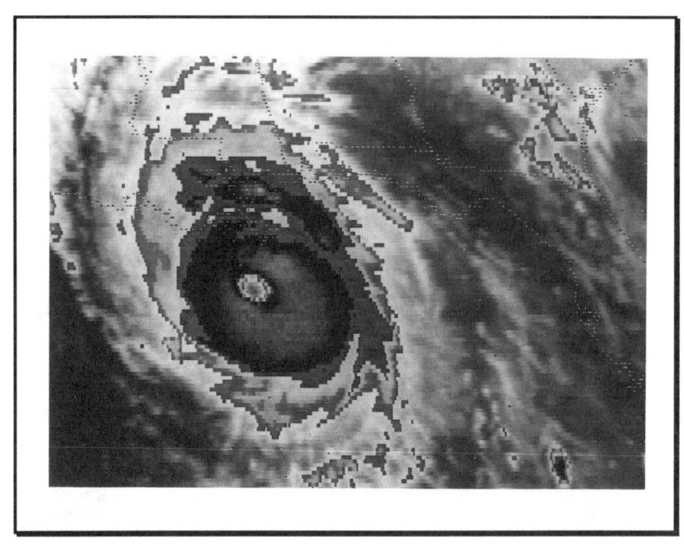

Figure 3.15. MB enhanced IR image of Hurricane Andrew (GOES IR, August 25, 1992)

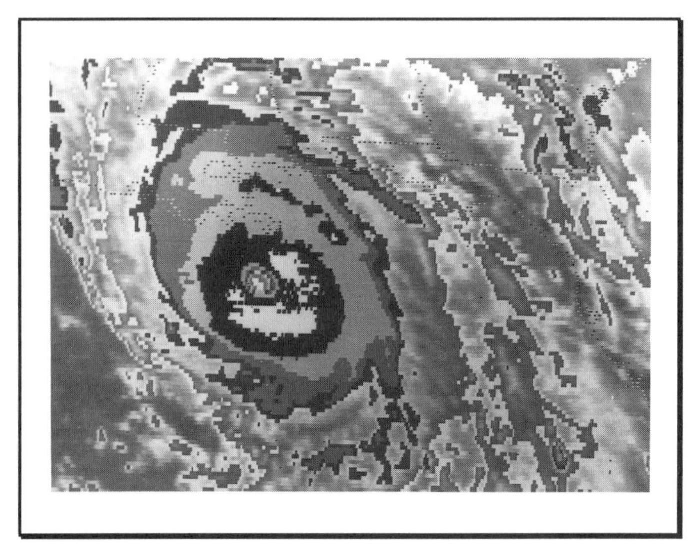

Figure 3.16. BD enhanced IR image of Hurricane Andrew (GOES IR, August 25, 1992)

that are readily absorbed by the atmosphere. Studying the IR energy at these wavelengths allows atmospheric gas concentrations to be studied without interference from surface features.

Two widely used applications of this concept are channel 9 (7.3 microns) and channel 10 (6.7 microns) on the GOES VISSR sensors. Energy emitted at these particular wavelengths is readily absorbed by water vapor in the atmosphere. Images that are taken in these channels are used to locate large concentrations of water vapor and water vapor gradients in the middle and upper **troposphere** (this is the lowest layer of the atmosphere and the location of the most significant weather). Figure 3.17 is a water vapor image from GOES 7. The darker regions are areas where very little water vapor exists in the middle and upper troposphere, and the lighter regions are very moist. Water vapor imagery has become a very valuable tool for weather analysis and prediction in the last ten years because water vapor imagery shows moisture in the atmosphere, not just cloud patterns. This

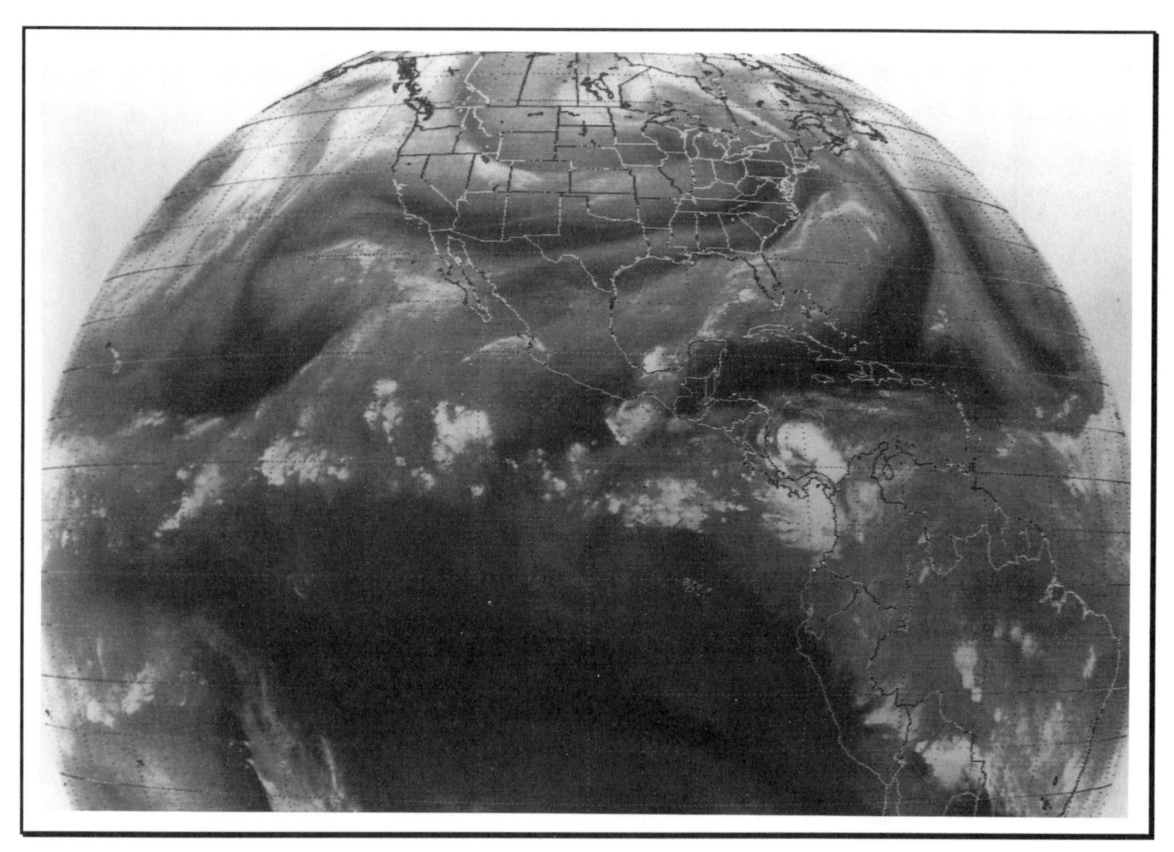

Figure 3.17. GOES water vapor image (GOES WV, November 2, 1991)

allows meteorologists to observe large-scale circulation patterns even when clouds are not present.

Errors in satellite imagery

When using satellite imagery to analyze the current state of the atmosphere, it is important to understand that the images that you see are not always accurate. Errors in satellite imagery can be caused by many factors, including limitations in the sensors, the angle between a sensor and the feature it is viewing, and parallax shifting. These errors often cause a certain object in a satellite image to appear to be displaced or to show brightness values that are not accurate. This is especially important to consider during severe storm and hurricane forecasting, in which a placement error of only a few kilometers can lead to incorrect interpretations and forecasts.

Cloud displacement

When interpreting GOES imagery it is important to remember that the satellite is viewing a curved Earth-atmosphere surface. Due to the angle of the observation, a very tall cloud might appear to be over a certain point even though it is actually several kilometers equatorward. This phenomenon is called **cloud displacement** and is illustrated in figure 3.18. From a different angle, the cloud might appear to be in a different location. The apparent shift in an object's position as a result of the viewing angle is called **parallax**. This is a very important problem, since a severe storm could be misinterpreted as being several kilometers from its actual location.

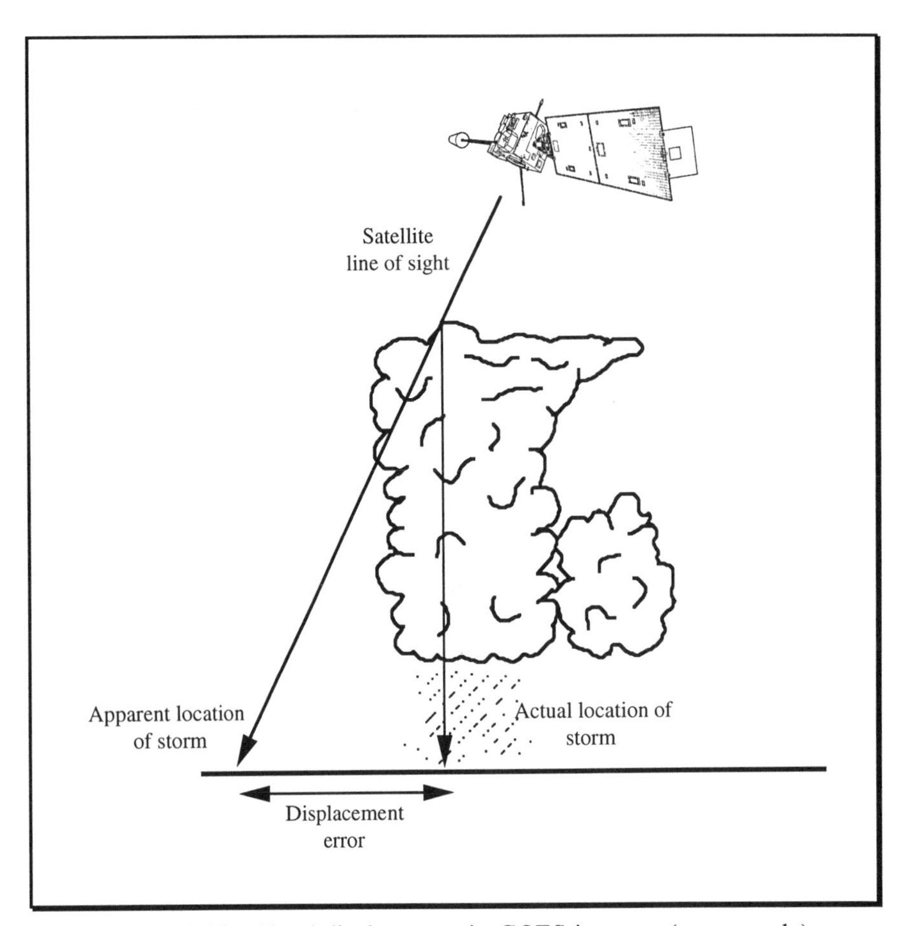

Figure 3.18. Cloud displacement in GOES imagery (not to scale)

A correction table based on trigonometric and geometric concepts, such as the one shown in figure 3.19, can be used to place the clouds where they belong. This correction chart can be used for any GOES image. To use this chart, align the central meridian on the chart with the longitude of the satellite subpoint. Locate the apparent location of the cloud on the satellite image; the true location of the cloud can be determined using the arrows and distances on the concentric rings on the chart. This chart assumes that the cloud tops have a height of 40,000 feet. Adjustments for other cloud heights are shown on the chart.

Limb darkening

On a satellite image, the data are most reliable and accurate directly below the satellite, at the satellite subpoint. Toward the outer edge of the image, the details become less accurate due to the scanning angle of the sensor. The radiation sensed at these locations has to travel a longer distance through the atmosphere, and more of this radiation is absorbed or scattered as it passes through the atmosphere. This affects brightness values, making points on Earth appear darker or lighter than they actually are. Such distortions in a satellite image are called **limb darkening**. In an IR image, for example, the cloud top temperature measurements will be most inaccurate on the outer edges of the image, often appearing much colder than they actually are.

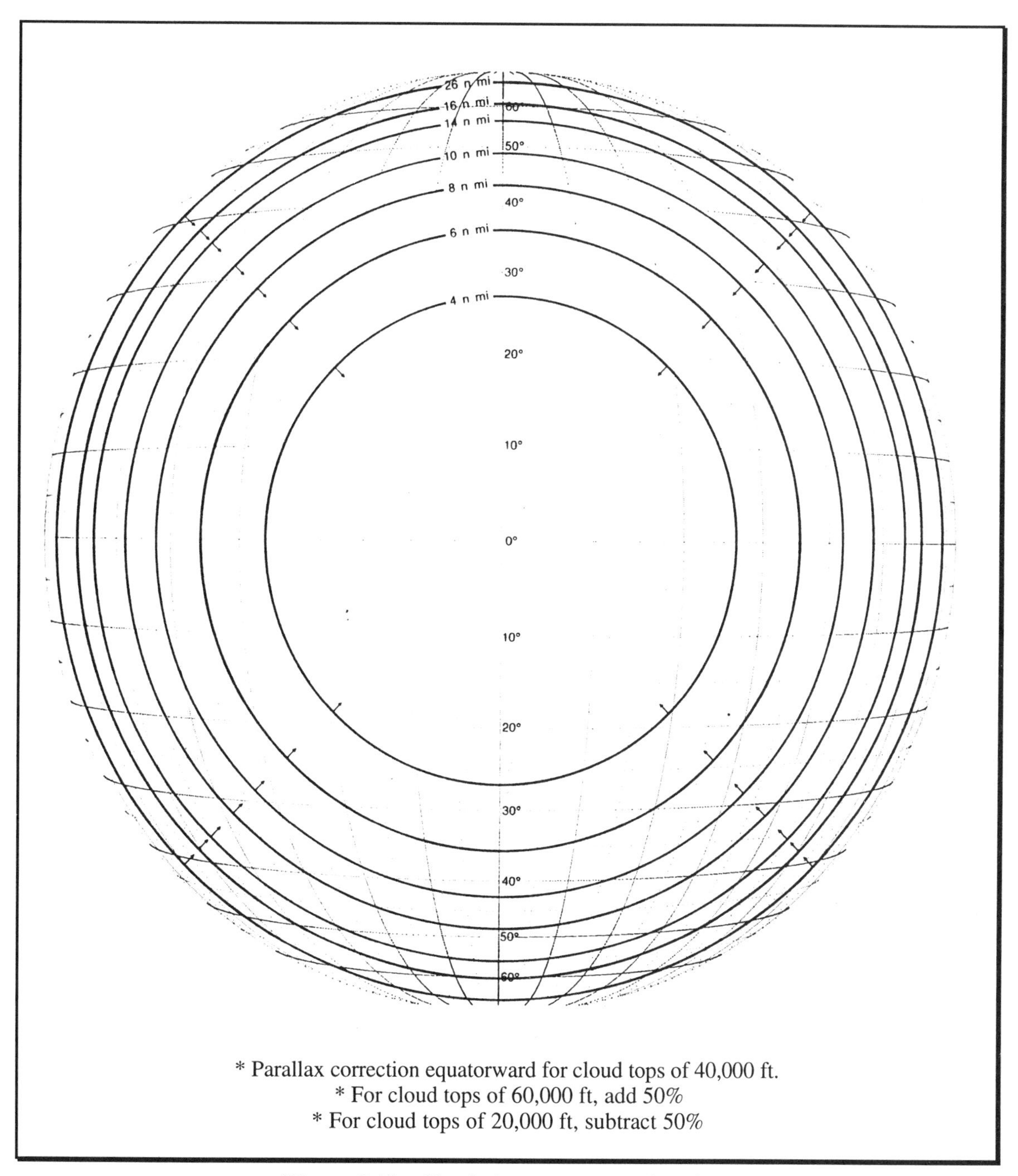

Figure 3.19. Cloud placement correction chart

Instrument response delay time

A final error that can be found in satellite imagery is known as **instrument response delay time.** As a satellite sensor scans across a portion of the Earth, it takes a certain amount of time to respond to changes in the brightness of the surface. When sharp changes in brightness occur, there is often a delay during which the sensor must adjust. This idea can be understood by considering an **anemometer**, which is a device that measures wind speed. As wind strikes the anemometer, the instrument takes time to respond. Initially, it spins slowly and reports a slower wind speed than is accurate. After a short period of time, it builds up to the proper speed to accurately measure the wind. As the gust of wind dies down, however, the anemometer takes a certain amount of time to slow down. During this period, the instrument reports the winds as being faster than they actually are. To a forecaster reading the output from the anemometer, it would seem as though the wind gust occurred a short time after it actually happened. This error occurs because a certain amount of delay takes place as the anemometer adjusted to an increase in wind speed.

Satellite sensors exhibit the same type of delayed response to sharp pixel-to-pixel changes in brightness. For example, around a thunderstorm a very sharp temperature gradient occurs. A thunderstorm cloud top can be extremely cold, while the ground temperature next to the cloud may be very warm. As an IR satellite sensor scans from west to east across the warm ground, it reports accurate temperatures. However, when it encounters the sharp drop in temperature associated with the cloud top, there is a delay while the sensor responds. During this delay, the temperatures that the sensor reports are warmer than the actual cloud top. As the sensor scans past the cold cloud top, it senses the warm Earth. There is another delay as it responds to a sharp increase in temperature. The temperatures reported during this time are colder than they should be. The result of this error is the apparent eastward displacement of the cold cloud tops in the image. This displacement is caused by the delay in the sensor's response to the sharp temperature gradient.

CHAPTER 4

Geographical Applications of Satellite Imagery

Locating geographical features

One of the most exciting aspects of satellite imagery is the variety of geographical features that can be seen on a satellite image. In a relatively cloud-free image, or when only thin clouds cover an area, it is possible to see a wide range of landforms, including lakes, rivers, cities, mountains, forests, snow cover, and ice. Coastlines, islands, and various ocean features can also be seen. Seeing these features from space offers a new perspective about the world. Becoming familiar with these features gives one a deeper appreciation and understanding of the planet. Additionally, since land features often influence weather patterns, it is very useful to know the geography of an area when using satellite imagery to assess weather conditions and make forecasts.

One of the first activities that should be practiced when learning how to interpret satellite imagery is locating geographical features. An essential tool for this activity is an atlas, preferably one with both geographical and political maps. Any satellite image can be used, although some images will show more features than others. The resolution of the image will determine what features are discernible. For example, if a satellite image has a resolution of 7 km (4 miles), objects smaller than 7 km will not show up in the image. As the resolution improves, the number of features visible in an image increases. In an image with a resolution of 1 km, more features will be seen than in an image with a resolution of 7 km.

In VIS imagery, the relative brightness of land surfaces is expressed as the albedo, or the percentage of visible radiation the surface reflects. Albedo is determined by surface characteristics such as soil composition, type of vegetation, ground cover, and the presence or absence of snow and ice (refer to table 1.1 for albedos of various Earth surfaces). The most highly reflective surfaces will appear bright in VIS imagery, while objects with low reflectivity will be dark. In IR imagery, the relative brightness of a feature is determined by its surface temperature. Differences in surface temperature make it possible to locate geographical features in IR imagery. Objects with warmer surface temperatures will appear darker in an IR image than will cooler objects.

Coastal features

Coastlines, bays, peninsulas, and small coastal islands are all very easy to identify in VIS and IR satellite imagery. The contrast between land and water is usually sharp in both types of imagery, allowing these features to stand out clearly. Figure 4.1 is a VIS HRPT image of the eastern United States. The Chesapeake Bay, Delaware Bay, Long Island, and Cape Cod are easily identified in this satellite image. If you examine the image closely, the long, thin barrier islands off the coasts of New Jersey, Maryland, Virginia, and North Carolina are visible, as are many small islands in the Chesapeake Bay and along the New England and Georgia coasts. In IR imagery, coastal features stand out best when the temperature difference between land and water is the greatest. This can be seen by comparing figures 3.9 and 3.10, where Baja California shows up best later in the day when the land is warmer.

Inland water features

Inland bodies of water are usually easy to identify, especially in VIS imagery. In satellite imagery it often helps to use several large landmarks to aid in locating smaller features that are more difficult to detect. The Great Lakes are a very striking feature in an image of the eastern United States. Other commonly used landmarks include Lake

Figure 4.1. The east coast of the United States (HRPT VIS, June 12, 1988)

Okeechobee (Florida), the Finger Lakes (upstate New York), the Chesapeake Bay, and the St. Lawrence Seaway. In figure 4.1, a VIS HRPT image, Lake Ontario and Lake Erie are seen in the upper left corner of the image. The Susquehanna River can be followed from the Chesapeake Bay to its headwaters in upper New York State, south of Lake Ontario. The Finger Lakes, formed by glacial scarring during the last ice age, are visible as a series of elongated lakes just south of Lake Ontario.

The low reflectivity of water contrasts with the varying reflectivity of different land surfaces, making rivers and lakes stand out well in VIS imagery. The only limitation is the resolution of the image that is being studied. Many rivers and small lakes will not show up simply because of their small size. Often, however, they can be detected by their influence on the surrounding vegetation. For example, when a river passes through a desert, the vegetated flood plain often will contrast with the desert. Even if the river itself does not show up in an image, the orientation of its flood plain will. Figure 4.2 is an HRPT image of the Middle East. The Nile River is easily seen as it flows northward through the deserts of northern Africa. Its flood plain is very heavily vegetated and appears much darker than the dry, sandy desert soils. The Nile River's delta is also very dark. It can be seen as a wedge-shaped dark region where the Nile flows into the Mediterranean Sea.

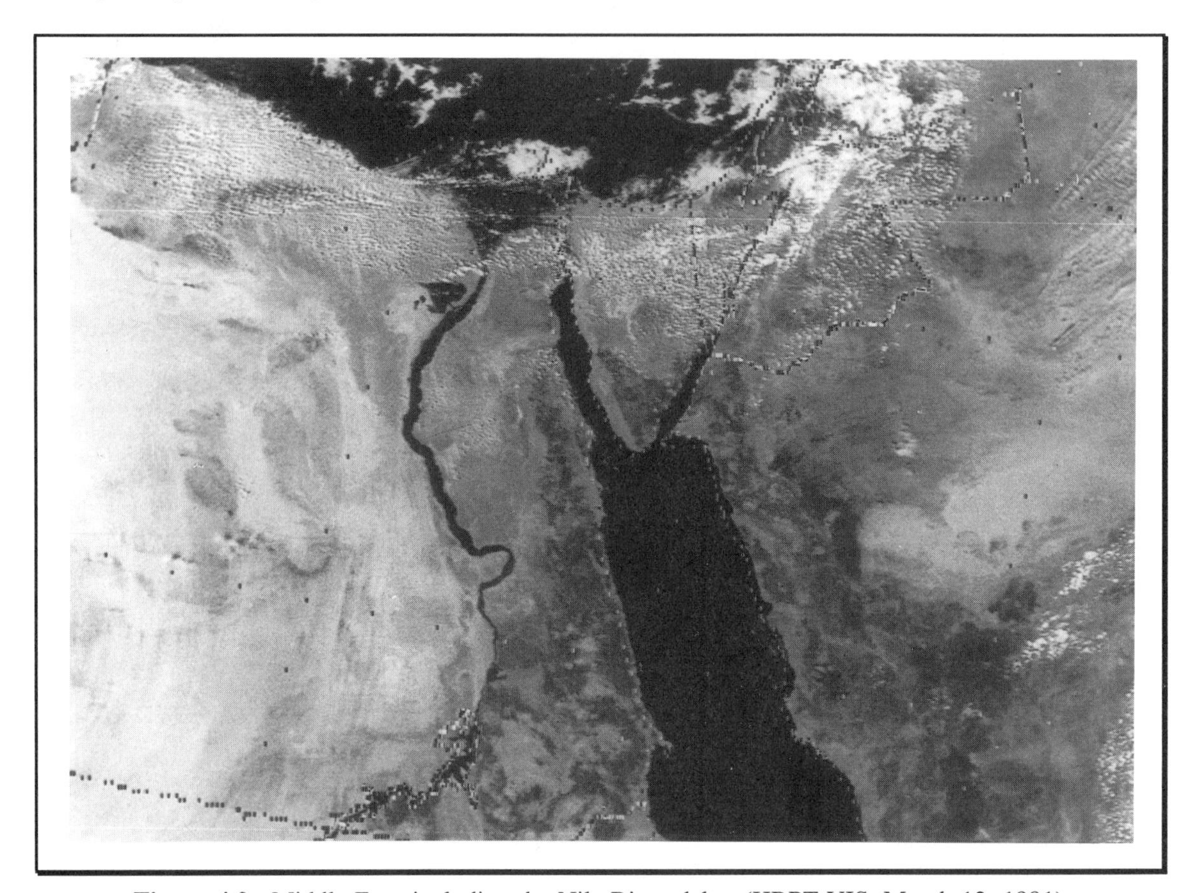

Figure 4.2. Middle East, including the Nile River delta (HRPT VIS, March 12, 1991)

Figure 4.3 is a VIS HRPT image of southeastern Canada. The dark ring-shaped feature north of the St. Lawrence Seaway is Lake Manicouagan (white arrow). This body of water is thought to have been formed as a result of the impact of a large meteorite several thousand years ago. This feature often stands out in satellite imagery because of its unique shape. Take some time with these images to identify other inland water features.

Terrain features

Many terrain features can be seen on satellite images. Heavily vegetated and wooded areas have a very low reflectivity and therefore appear darker in VIS imagery. Mountain ranges are often identified on imagery by the patterns created by wooded slopes (which appear dark in an image) and farmland in the valleys (which appears lighter). Snow-covered mountain peaks can appear white in VIS imagery. Large river flood plains that are relatively treeless often mark out the area along a river that is not wide enough to be seen in the image. Alkali or salt flats and the sparsely vegetated soils of desert regions also appear bright in VIS images due to the very high reflectivity of the light-colored sands and soils.

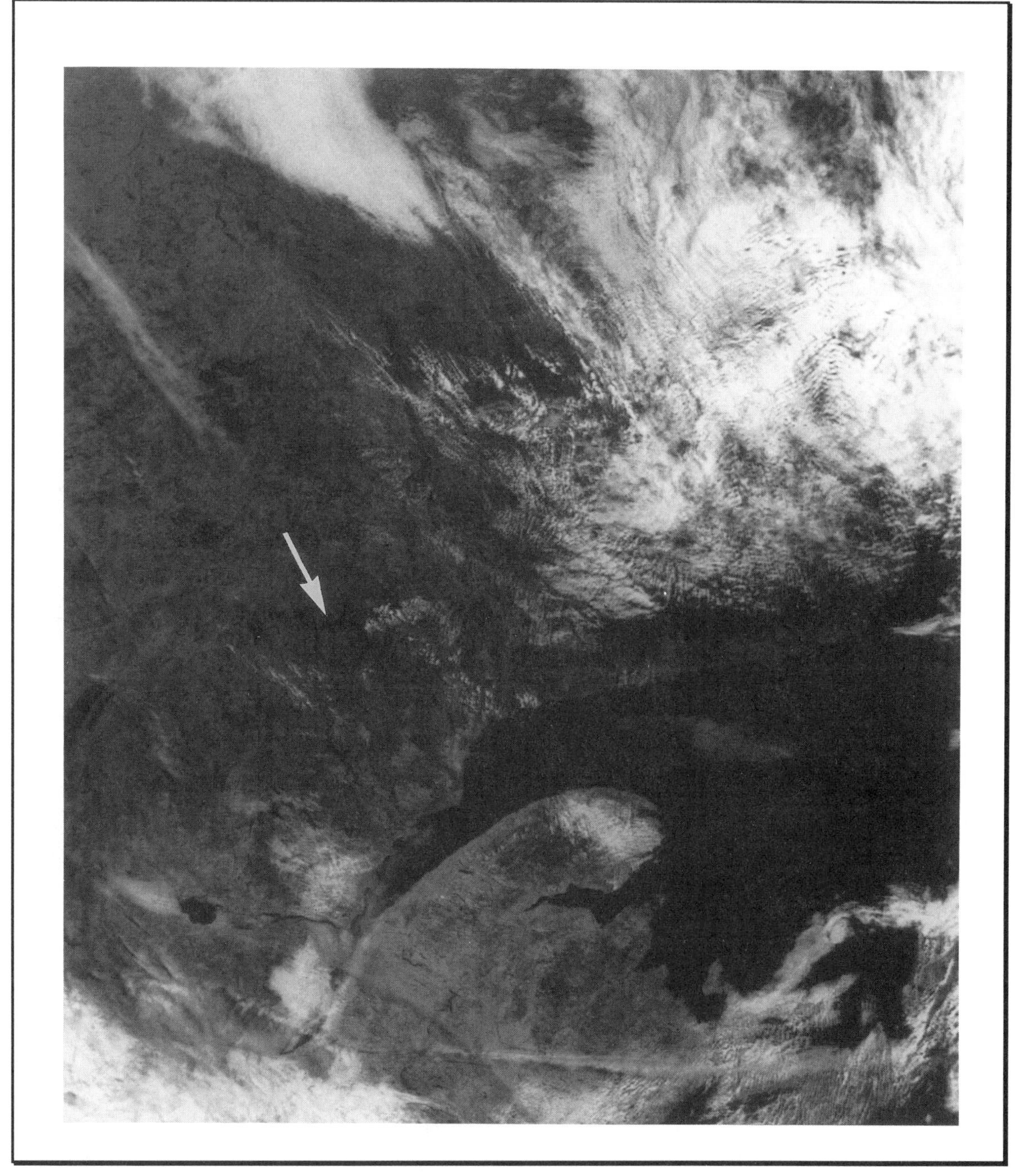

Figure 4.3. Southeastern Canada (HRPT VIS, June 25, 1987)

Figure 4.4 is an HRPT image of the southwestern United States. White Sands National Park, famous for its broad expanses of very light-colored sands, shows up as a bright white spot in southern New Mexico. Compare White Sands to the Sacramento Mountains immediately to the east. These mountains show up as a dark patch due to the forests on the mountain slopes. The Rio Grande is also visible as a dark line running in a north-south direction through central New Mexico. Although this river cuts through much desert, its flood plain is very fertile and vegetated. This gives it a darker tone than the surrounding desert. In Utah, the Great Salt Lake is visible in the northwestern corner of the state. The surrounding lands, especially to the west and the south of the lake, are primarily desert and appear relatively bright in this image. In southwest Utah, the San Rafael Desert appears as

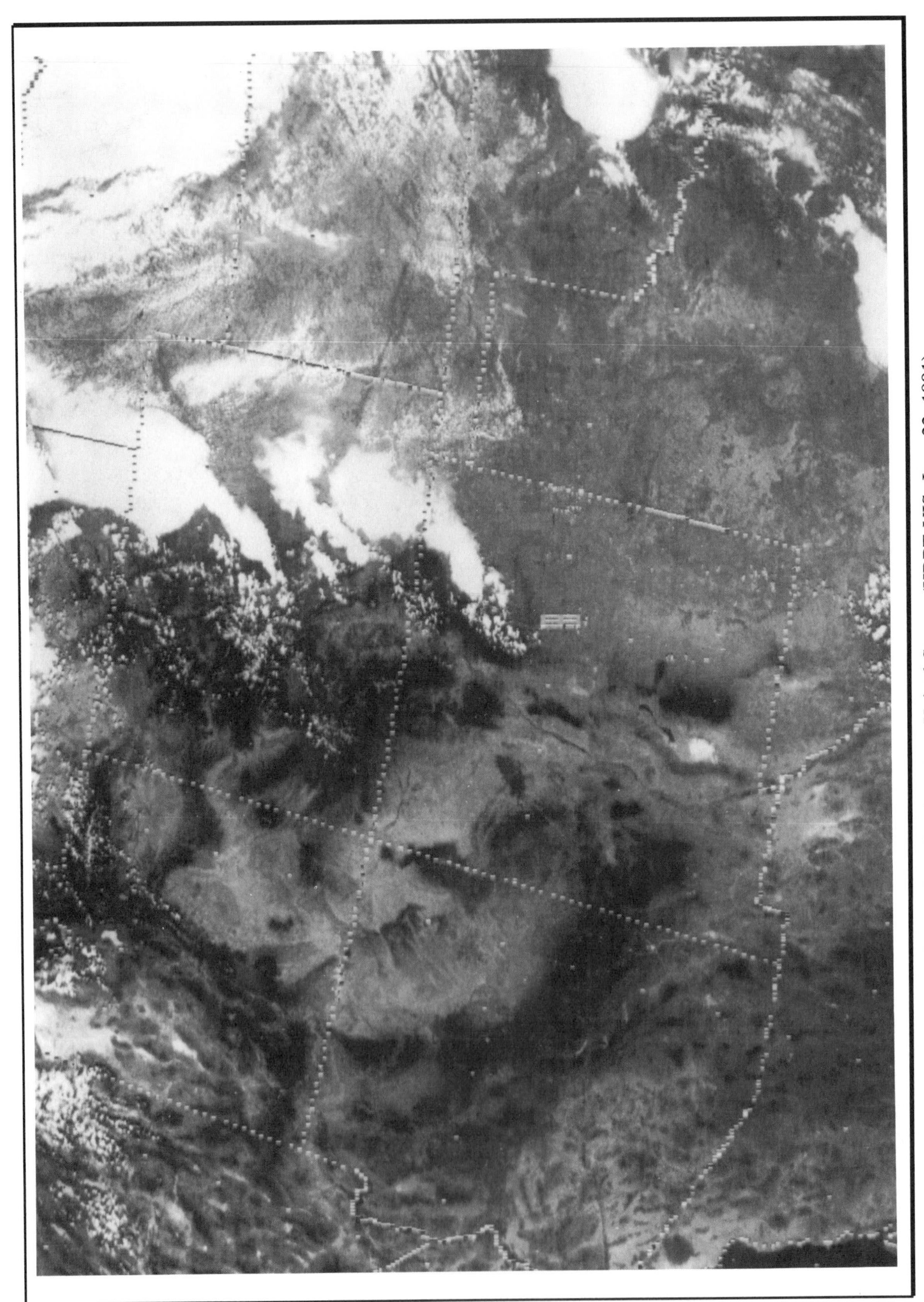

Figure 4.4. Southwestern United States (HRPT VIS, June 22, 1991)

a large bright area, as does the Painted Desert in northeastern Arizona and the Sonora Desert in southwestern Arizona. Compare these deserts to the San Juan Mountains in southwestern Colorado or the broad, flat farmlands of central Texas.

Urban heat islands

A **heat island** is an area characterized by warmer temperatures than the surrounding region. The central portions of cities are normally warmer than the surrounding countryside; therefore, cities are often called **urban heat islands**. One reason cities are warmer than the surrounding regions is that the structural materials in a city (concrete, asphalt, brick) become hotter during the daytime than do nearby areas where solar energy is used in evaporation and transpiration from vegetative cover. At night, the cities retain heat longer, thus decreasing the cooling rate overnight. Pollution and haze in the cities act as a **thermal blanket** (they hold in the heat as a blanket would), further slowing the cooling process both day and night. The temperature difference between the city and the countryside can often be 5° or more. The magnitude of an urban heat island depends on the size of the urban area and on atmospheric conditions such as the presence of an inversion layer, the type of winds present, and the geography of the area.

In IR satellite imagery with little to no cloud cover, urban heat islands show up as areas darker than the surrounding regions. Temperatures in each city can be measured and compared to the surrounding areas. Most major cities in a satellite image will show this effect. In figure 4.5, an IR image of the northeastern United States, Washington D.C., Baltimore, Philadelphia, New York, and Boston all show up clearly as urban heat islands.

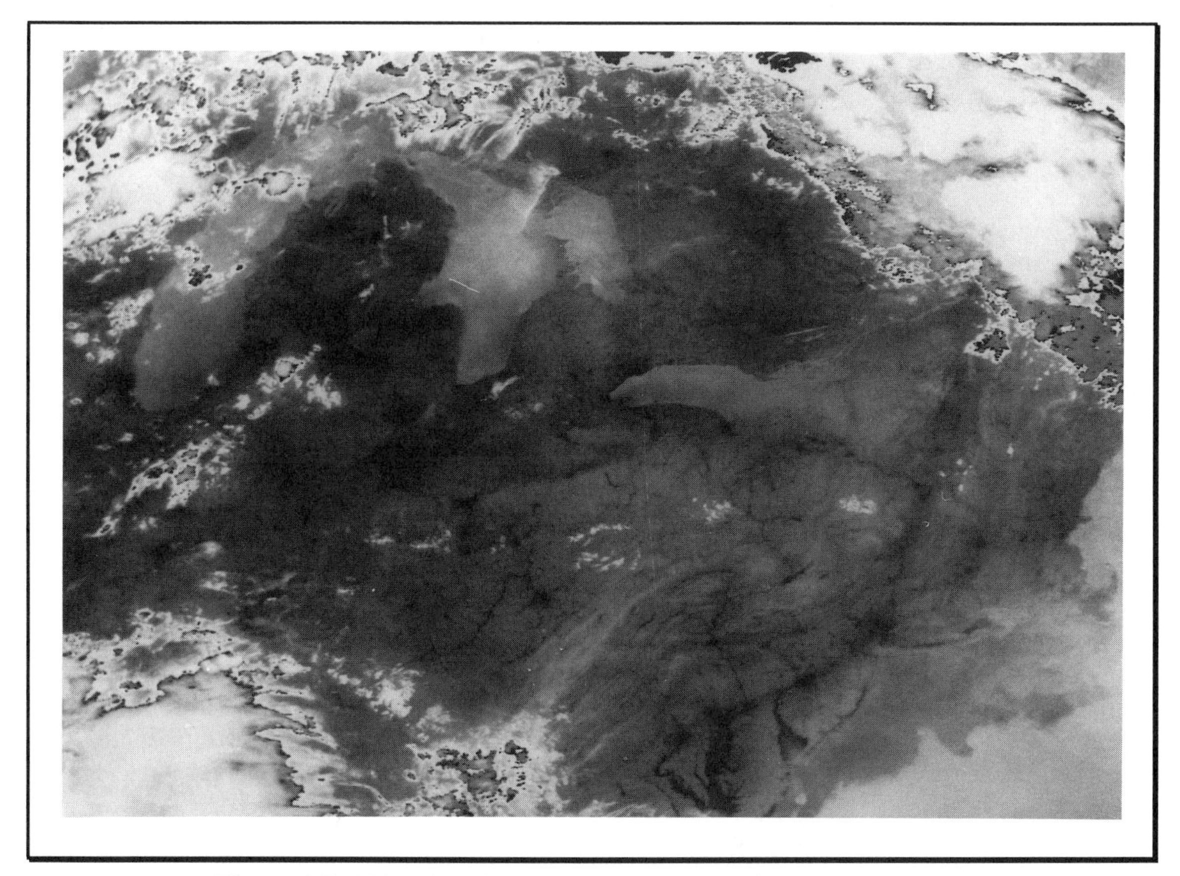

Figure 4.5. Urban heat islands (HRPT, enhanced IR, June 25, 1987)

These cities form a nearly continuous dark line up the east coast of the United States. Many other cities, including Chicago, Detroit, Buffalo, Rochester, Columbus, and Toronto, show up as dark patches on this image.

Understanding urban heat islands is important because the warmer temperatures affect the climate of the city. This can lead to many energy-related and biological implications. For example, a home in the city may use less energy in the winter for heating than a comparable home in the suburbs. The reverse may happen in the summer due to increased air conditioning use. Plants in the city may respond to the differences in temperature by blooming sooner in the spring; leaves on deciduous trees may change colors later in the fall. In addition, warmer city temperatures may allow for a city rain event while nearby suburban or rural areas receive snow and ice. Decreased air quality and visibility can result should the heat island act like a miniature low-pressure system, drawing **country breezes** into the center of the city. This process carries more pollution toward the central part of the city and further degrades the quality of the city air.

Detecting snow cover

Snow cover on land surfaces plays an important role in the climate and weather patterns across the Earth. Fresh dry snow has a very high albedo; thus it reflects almost all of the sunlight that strikes it. Since solar heating is the mechanism that drives most atmospheric processes, snow cover has a direct influence on the Earth's weather patterns. By reflecting away incident radiation, snow cover limits the amount of heating that will occur in the atmosphere. This can affect surface and air temperature, atmospheric circulation patterns, storm tracks, cloudiness, evaporation, and precipitation. Additionally, snow that is resting on the surface of the Earth represents a large amount of stored fresh water. This affects soil moisture and fresh water availability. When this snow melts, it can provide a vital source of fresh water to dry regions. It can also lead to heavy flooding if it melts very quickly.

Since 1966, NOAA has produced Northern Hemisphere Weekly Snow and Ice Cover Charts (figure 4.6). NOAA meteorologists use VIS satellite data from many satellites to map out the extent of snow and ice cover in the Northern Hemisphere. Then they examine surface data received during the current week, compare the snow and ice cover with that on the previous week's chart, and adjust the position of the **snow line** (the line that marks the most equatorward occurrences of snow). These charts show the areal extent and brightness of the snow but do not show the snow depth. They are used to determine the relationships that may exist between the amount of surface snow cover and other atmospheric phenomena. They are also used in numerical prediction models. This data can also be used to calculate the frequency of snow cover for a given area over a certain period of time or the average amount of land covered by snow during a given period.

Using data from meteorological satellites, you can create snow cover maps for North America. As you collect satellite images, draw in the extent of snow cover on a blank base map of the North American continent. Once you identify the snow in the image, you have located the snow line. The area of the snow cover (in square kilometers) can be estimated with this chart. These data can then be stored and compared with data for different months, years, and locations. Seasonal changes in snow cover can be determined, and yearly snow cover data can be examined for evidence of global climate change.

When producing such a map using satellite imagery, it is important to distinguish snow cover from other features. Snow on the surface appears bright white to light gray in VIS imagery, depending on the nature of the terrain, the vegetation, the age of the snow, and

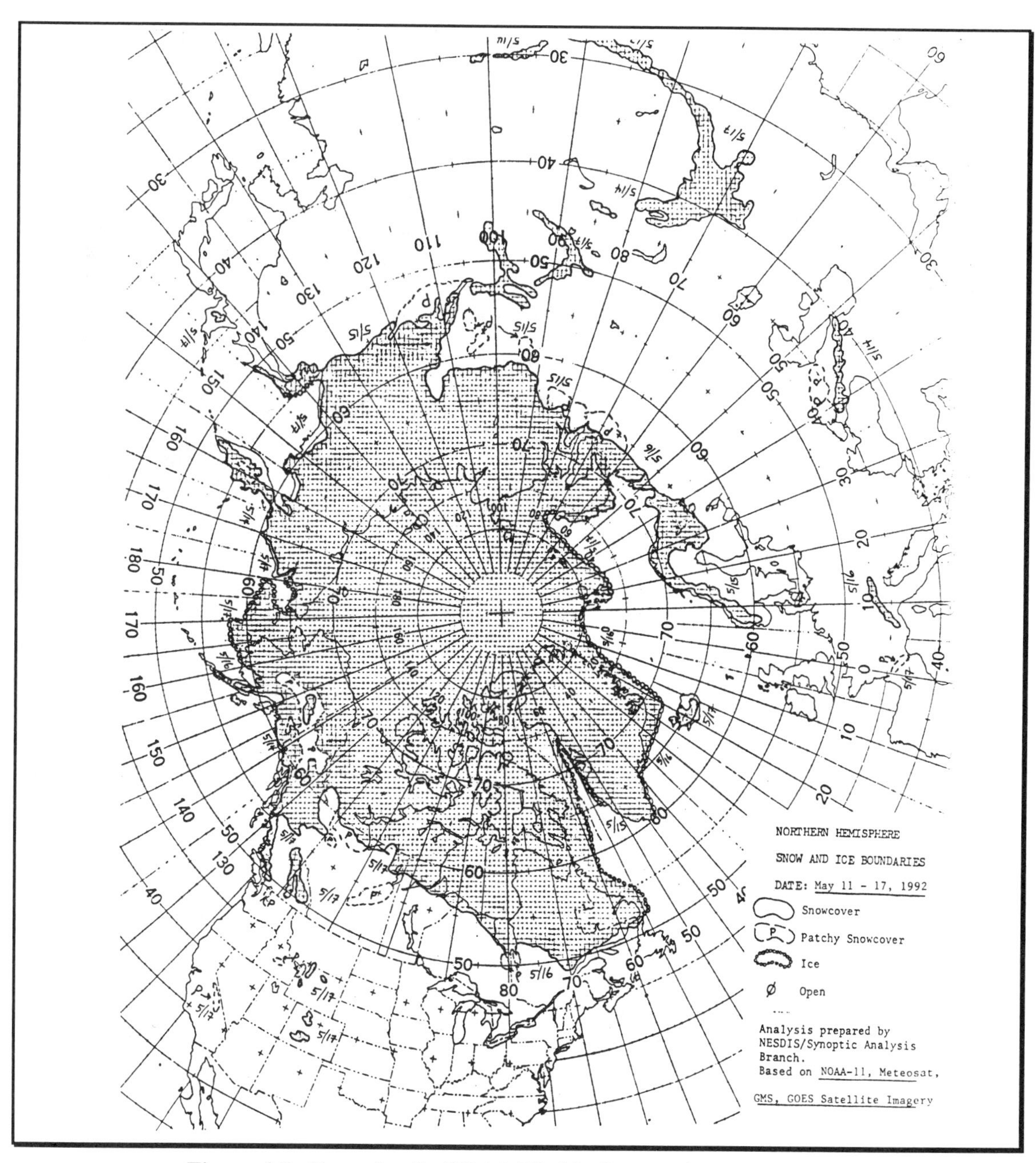

Figure 4.6. Example of a NOAA Weekly Snow and Ice Cover Chart

the illumination angle. Since fresh snow can appear very white and resemble cloud cover, it is important to be able to distinguish between the two. It is helpful to examine a series of image shown in motion; the clouds usually move, while snow does not.

Knowledge of the terrain can also help identify snow. Snow cover ends at unfrozen rivers or lakes, while clouds would likely cover these features. Snow on broad, flat, treeless areas (such as tundra) appears uniformly white, resembling clouds. Without clues such as rivers or lakes, it may not be possible to distinguish between snow and cloud cover. In areas covered by extensive forests, a snow-covered surface appears spotted, with lighter shades in treeless areas, and darker shades in wooded areas where the trees are thick and obscure the snow. Mountain snow can often be easily recognized by the dendritic pattern that forms due to lighter, snow-covered ridges surrounded by darker, tree-filled valleys and/or hillsides. Figure 4.7 is an HRPT image of the far eastern Soviet Union (the Sea of Okhotsk is on the righthand side of the image). The snow-covered mountaintops are easily identified because of their white dendritic pattern. The tree-lined river valleys and mountainsides are not snow-covered and therefore appear dark.

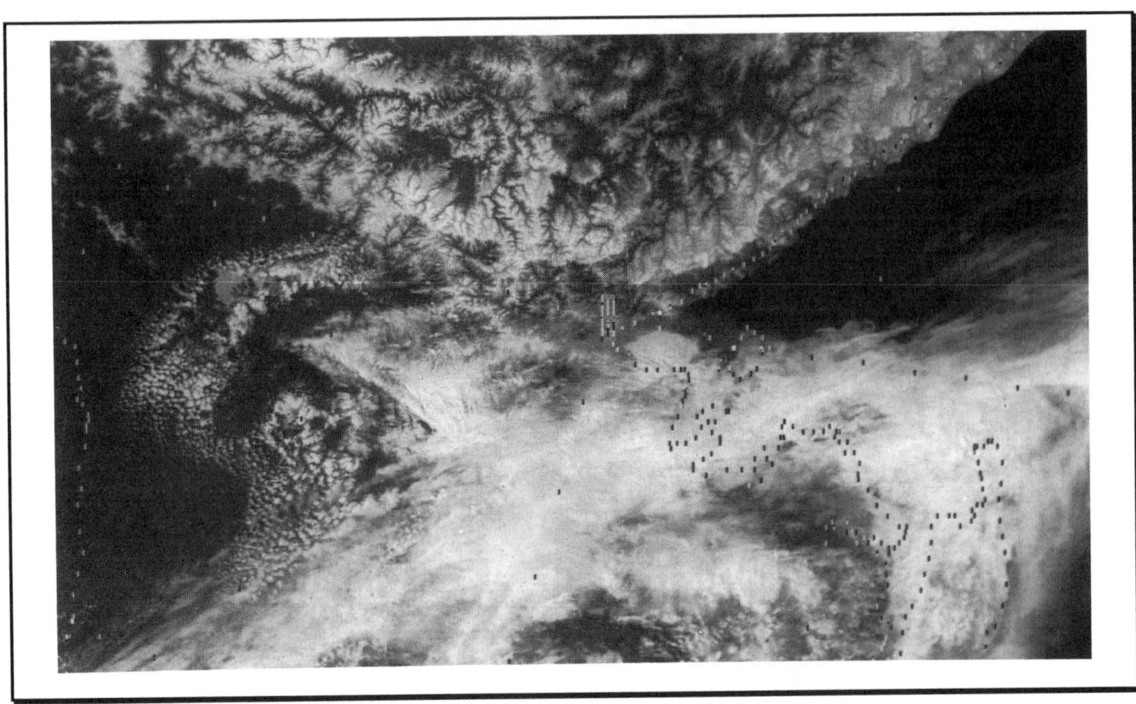

Figure 4.7. Snow-capped mountain peaks in far eastern Russia (HRPT VIS, May 15, 1991)

Figure 4.8 is a GOES VIS image of the central United States following a December snowstorm. The southern boundary of the snow line can be seen most clearly in Kansas where it runs diagonally across the state from southwest to northeast. Fresh snow can be seen covering most of North and South Dakota, Nebraska, Iowa, and Minnesota. The Missouri River can be seen in North and South Dakota, and the Mississippi river can be seen in Minnesota. Since these rivers remain unfrozen, they are not covered with snow and appear darker than the surrounding land. In northern Minnesota, many large lakes appear bright white, since they are frozen and covered with snow. The land surrounding these lakes is heavily forested, and the trees block the snow-covered ground from the satellite's view; therefore, it appears darker than the treeless farmlands to the south. In the extreme western portion of this image, the snow-covered peaks of the Colorado Rocky Mountains can be seen.

Shadows in VIS imagery can also aid in separating cloud cover from snow. In figure 4.8, clouds can be seen casting shadows on the snow-covered surface in southeastern Wyoming and Nebraska. Figure 4.9 is an enhanced IR image of the same scene. The MB enhancement curve highlights the colder cloud tops and helps distinguish them from the snow-covered surface. Notice also that the wider portions of the Missouri River can still be seen in the IR image, since the water is warmer than the surrounding snow-covered land. IR imagery can also be enhanced to highlight the temperatures associated with snow and ice on the Earth's surface in order to locate these conditions in nighttime IR satellite imagery.

Figure 4.10 is an image of the eastern United States following the Blizzard of 1979. This very strong winter storm even had a hurricane-like eye and dumped several feet of

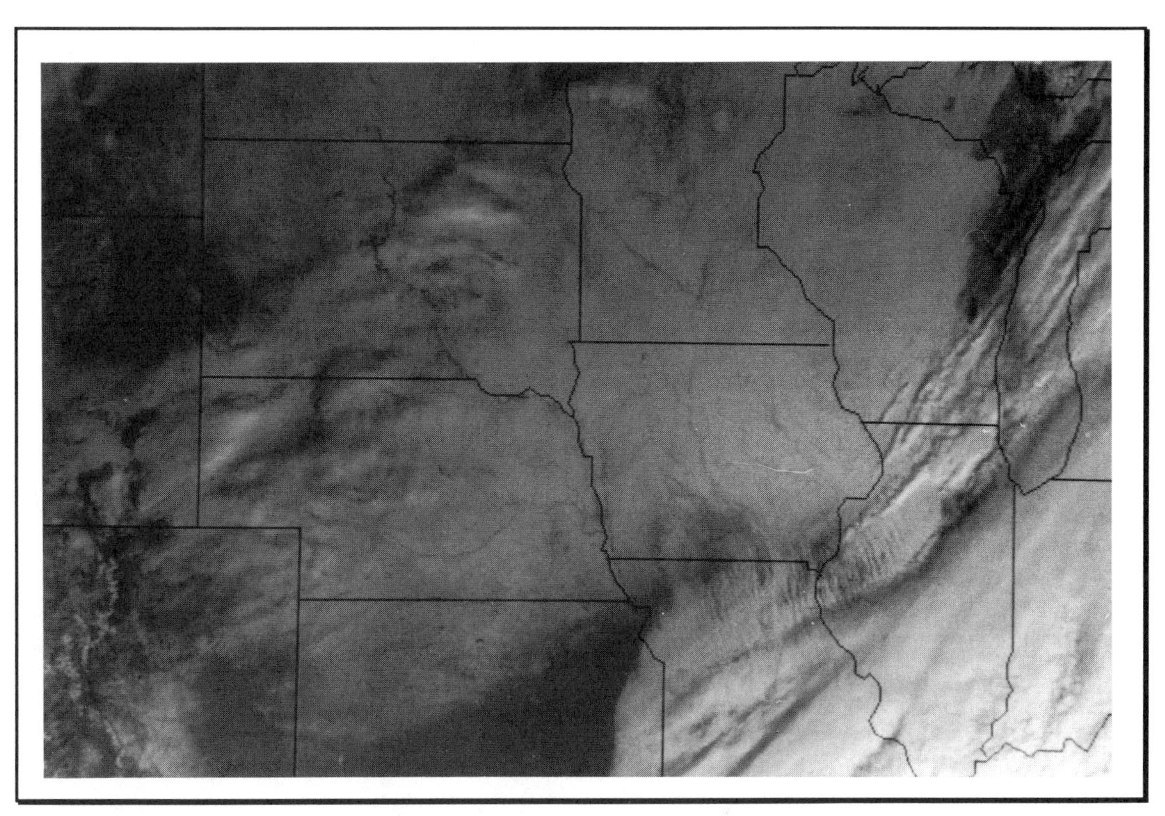

Figure 4.8. Midwestern snow cover (GOES VIS, December 9, 1994)

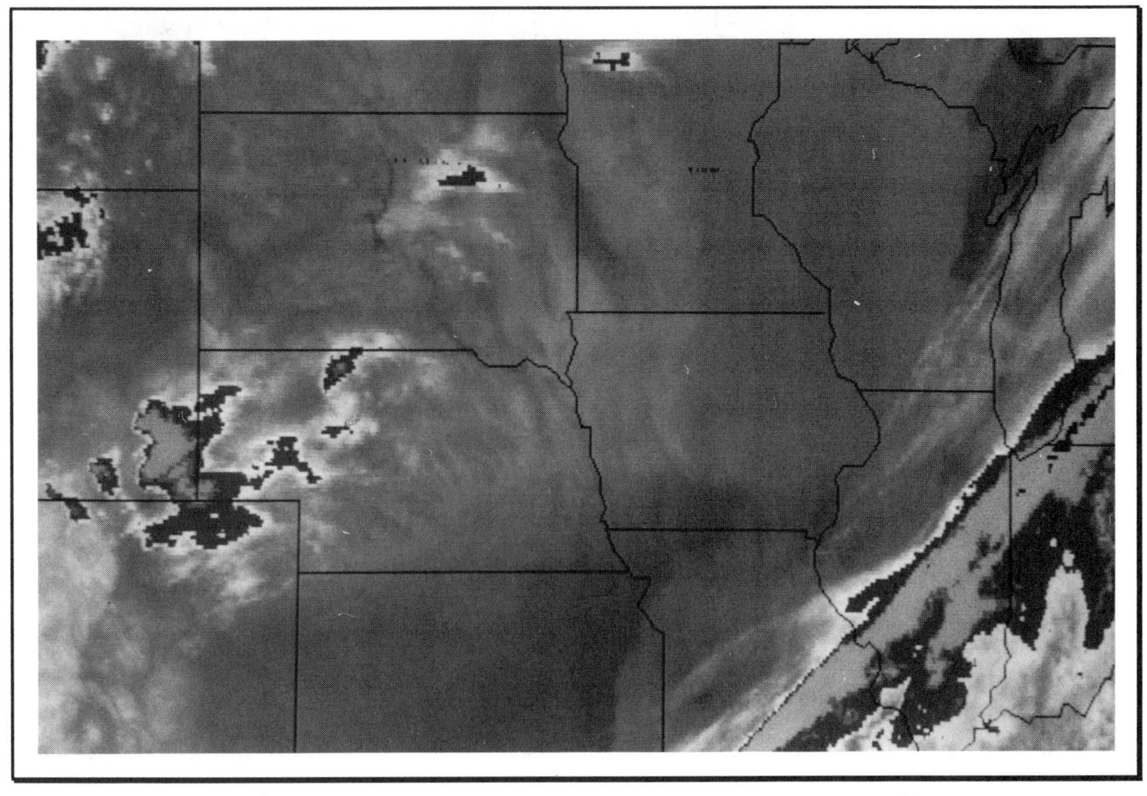

Figure 4.9. Midwestern snow cover (GOES IR, December 9, 1994)

snow on the midwest and along the East Coast. The snow cover on the ground is easily recognized, since it follows many geographical features. In the Appalachian Mountains of Pennsylvania and West Virginia, the snow follows the ridge and valley pattern. Along the Atlantic coast, many rivers cut through the snow-covered land. Other river valleys can be seen in the Midwest. Notice that snow-covered land in Canada shows up as a darker tone. This is because extensive coniferous forests block the snow from the satellite sensor. Farther north, in the tundra, the snow-covered land shows up as white, since there are no trees to block the snow. The boundary between the northern coniferous forests and the relatively treeless tundra is known as the **tree line**. This boundary is often clearly visible in VIS satellite imagery. Note also that the Hudson Bay is covered with ice and snow, and its shape is clearly visible next to the tundra zone surrounding it.

Figure 4.10. Snow cover following the Blizzard of 1979 (GOES VIS, February 19, 1979)

CHAPTER 5

The Atmosphere

Introduction to the atmosphere

The **atmosphere** is an amazingly thin shell of gases surrounding the Earth, extending to about 800 km (500 miles) above the Earth's surface. Relative to the Earth's radius of 6,600 km (4,000 miles), this is a very thin layer. In fact, more than half of the gases that make up the atmosphere are contained in a layer extending upward only 6 km (3.7 miles) from the Earth.

It is important to study the atmosphere, since it affects everyone's life. Natural and manmade atmospheric pollution affects the health of all our planet's inhabitants, causing poor air quality and possibly altering the Earth's climate. Changes in climate and weather patterns can affect the entire world's food supply, which is dependent on rainfall, sunlight, and temperature. Severe storms can destroy property, create floods, and kill or injure plants, animals, and humans; and can disrupt a national, regional, or local economy by knocking out power, disrupting transportation, or destroying industries, (e.g., agriculture and recreation). Cold winters or hot summers cause us to increase our energy consumption and use up vital natural resources. These are just a few of the ways in which our atmosphere affects our lives; it therefore is vital for us to study and attempt to understand these phenomena and forecast their occurrence.

However, understanding and predicting the atmosphere is difficult, since there are complex interactions between incoming and outgoing energy, the Earth's surface, and the gases of the atmosphere. Differences in solar heating and cooling cause temperature variations in the atmosphere, which, in turn, cause volume changes that result in pressure changes. These pressure changes cause air to move horizontally and vertically, which can create or modify circulation patterns. These circulation patterns then affect temperatures, wind patterns, and precipitation, resulting in a very complex system that is in constant transition. Only through careful study and observation can we unwrap the mysteries of the processes that operate within our atmosphere. This chapter will review some of the basic characteristics of our atmosphere in order to prepare you for understanding the complex processes that can be observed in satellite imagery. If further information is needed for any of the topics discussed here, you may find an introductory-level meteorology textbook to be helpful.

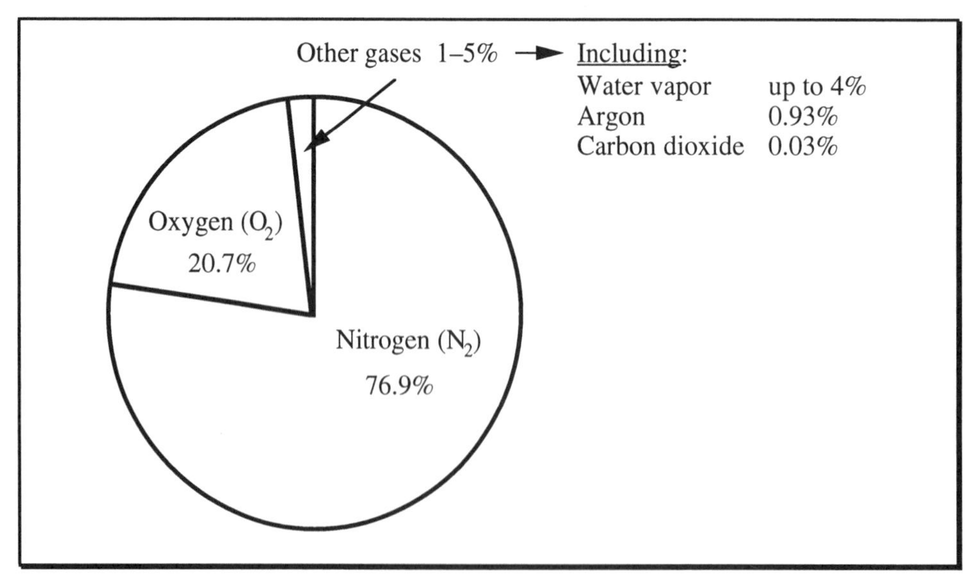

Figure 5.1. The average composition of the atmosphere (by volume)

Atmospheric composition

The atmosphere is composed mainly of nitrogen and oxygen, with smaller amounts of other gases mixed in (figure 5.1). For weather studies, a "minor" gas that is of great importance is **water vapor**. The atmosphere contains a reservoir of fresh water for the Earth, since, depending on environmental conditions, up to 4% of the atmosphere can be water vapor. Winds transport water vapor around the planet, bringing water to areas that are not necessarily near large water sources.

The water cycle

Water vapor enters the air when liquid water absorbs energy from the sun and **evaporates** (changes from a liquid to a gas) or when ice **sublimates** (changes from a solid to a gas). Air can only hold a specific amount of water vapor at a particular temperature. When the air is holding the maximum amount of water vapor that it can, it is said to be **saturated**. If more water vapor is added to saturated air, or if the air is cooled, **condensation** of the excess water vapor into liquid will occur. If the temperature is cold enough, deposition of water as ice crystals will occur. The temperature at which the water vapor begins to condense is known as the **dew point**. When air is cooled to the dew point and water vapor condenses, clouds often form, or dew or frost appears.

The process of condensation releases the energy originally absorbed from the sun during evaporation. Each time a droplet of water condenses out of the air, a small quantity of heat is released into the atmosphere. This energy, known as **latent heat**, heats the surrounding air, causing it to become more buoyant. As this newly warmed air rises, it can enhance vertical airflow within a storm, making it stronger. In fact, much of the powerful and destructive force created in a hurricane is due to the latent heat of condensation released over warm tropical seas.

When raindrops, hailstones, and snowflakes become too heavy to remain suspended in the air, they fall as **precipitation**, bringing fresh water to the Earth's surface. The water on the land runs off into bodies of water and evaporates into the air as water vapor. In this manner the cycle continues. This repeating pattern of condensation, precipitation, and evaporation is known as the **water cycle**; it is illustrated in figure 5.2.

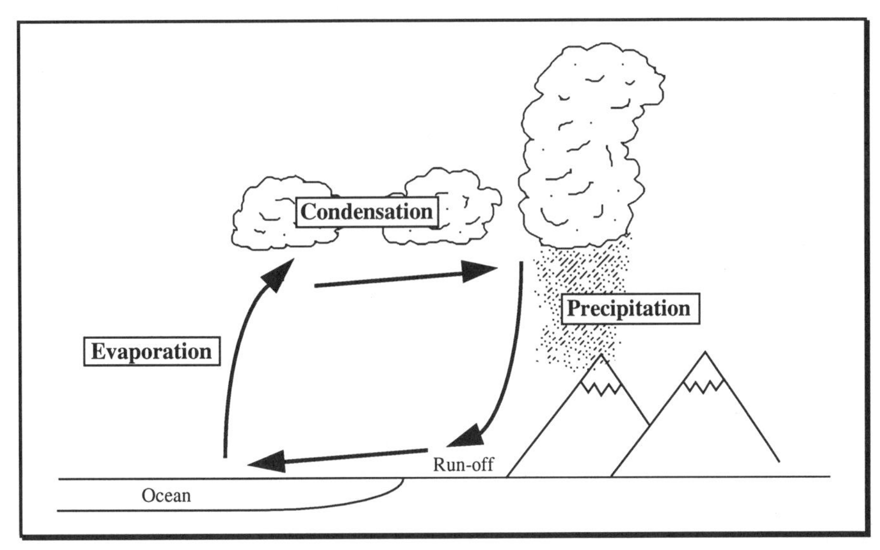

Figure 5.2. The water cycle

Heating of the atmosphere

Radiation from the sun supplies the energy for all processes that occur in the atmosphere. The majority of the solar energy that reaches the Earth passes through the atmosphere without heating it. This radiation is transmitted through the atmosphere, until it reaches, and is partially absorbed by, the surface of the Earth, thus heating it. Air in contact with the warmed Earth is then heated by a process known as **conduction** (heating through contact). As the air is heated, it becomes less dense (and therefore more buoyant), and it rises. As it rises, cooler, denser air sinks to replace it. This pattern of rising warm air and sinking cool air is known as **convection**. Convection is a very important process in the atmosphere, since it carries excess heat away from the Earth's surface and distributes it throughout the atmosphere. Convection is also part of very dynamic weather patterns, on both a small scale and a large scale. Thunderstorms and sea breezes are examples of small-scale convective weather patterns. On a larger scale, convective processes create high- and low-pressure systems that are responsible for the weather systems described on television weather reports. Figure 5.3 summarizes the main processes that result in the transfer of heat energy throughout the atmosphere.

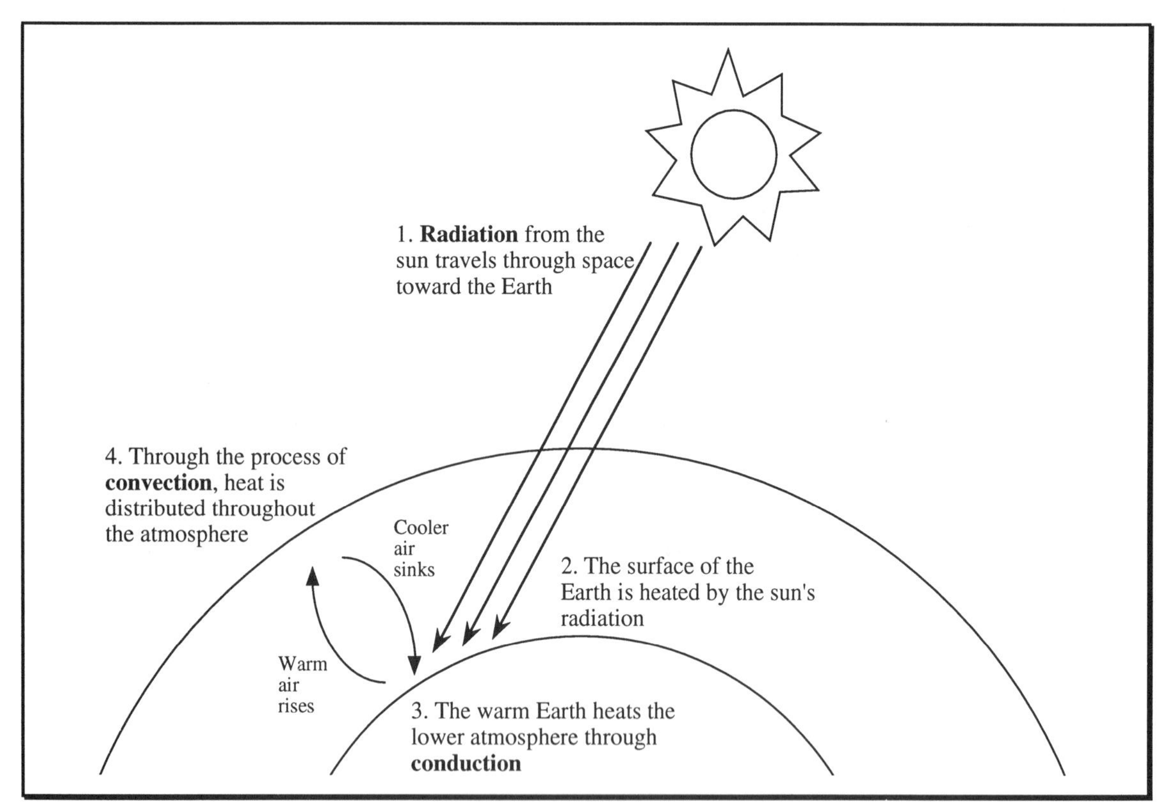

Figure 5.3. The heating of the Earth's atmosphere (not drawn to scale)

Seasonal variations in atmospheric heating

As the Earth orbits the sun on a yearly basis, the tilt of the Earth's axis changes with respect to the sun. This causes the location of the sun's most direct (and hottest) rays to migrate between the Tropic of Cancer (23.5° N) and the Tropic of Capricorn (23.5° S). The greatest heating of the atmosphere occurs here, where the sun's most direct rays are located throughout the year. Temperatures between these latitudes, therefore, remain warm throughout the year. In higher latitudes, temperatures change seasonally as the sun's direct

Figure 5.4. Summer solstice (GOES VIS, June 21, 1982)

rays migrate closer and farther away. The changing location of the sun's direct rays causes changes in temperature gradients in each hemisphere, which, in turn, affect weather patterns.

When the Northern Hemisphere is tilted toward the sun, the most direct rays are located north of the equator. When the sun's direct rays reach their northernmost latitude at the Tropic of Cancer (23.5° N), this is known as the **summer solstice**. The summer solstice occurs each year on June 21 or 22, marking the first day of summer in the Northern Hemisphere and the first day of winter in the Southern Hemisphere. During the period around the summer solstice, the greatest amount of heating takes place in the Northern Hemisphere, and temperatures are much warmer than in the Southern Hemisphere. Figure 5.4 is a GOES VIS image during the 1982 summer solstice. Notice that the terminator shows the Northern Hemisphere covered by more sunlight than the Southern Hemisphere. Since a larger portion of the Northern Hemisphere is covered in sunlight, the days there are longer. At the time of the summer solstice, everywhere above the Arctic

Figure 5.5. Autumnal equinox (GOES VIS, September 21, 1982)

Circle (66.5° N) experiences 24 hours of daylight, while in the Southern Hemisphere everywhere below the Antarctic Circle (66.5° S) experiences 24 hours of darkness.

After the summer solstice, the sun's most direct rays migrate southward and reach the equator on September 21 or 22. The point at which the sun's direct rays are over the equator (0° latitude) is called the **autumnal equinox**. This marks the first day of autumn in the Northern Hemisphere and the first day of spring in the Southern Hemisphere. During the autumnal equinox, the greatest amount of heating occurs in the equatorial region, and mid-latitude temperatures are mild in both the Northern and Southern Hemispheres. Figure 5.5 is an image taken during the 1982 autumnal equinox. In this image, the terminator is oriented in a north-to-south line, and both hemispheres are covered equally by sunlight. All points on Earth experience about 12 hours of daylight and 12 hours of darkness at this time.

Figure 5.6. Winter solstice (GOES VIS, December 21, 1982)

The sun's most direct rays reach their southernmost location on December 22 or 23, when they reach the Tropic of Capricorn (23.5° S). This occurrence is known as the **winter solstice**, and it marks the first day of winter in the Northern Hemisphere and the first day of summer in the Southern Hemisphere. During the winter solstice the greatest amount of heating occurs in the Southern Hemisphere and temperatures in the Northern Hemisphere are much colder. Since a larger portion of the Southern Hemisphere is covered by sunlight (figure 5.6), days are longer than in the Northern Hemisphere. The south polar regions below the Antarctic Circle experience 24 hours of daylight, while the regions above the Arctic Circle experience 24 hours of darkness.

Following the winter solstice, the location of the sun's direct rays begins to migrate northward again. They are over the equator on March 21 or 22 during the **vernal equinox**. This marks the first day of spring in the Northern Hemisphere and the first day of autumn in the Southern Hemisphere. As with the autumnal equinox, the greatest amount of heating occurs around the equator and mild temperatures occur in the Northern

Figure 5.7. Vernal equinox (GOES VIS, March 21, 1982)

and Southern Hemispheres. Figure 5.7 is an image of the vernal equinox. Since each hemisphere receives equal sunlight, day and night are equal across the Earth.

Structure of the atmosphere

When studying a satellite image, it is easy to think of the atmosphere as a flat, two-dimensional layer of gases. It is important, however, to understand that processes in the atmosphere operate differently at various levels. For example, the surface winds at a place may be from the southwest while, at the same time, the winds 5 km above the ground are from the northwest. To fully understand the weather at any given location one must know something about both the upper-level and the near-ground weather conditions.

Therefore, **vertical profiles** of the atmosphere are studied. Important profiles include temperature, pressure, wind, and dew point. Profiles can be constructed by a variety of methods. Scientists can use **radiosondes** to probe the atmosphere at different altitudes. A radiosonde is carried on a weather balloon, and it measures weather variables as it rises to high altitudes. The balloon is also tracked by radar to obtain wind information. Radar and satellite sensors can take **soundings**, which are a vertical series of measurements taken at different heights above the ground.

One of the vertical profiles of the atmosphere that is important to understand is the temperature profile. Figure 5.8 is a graph that illustrates the temperature profile of what is known as the **standard atmosphere**. This generalized diagram represents the average temperature at different altitudes and can be used as a basis for comparison with observed temperatures. Using this diagram, the atmosphere can be divided into four separate layers on the basis of how the temperature changes with increasing altitude.

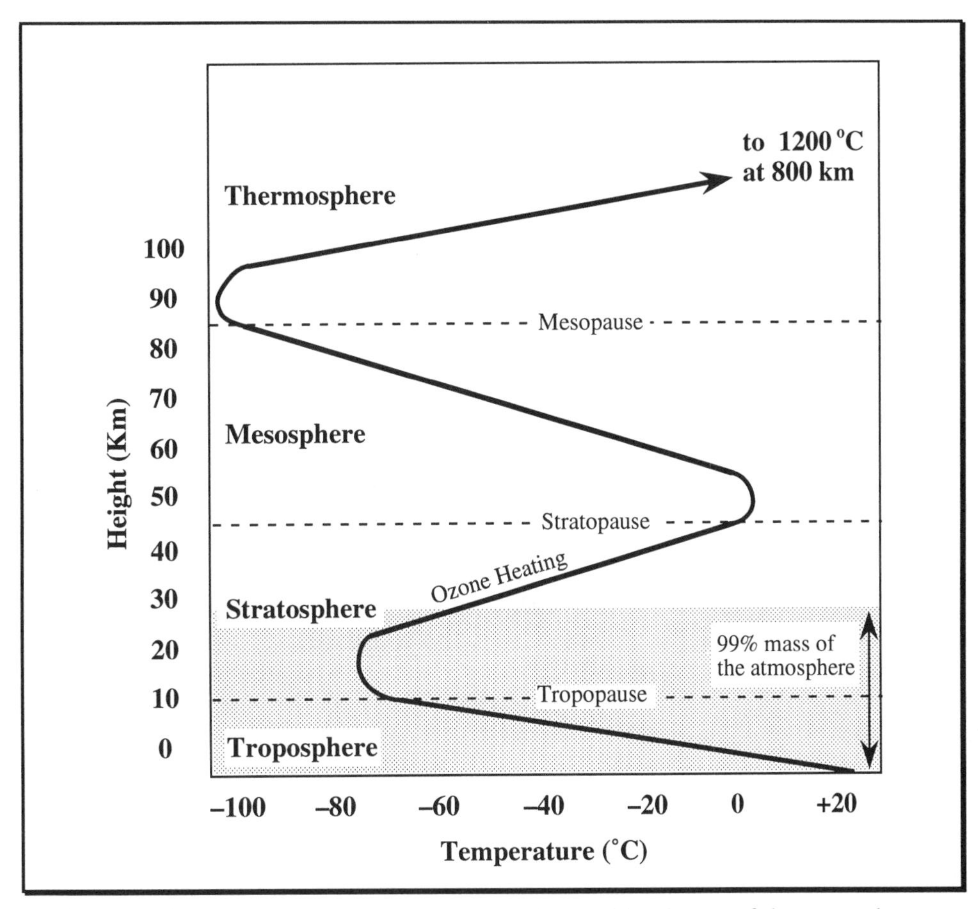

Figure 5.8. The temperature structure and the resulting layers of the atmosphere

Close to the ground and up to a height of about 12 km (7.2 miles), the temperature cools fairly rapidly with increasing altitude. This layer of air is heated by the Earth; therefore, the farther away from the Earth, the colder the temperatures. This is a layer of the atmosphere known as the **troposphere**. Most weather phenomena take place in the troposphere, so this is the layer of the atmosphere that is studied most.

The vertical variation of temperature in the troposphere is known as the **lapse rate**. A **normal lapse rate** occurs when temperature drops with increasing altitude. When an increase in temperature with height exists, this is called an **inversion**. Inversions can develop at different levels in the troposphere, and they are important because cold, heavier air is present beneath warmer air. Inversions produce an atmosphere with very little vertical air motion and very little mixing. When the atmosphere is in this state it is said to be a **stable atmosphere**. Atmospheric conditions that are characterized by rising and falling air create an **unstable atmosphere**. The stability or instability of the atmosphere directly determines the types of clouds and weather that occur in any area.

The **tropopause** is the point at which air begins to warm with increasing altitude. This boundary marks the maximum extent of most cloud tops, since the air is warmer above it, and conditions for convective cloud formation are not present. The tropopause also marks the boundary between the troposphere and the second layer of the atmosphere: the **stratosphere**. The air warms with altitude in the stratosphere because the ozone layer is concentrated here. Since ozone absorbs ultraviolet radiation from the sun, it heats this layer of the atmosphere. The stratosphere is also very important to study, since the upper-level winds in the lower stratosphere may have an influence on the movement of large-scale weather systems. Recently, interest in stratospheric ozone (or the lack thereof), volcanic dust, water vapor, and other atmospheric constituents is focusing more attention on this atmospheric layer.

Beyond the influence of the ozone layer in the stratosphere, temperatures begin to cool again as altitude increases. The altitude at which temperatures begin to decrease with altitude is known as the **stratopause**. This marks the boundary between the stratosphere and the third layer of the atmosphere, known as the **mesosphere**. At this level, the air begins to warm up with increasing altitude. This altitude is called the **mesopause**; it separates the mesosphere from the outermost layer of the atmosphere, known as the **thermosphere**.

The thermosphere extends into outer space. Contrary to what many people believe, the temperatures in the thermosphere are extremely high, often over 1200° C (2192° F). This is better understood when one recalls that temperature is a measure of how rapidly air molecules are moving, not how hot the air feels to the touch. While the density of the air in the thermosphere is so low that it would not feel hot to an observer, the actual temperatures in this layer are extremely high, since the air molecules are excited by high-energy particles coming from space. As these electrically charged particles (mostly from the sun) travel through this layer, they excite the air molecules and cause them to give off light, sometimes creating brilliant displays known as the **aurora borealis** or the **northern lights** in the Northern Hemisphere and the **aurora australis** in the Southern Hemisphere. It should also be noted that most polar orbiting satellites orbit the Earth in the thermosphere.

Atmospheric pressure

Generally, we do not think of the air in our atmosphere as having weight. Air, however, is nothing but gas molecules. These molecules have a definite mass, and because gravity attracts them to Earth, they have weight. We do not feel this since our bodies are adapted to the weight of the air above us.

The force exerted by the mass of the atmospheric gases on a unit area of a surface is known as **atmospheric pressure** (or air pressure). At sea level, air pressure averages 14.7 lb/sq in. This means that at sea level, every square inch of surface area has a column of air above it, extending to space and weighing 14.7 lb. Pressure at sea level is used as the standard atmospheric pressure for comparison with observed measurements of air pressure. A more common unit for measuring air pressure today is a metric unit known as the **millibar (mb)**. Average atmospheric pressure at sea level is 1013.25 mb.

Atmospheric pressure decreases with altitude, since gravity concentrates more mass of air close to the Earth's surface. This explains why it is more difficult to breath in areas on the Earth that are at very high elevations. At these heights, there is less air available, since the atmosphere is less dense than it is near sea level. Figure 5.9 is a profile showing how atmospheric pressure varies as altitude increases.

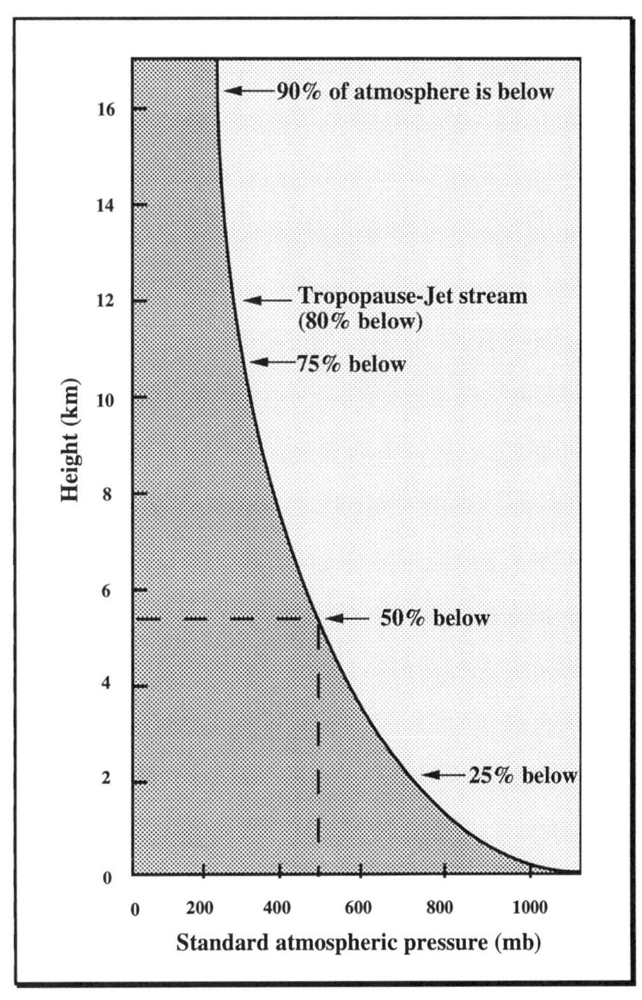

Figure 5.9. The relationship between atmospheric pressure and altitude

Note that 50% of the mass of the atmosphere exists below about 6 km (3.7 miles). At this point the atmospheric pressure is one half of standard atmospheric pressure, or 500 mb. Heights in the atmosphere are often referred to by their associated atmospheric pressure. For example, a map of the wind speeds at a height of about 6 km would be referred to as a 500 mb chart. Scientists also study 200 and 250 mb analysis charts quite often. At these levels, it is easy to detect the strong bands of high-velocity winds known as **jet streams**. This is because of the large number of aircraft reports of wind speed and direction which are available. Jet streams have a strong influence on the movement and development of weather systems, the occurrence of severe weather, and the creation of high-altitude turbulence. Thus, scientists look at these levels often as they predict and research weather.

CHAPTER 6

Identifying Cloud Types in Satellite Imagery

Introduction to cloud identification

When the world's first satellite pictures of the atmosphere were viewed, the most striking feature was the extensive cloud cover over large parts of the Earth. Even today, the first feature that one usually notices in a satellite image is the clouds. At first glance, these clouds may seem to be random in shape and distribution; however, they form as the result of very specific interactions between many different meteorological factors. When these factors interact in certain ways, different cloud types are formed. Clouds that form under similar conditions can be classified into individual categories on the basis of their appearance from the ground, and each cloud type will exhibit unique patterns in satellite imagery. Once the patterns are known, it is possible to identify the cloud types present in a satellite image. This can be useful, since the ability to recognize cloud types on satellite imagery can give you clues about the state of the atmosphere and the types of weather to look for. Identification of cloud types can also help in locating hazardous weather conditions such as severe thunderstorms, snow and ice storms, upper-level air turbulence, and ground fog. In this chapter, you will learn how clouds form and how meteorologists classify clouds. How each cloud type appears when seen from space will be discussed, to allow the reader to identify cloud types in satellite imagery.

Cloud formation

Clouds are formed when air is cooled to its dew point. At this temperature, water vapor in the air condenses onto particles in the air known as **condensation nuclei**. Condensation nuclei can be sea salt, dust, or other minute particles in the atmosphere. A cloud is just a collection of condensed water droplets or ice crystals that are small enough to remain suspended in the air.

Air is most commonly cooled through lifting. If there is enough lifting and the air is moist enough, clouds can form. Air can be lifted by convective processes, along frontal boundaries, or by flowing over a mountain. Most of the clouds that are seen on a satellite image have been formed as a result of one or more of nature's lifting mechanisms. Another way in which air can be cooled is by coming into contact with a cold surface. For example, when warmer air comes into contact with the colder Earth or water, **fog** may form. Fog is nothing but a cloud that has formed at ground level.

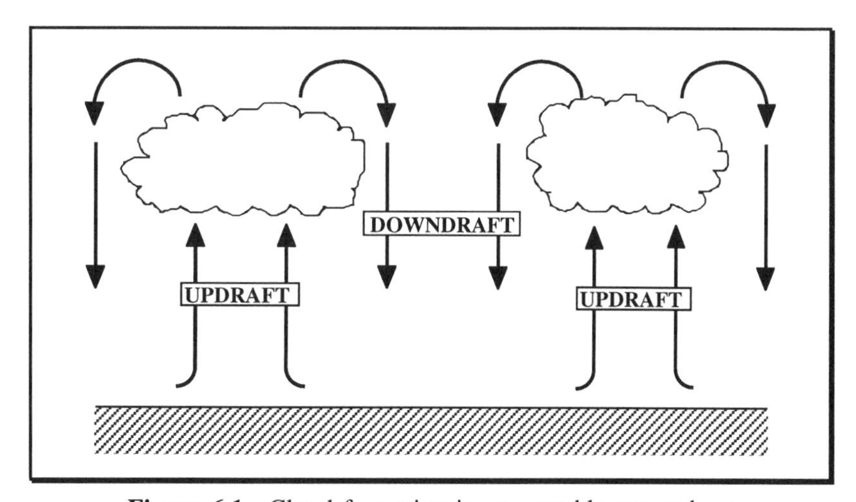

Figure 6.1. Cloud formation in an unstable atmosphere

Cloud formation is closely related to the stability of the atmosphere. In unstable air, uneven heating of the Earth causes convective currents to form. Pockets of warm air rise and create **updrafts**. As the rising air cools, clouds form in the updraft regions (figure 6.1). These clouds are characterized by vertical development, and they may grow very tall. Clouds that form in unstable air are associated with thunderstorms, showery precipitation, and gusty winds. Between the updrafts, **downdrafts** of sinking air are present. Where there are downdrafts, clouds usually are not present.

In a stable atmosphere, there is very little vertical motion. Clouds forming in a stable atmosphere do not exhibit much vertical growth, since the air is unable to rise. These clouds are characterized by a flat, layered, or sheetlike appearance. Low cloud ceilings, poor visibility, and steady precipitation are associated with stable air masses at low levels in the atmosphere.

Large-scale cloud patterns

On a satellite image, large-scale weather events form cloud systems with recognizable patterns. These offer clues about the state of the atmosphere. It is often helpful to recognize some of the large-scale cloud patterns that form. Some of the most common cloud patterns are listed below (and shown in figure 6.2).

Cloud shield. A cloud shield is a broad cloud pattern that is no more than four times as long in one direction as it is wide in the other direction. In figure 6.2 a cloud shield can be seen north of Lake Ontario.

Cloud band. A cloud band is a nearly continuous cloud formation with a distinct long axis; the ratio of length to width is at least 4 to 1, and the width of the band is greater than 1° of latitude. A large cloud band associated with a cold front can be seen off the east coast of the United States in figure 6.2.

Cloud line. A cloud line is a narrow cloud band in which the individual cloud cells are connected and the line width is less than 1° of latitude. Several small cloud lines can be seen in figure 6.2 ahead of the frontal band in the subtropical Atlantic Ocean.

Cloud street. A cloud street is a narrow cloud band in which the individual cells are not connected. Several streets generally line up parallel to each other and along the low-level wind flow. Each street is less than 1° of latitude in width. Cloud streets can be seen extending from northeast New York State into Vermont in figure 6.2.

Cloud finger. A cloud finger is an extension from the forward side of a frontal cloud. A cloud finger usually extends in a southerly direction from a cloud band associated with a cold front.

Cloud element. A cloud element is the smallest cloud form that can be resolved in a satellite image. Many small cloud elements can be seen over Georgia, Tennessee, Pennsylvania, New England, and eastern Canada in figure 6.2.

Comma cloud. A vortex that contains one or more spiral cloud bands that converge toward a common center. In figure 6.2 a comma cloud is located in southern Canada just north of North Dakota.

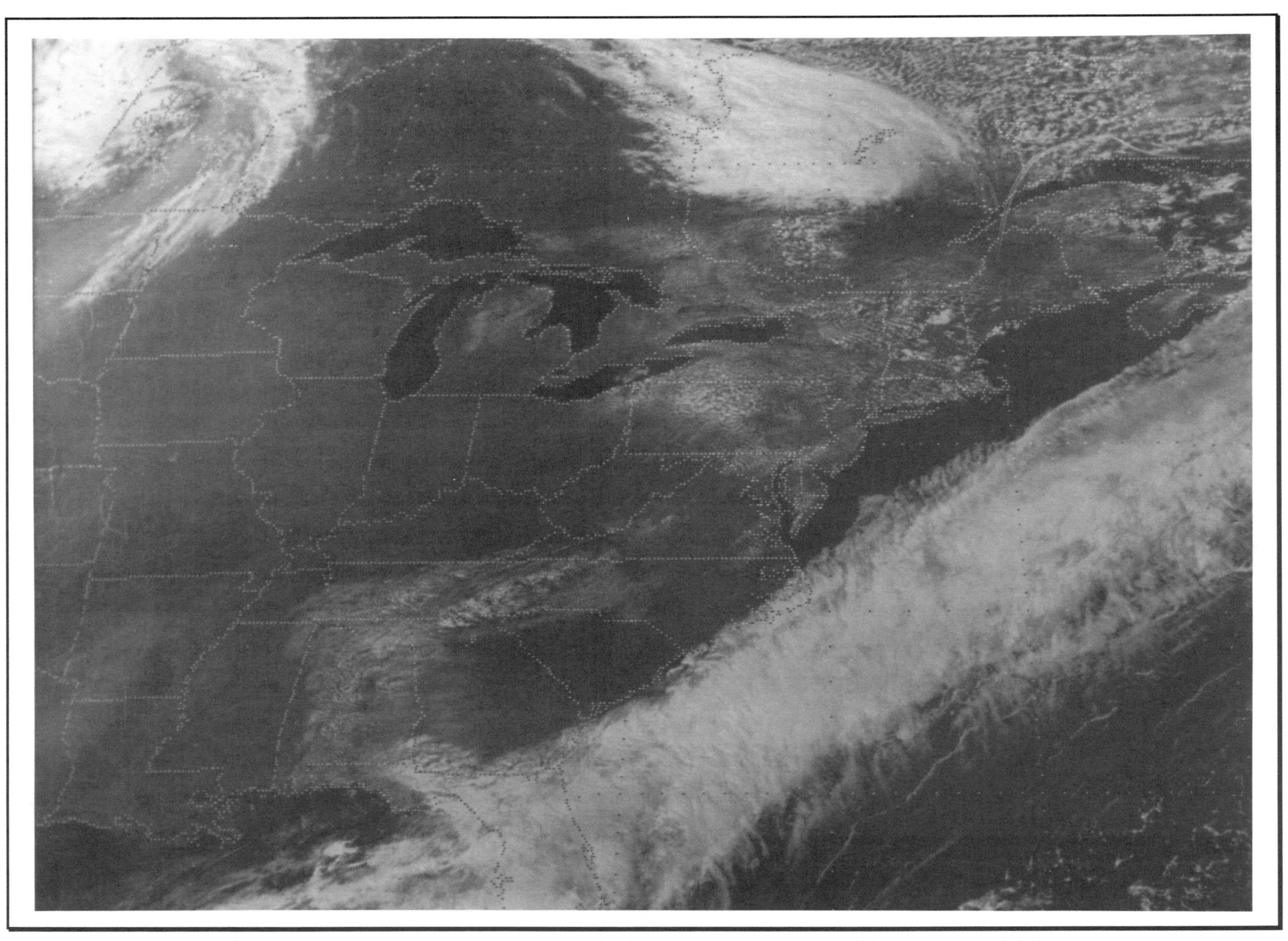

Figure 6.2. Examples of various large-scale cloud formations (GOES VIS, September 7, 1988)

Features that aid in cloud identification

Individual cloud types can be identified by observing various cloud features in a VIS satellite image. These features include:

—brightness
—texture
—organizational pattern
—edge definition
—size
—individual shape

The **brightness** of a cloud in a satellite image is one of the best clues to use when identifying its characteristics. Brightness can often be used to indirectly determine the thickness and the height of a cloud. In general, high brightness values in VIS imagery are associated with thick clouds, which tend to reflect much of the sun's light. Therefore, thick clouds will appear white or very light gray in a VIS satellite image. Thin clouds will appear darker or even be transparent in VIS imagery. In IR imagery, high brightness values are associated with the coldest cloud top temperatures. Therefore, very high, cold cloud tops will appear white or very light gray in an IR satellite image. Low, warm clouds will appear as a darker shade of gray or may even blend in with the ground or water surface.

Cloud **texture** is a second feature of a cloud that may aid in identification. This is a feature that can only be seen in VIS imagery, since it is a function of the amount of shadow that is cast on parts of clouds. Clouds that have a bumpy shape and grow vertically will exhibit a great deal of shadowing. This is often referred to as a lumpy texture. Clouds with a smooth, flat upper surface do not generally have many shadows on their tops. These clouds are said to have a smooth texture. Clouds with a smooth texture may cast shadows on lower-level clouds and/or the ground, but only at their edges. Some high-level clouds get stretched out as upper-level winds cause them to streak. These clouds often have a fibrous texture in satellite imagery.

Other features can also aid in cloud identification. Clouds can be observed to have different **organizational patterns**, including banded, linear, circular, and cellular patterns. They can often be identified by their **edge definition**, which can be ragged or well-defined. The size and shape of the individual cloud elements can also be useful in determining what types of clouds are present in an image.

In general, the most effective method for identifying individual cloud types is to obtain a VIS and an IR image of the same scene. The VIS image can be used to identify cloud shapes, textures, organizational patterns, and thicknesses. These data can then be compared to an IR image in order to determine the height of the clouds. When all this information is put together, it is usually possible to make a reliable assessment of what types of clouds are present in the image and the weather that is associated with each. The remainder of this chapter will focus on the various cloud types and their associated patterns in satellite imagery. At the end of the chapter, examples of this method for cloud identification from satellite imagery will be shown.

Stratiform Clouds

Stratiform clouds form in a stable atmosphere and are characterized by a very flat or layered appearance. **Stratus** clouds are low-level clouds that often cover the entire sky and create a dull gray or overcast appearance. Stratus clouds may be accompanied by steady light rain, drizzle, or snow grains.

In satellite imagery, stratiform clouds are characterized by smooth, flat tops and lack an organized pattern (table 6.1). The boundaries of these clouds are often sharp and defined by topography. Since these clouds develop at a low altitude, their temperatures tend to be warm; therefore, they usually appear as dark to medium gray in IR imagery. Low stratus clouds are often difficult to distinguish from land surfaces when the temperature contrast between land and cloud is small. In VIS imagery, however, these clouds can be very bright when they are thick. Shadows will not generally appear on the cloud tops; however, they may be visible on the ground next to the edge of a stratus cloud.

Basic Cloud Identification: Stratiform Cloud Types

Type	Content	Base Height	Characteristic Shape or Pattern	Tone	
				VIS	IR
Fog and stratus combination	Water (can be ice in winter or high latitudes)	less than 1 km (3000 ft)	Smooth tops. Boundaries often sharp and defined by topography.	Bright white to medium gray when thick. When thin, may have a mottled appearance.	Uniform dark to medium gray, varying with seasonal temperature changes. Difficult to detect when land-cloud temperature difference is small. Can be darker than ground when ground temperatures are very cold.
Altocumulus and altostratus	Water and/or ice	2 to 4 km (6500–12,000 ft)	Smooth tops, often in layers; boundaries can be ragged or smooth. Cellular structure of altocumulus too small to be distinguished from altostratus. Often associated with high-level cirrus.	Light gray. Appears mottled or striated, depending on thickness or layered structure.	Uniform medium gray, depending on height.

Table 6.1

Fog identification

Identification of fog is a very important aspect of satellite meteorology. Fog is simply a stratus cloud that has formed at ground level. Fog locations must be noted to warn pilots, boaters, and motorists about poor visibility. Identifying fog on satellite images can improve this warning system; however, it can be very tricky.

In VIS imagery, fog appears to have a smooth flat texture resembling stratus cloud layers. This makes it difficult to distinguish fog from stratus clouds. In IR imagery, fog appears as a dull shade of gray, if it can be seen at all. If the temperature of the land surface is about the same as the temperature at the top of the fog it becomes nearly impossible to see fog in an IR image because of the poor land-cloud contrast. If this occurs at night, when there is no VIS imagery to aid interpretation, fog-covered areas may go unnoticed.

Generally, VIS and IR images are used in combination to locate fog. When multiple or time-lapse images are available, a good rule of thumb to follow is that if a smooth, white cloud does not move it is usually fog. Also, as fog evaporates, it usually dissipates from its outside, thinner edge and works its way inward. Fog can also be identified at times by sharp-edged boundaries, such as mountains or river valleys, that limit the area the fog can cover. Ground fog can be identified because it is often restricted to valley areas. For example, in figure 6.3, an HRPT image of the west coast of the United States, the San Joaquin Valley is covered by an extensive fog bank. The fog is bounded on the west by the Diablo mountain range and on the east by the Sierra Nevada range. Notice the many notches in the edge of the fog that are located where rivers flow into the valley.

In mountainous regions, the valleys often fill up with fog, while the mountain ridges remain clear. This produces an image in which the valleys show up as white lines that are arranged in the drainage pattern that the mountain rivers follow. If the mountain valleys form a **dendritic pattern** (shaped like a tree), a dendritic pattern of white lines will be seen in the image. If the mountain valleys are a series of parallel valleys **(trellis arrangement)**, the fog will show up as a series of parallel white lines on the image. This can be seen in figure 6.4, an HRPT image of the eastern United States. The eastern portion of the Appalachian Mountain range is characterized by a series of parallel ridges and valleys. In this image, the valleys are covered with fog and they appear as a series of parallel white streaks that extend from northeastern Alabama through eastern Tennessee and into the southwest portion of Virginia. The plateau region of the Appalachians lies to the west of the ridge and valley region. These plateaus, including the Allegheny Plateau in West Virginia and the Cumberland Plateau in Kentucky and Tennessee, are characterized by rivers that flow in a dendritic pattern. As these valleys fill with fog, a branching pattern of white, fog-filled valleys is evident.

Figure 6.3. Fog bank over the San Joaquin Valley, California (HRPT VIS, January 9, 1992)

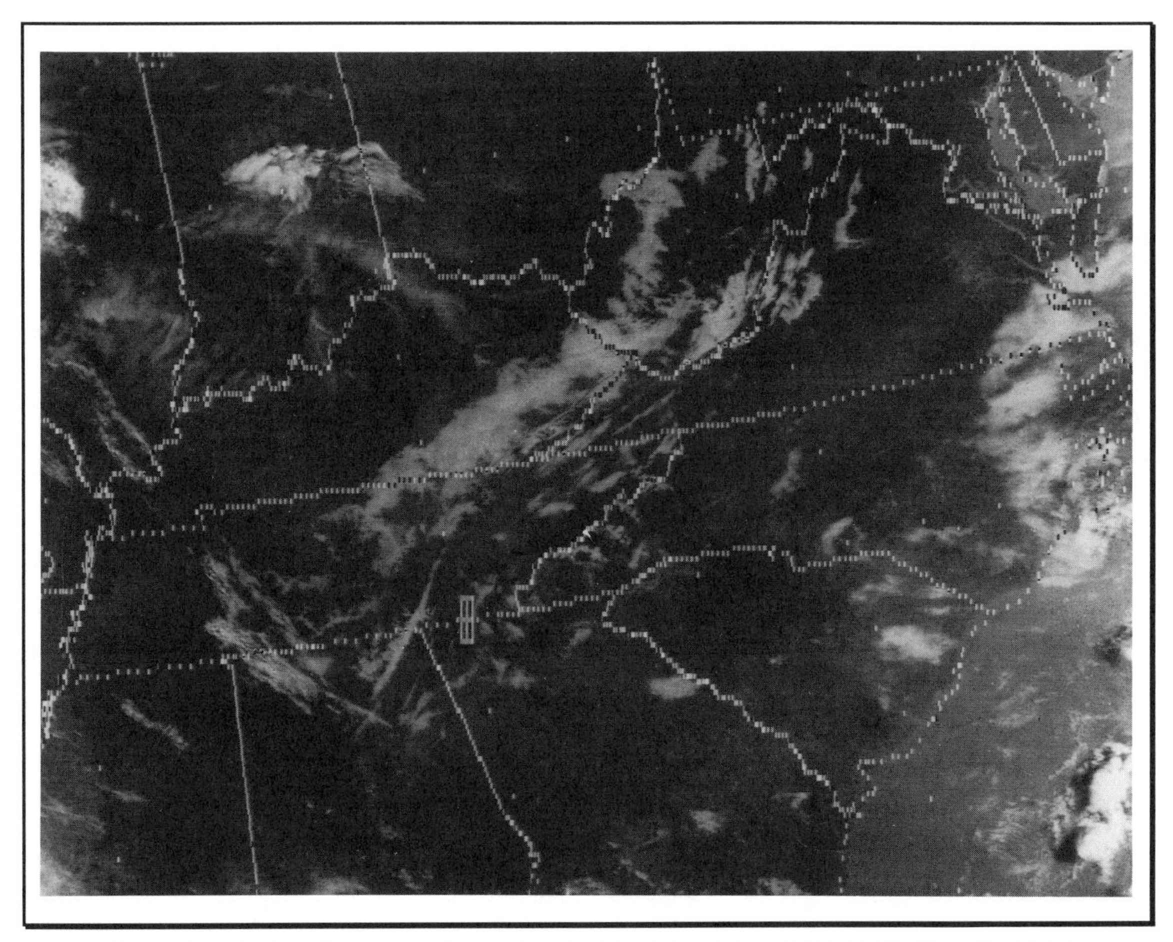

Figure 6.4. Valley fog pattern in the Appalachian Mountains (HRPT VIS, May 15, 1991)

Cumuliform clouds

Cumuliform clouds form in an unstable atmosphere in which air is rising and sinking, giving the clouds vertical development. **Cumulus** clouds form in an updraft, where the air is rising. They tend to be irregular in shape and appear puffy or cottonlike. Between cumulus clouds (in cloud-free areas) the air is sinking. Cumulus clouds are generally associated with fair weather. From space, low cumulus clouds will appear as irregularly shaped cloud elements of various sizes. The ground will often be visible between individual clouds or cloud clusters. These clouds can also be banded or cellular, or exhibit wavelike patterns. They will be very lumpy in VIS images due to shadowing on the irregularly shaped cloud tops. In IR imagery cumulus clouds exhibit gray tones ranging from dark to medium gray. Table 6.2 summarizes the characteristics of cumulus clouds in satellite imagery.

Stratocumulus clouds

In some instances, cumulus clouds begin to develop over an area but an inversion layer prevents the clouds from developing further vertically. This causes the cumulus clouds to spread out and develop a layered appearance, similar to that of stratiform clouds. These clouds, having characteristics of both stratus and cumulus clouds, are called **stratocumulus** clouds. In satellite imagery, they are often arranged in sheets, lines, or

Basic Cloud Identification: Cumuliform Cloud Types

Type	Content	Base Height	Characteristic Shape or Pattern	Tone	
				VIS	IR
Cumulus	Water (can be ice in winter or high latitudes)	less than 2 km (6500 ft)	Small individual elements, irregularly shaped. Shadows when sun angle is low. Lumpy texture.	Medium bright	Dark to medium gray. Small clouds and individual elements can be difficult to detect.
Cumulonimbus	Water and ice mixed	base less than 2 km (6500 ft) and often less than 4000 ft	Globular or carrot-shaped, depending on upper-level winds. Tops can be lumpy in texture. Tops typically 35,000 ft or more; can reach 60,000 ft in severe storms.	Very bright; shadows distinct when sun angle is low.	Very bright, especially in areas of active cumulus cloud top growth.
Stratocumulus	Water (can be ice in winter or high latitudes)	less than 2 km (6500 ft)	Irregularly shaped, globular or cellular pattern. May be arranged in lines or groups. Edges of clouds often touch each other. Texture can be lumpy.	Bright in center; edges are gray where cloud thins.	Uniform dark gray. Cellular structure not as evident. Difficult to detect when cloud-land or cloud-water contrast is poor.

Table 6.2

street patterns, especially over water in the winter. They will appear medium gray in IR imagery, while in VIS imagery they can be very bright and lumpy in appearance. In figure 6.5, a stratocumulus formation can be seen over the Pacific Ocean off the west coast of Mexico.

Towering cumulus and cumulonimbus clouds

When cumuliform clouds form in an unstable atmosphere, rapidly rising air causes the clouds to grow very tall and develop into **towering cumulus** clouds. In extreme cases, when a towering cumulus cloud reaches high altitudes, upper-level winds will often cause the top of the cloud to be spread downwind, far away from the cloud base. The cloud develops a flat top with an anvil-like appearance. This type of cloud formation is known as a **cumulonimbus** (or thunderstorm) cloud and may be associated with high winds, hail, heavy rainfall, and tornadoes. In satellite imagery these clouds can appear globular, carrot shaped, or triangular, depending on the strength of the upper-level winds. A cumulonimbus cloud will appear very bright in both VIS and IR imagery, since it is characterized by very thick, tall cloud development and very cold, high cloud tops. In VIS imagery the tops of these clouds are often lumpy and shadowed where clouds shoot above the anvil. These regions of lumpy cloud top texture in a thunderstorm are known as **overshooting tops**.

Figure 6.6 is a VIS GOES image of Kansas and Oklahoma. Several cumulonimbus clouds can be seen in the center and lower left-hand portion of this image. The tops of these clouds appear very lumpy due to many overshooting tops embedded in the storm

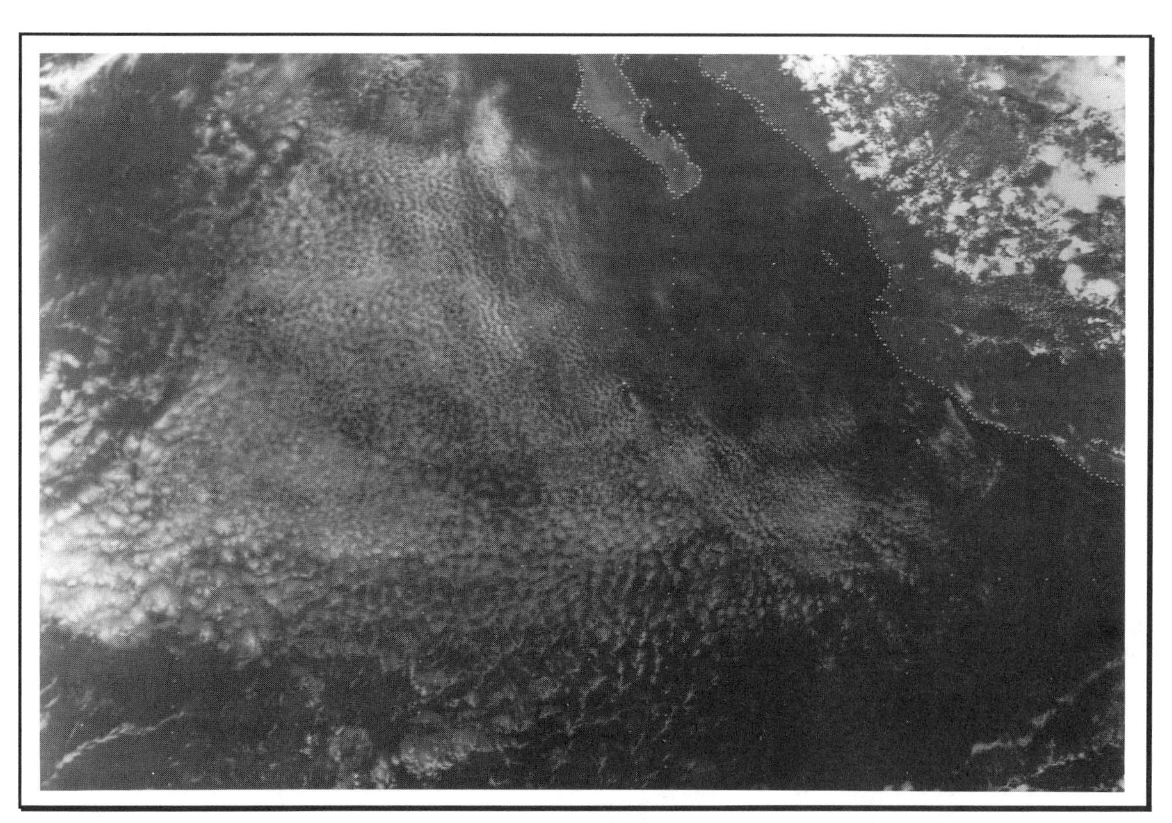

Figure 6.5. Stratocumulus clouds over the Pacific Ocean (GOES VIS, May 25, 1993)

Figure 6.6. Cumulonimbus clouds over Kansas and Oklahoma (GOES VIS, July 11, 1993)

clouds. Some of the tallest clouds have cast shadows on lower clouds and the ground. A line of mature cumulonimbus clouds extends across central Kansas. These clouds have grown into upper-level winds, and their tops have been spread downwind. Where a cloud is being spread out its edges are thin and therefore not as bright. To the immediate south of this line is a younger, growing cumulonimbus cloud centered over Wichita, Kansas. Other young cumulonimbus clouds can be seen in the Oklahoma panhandle and in northern Texas. These clouds exhibit a shape commonly seen in cumulonimbus clouds, with an apex in one corner of the storms and a pronounced spreading of the cloud top. The younger cumulonimbus clouds have sharper edges, since the upper-level winds have not had as much time to spread out the anvil.

Middle-level (alto) clouds

Middle-level clouds are usually given the prefix "alto." These clouds can have a cellular or cumulus appearance (in which case they are called **altocumulus** clouds), or they can have a flat stratiform shape (in which case they are called **altostratus** clouds). Altocumulus clouds can exhibit wave patterns like stratocumulus clouds, only they are higher and colder. These clouds can be difficult to identify in satellite imagery because they usually occur along frontal boundaries in conjunction with higher cirriform clouds that tend to block them from the satellite. In VIS imagery they can be relatively bright and flat and can often be identified by the shadow they cast on lower cloud layers, especially when the sun angle is low. In IR imagery they are usually a medium gray color (see table 6.1).

High-level (cirro) clouds

Cirriform (ice crystal) clouds form at higher altitudes where temperatures are very cold. **Cirrus** clouds often have a wispy, windswept appearance caused by high-altitude winds spreading the ice crystals across the sky. These clouds form at high altitudes where

Basic Cloud Identification: Cirriform Cloud Types

Type	Content	Base Height	Characteristic Shape or Pattern	Tone	
				VIS	IR
Cirrus	Ice	6 to 18 km (20,000 to 60,000 ft)	Banded fibrous structure. Very thin. Underlying terrain and lower clouds often visible through cirrus.	Dark to medium gray, depending on the underlying surface.	Light gray in tone. Fibrous structure not as evident.
Cirrostratus and cirrocumulus	Ice	6 to 18 km (20,000 to 60,000 ft)	Generally smooth, uniform tops. Can be fibrous, in long bands, or an extensive sheet.	Light gray when thin; whiter as thickness increases.	White to light gray. Can be difficult to separate from middle clouds.
Anvil cirrus	Ice	6 to 18 km (20,000 to 60,000 ft)	Smooth texture, except where overshooting tops cast shadows. Sharp edge upwind; downwind edge filmy, fibrous, and indistinct.	Very bright over most active part of thunderstorm cell; brightness decreases downwind.	Very bright over most active part of thunderstorm cell; brightness decreases downwind.

Table 6.3

very little water vapor exists; therefore, they also tend to be very thin. The wispy appearance of these clouds gives them the popular name "mare's tails." In VIS satellite imagery they typically have a fibrous appearance and the ground is often visible through the clouds. In IR imagery they appear very bright due to their low temperatures (table 6.3). In IR imagery, cirrus clouds often appear to be more extensive than they are in VIS imagery as a result of the "smearing" effect that is caused by the lower resolution of IR sensors.

High clouds with a stratiform appearance are known as **cirrostratus** clouds. These clouds are usually smooth in appearance with uniform tops. They often form long bands or sheets. In IR imagery they are bright, and in VIS imagery they tend to be a light gray color. High clouds with a cellular pattern are called **cirrocumulus**. Their cellular structure is too small to observe from satellite imagery, so they are often difficult to distinguish from cirrostratus clouds.

Examples of cloud identification

Figure 6.7 is a VIS image of the Pacific Ocean and the west coast of the United States. Figure 6.8 is an IR image of the same sector that was taken only 30 minutes later. Although the clouds have moved slightly over this half-hour period, these two images can be used together to identify cloud types.

The first clouds that will be examined here will be the broad cloud shield that is centered over South Dakota, Nebraska, and Iowa. In the VIS image, these clouds appear relatively bright, indicating that they are thick. They have a smooth texture and fairly well defined edges. They also appear to be arranged in a sheetlike pattern. When viewed in the IR image, they are a dull gray, indicating warmer temperatures and lower altitudes. These observations lead to the conclusion that these clouds are layered stratus clouds.

Off the coast of California there is a group of clouds that are arranged in a cellular pattern. These clouds have a lumpy texture, and they appear to be cumuliform. In the IR image, they are very dull gray, indicating that they are low and warm clouds. The cloud top temperatures for the clouds in this group are all approximately the same, indicating that there is a limit to how high these clouds can grow. This is a characteristic of stratiform clouds. These low clouds, which exhibit characteristics of cumuliform and stratiform cloud types, are stratocumulus clouds.

Far out in the North Pacific Ocean, there is a comma-shaped cloud that is part of a developing storm system. This comma can be said to have a head, and a tail extending away from the head. Storm systems like this are characterized by a mixture of many different cloud types. In the VIS image, most of the clouds have the same brightness, making it difficult to tell the lower clouds from the higher clouds. In the head of the comma, just below the center of rotation, there are many clouds that are fairly bright and linear. These clouds are barely visible in the IR image, indicating that they are very low and warm clouds. These are cumulus clouds that are arranged in lines oriented parallel to the wind direction in the developing storm. In the long tail that extends from the comma, the clouds are very bright; however, this only tells us that there is a thick layer of clouds present. When the tail is observed in the IR image, however, the cold, high cirrus clouds appear very bright while the warmer, lower clouds appear dull gray. Notice that the IR image clearly shows that the cirrus shield does not curve around the storm system.

Finally, along the eastern edge of the Rocky Mountains there is a series of triangular-shaped clouds. In the VIS image, these clouds can be observed as very narrow, bright points that spread out to the east. The tops exhibit a smooth texture. In the IR image,

Figure 6.7. Pacific Ocean cloud features (GOES VIS, 2001 UTC, June 2, 1993)

Figure 6.8. Pacific Ocean cloud features (GOES IR, 2031 UTC, June 2, 1993)

the cloud tops are very bright, indicating that they are cold and high. These are cumulonimbus clouds. The tops of these clouds are being blown to the east by strong upper-level winds. These smooth tops are often referred to as anvil cirrus clouds. The points, or the apexes, of the triangles point upstream with regard to the upper-level winds. It is very common to see such clouds and their associated thunderstorms form on the mountains, especially in the summer months when heating of the land is greatest.

Figures 6.9 and 6.10 are a VIS and IR pair depicting a storm system in the southeastern United States. The center of the storm is located near the Kentucky and Tennessee border. One of the striking differences between the two images is the cloud cover over southern Texas and the Gulf of Mexico. In the VIS image, an extensive cloud shield is easily visible in the region; however, it is barely detectable in the IR image, indicating a very low, warm cloud layer. The smooth texture in the VIS image indicates that this is a layer of very low stratus clouds and possibly fog. When compared to a physical map of this region, it is also evident that this low layer of clouds is banked up against mountain topography along its western edge, a further indication that these are low-lying clouds.

In mid-latitude storm systems, there are often many different layers of clouds at many different heights. Distinguishing between different cloud types can aid in locating the boundaries between the different segments of a storm. In the VIS image, a narrow band of very bright, lumpy clouds can be seen running roughly from southwest to northeast. It starts in the Gulf of Mexico, along the eastern edge of the stratus shield, and extends into southern Louisiana. In the IR image, this line of clouds appears very bright and exhibits the windswept appearance of anvil clouds, indicating a line of cumulonimbus clouds. This line of storms marks the cold frontal segment of this mid-latitude storm system, which is a boundary between colder air to the west and warmer air to the east. Cold fronts are often associated with cumuliform clouds and thunderstorms. Another region of cumulonimbus clouds can be seen behind the leading line of storms. In the VIS image, they can be distinguished by the shadows on the lower altostratus and stratus layers. In the IR image, these storms have a wedge, or V, shape that points to the southwest.

In the IR image, a region of warmer clouds can be seen in mid-Tennessee. These clouds appear relatively smooth and slightly less bright in the VIS image, indicating a layer of clouds that is thinner than the surrounding cloud layers. This section of clouds is a layer of altostratus clouds that is embedded in the storm system just ahead of a warm frontal boundary . A warm front separates warmer air to the south from colder air to the north and is often characterized by many layers of clouds that are vertically stacked. In this storm the warm front extends eastward from the center of the storm, through northern North Carolina toward the Atlantic Ocean.

Other middle-level clouds can be seen in Arkansas and northeast Texas. In the IR image there is a sharp line running north to south in eastern Arkansas that separates a very high, cold layer of clouds to the east from a lower, warmer layer to the west. These clouds are not at ground level, since they still appear considerably colder than ground temperature and a layer of stratus clouds to their north and south, indicating a middle level cloud. In the VIS image, this boundary line is marked by a line of shadows cast by the higher clouds on the altostratus layer. The altostratus clouds extend across the edge of the low cloud layer in western Arkansas, providing further evidence of the multi-layer structure.

There is also a thick layer of altostratus clouds along the southern Texas coast. The clouds are brighter than the nearby stratus clouds in IR imagery and cast shadows on the stratus clouds in the VIS imagery.

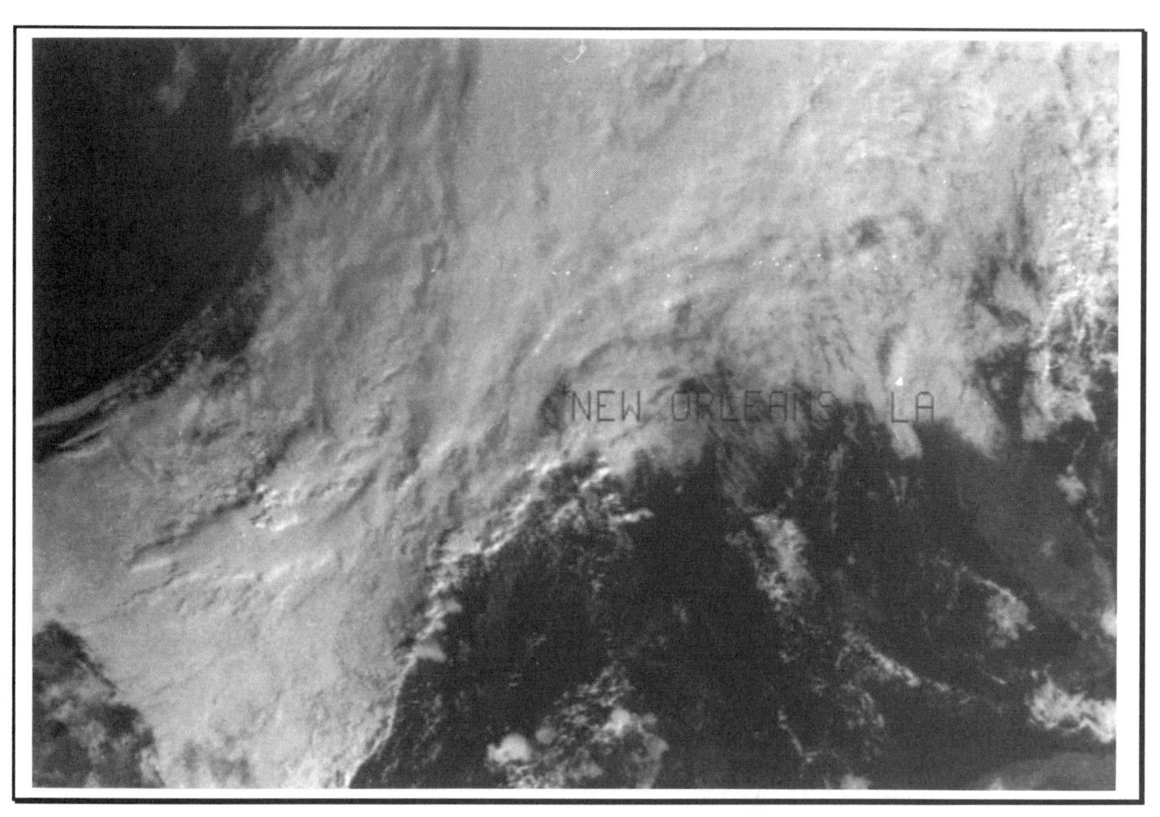

Figure 6.9. Southeastern U.S. storm system (GOES VIS, 1501 UTC, December 10, 1994)

Figure 6.10. Southeastern U.S. storm system (GOES IR, 1501 UTC, December 10, 1994)

Detection of small and thin clouds

Detection of very small clouds or thin clouds may be difficult in VIS imagery because of the sensor's spatial resolution. In an image with 1 km (0.6 mile) resolution, each pixel represents the average brightness over an area of 1 km x 1 km. In an area of clouds smaller than 1 km, or in an area of very thin clouds that transmit large amounts of visible light, part of the reflected light sensed by the satellite is reflected from clouds and part is reflected from the ground (figure 6.11). The resulting gray shade depicts the average reflectivity over the area, and is darker than the cloud area but lighter than the surface. Because of this, small, thin clouds are often difficult to distinguish from the surface of the Earth in VIS imagery.

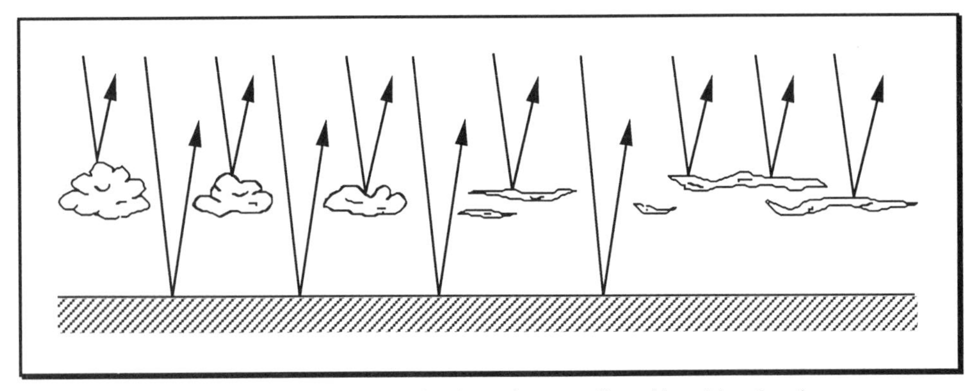

Figure 6.11. Problems in detecting small and/or thin clouds

In IR imagery, smaller cloud features are especially difficult to detect, since IR sensors often have lower resolution than VIS sensors. IR sensors on the GOES 7 satellites have a resolution of 7 km, which means that all brightness values over a 7 km x 7 km area are averaged into one pixel. This causes small or thin cloud features to be smeared on an IR image and to appear to be more extensive than they actually are.

Figures 6.7 and 6.8 illustrate the problem with detecting small clouds in satellite imagery. On the western coast of Mexico, there is an area of cumulus clouds that have formed over the Sierra Madre mountain range. In the VIS image, which has a resolution of 1 km, many small cloud elements can be seen. In the IR image, with a resolution of only 7 km, the smallest cloud elements are not visible. Additionally, the IR sensor has averaged the brightness values. The effect is that the clouds that can be seen appear larger in the IR image than they actually are.

Figures 6.12 and 6.13 also illustrate the problems associated with detecting small or thin clouds. In figure 6.12, a VIS image of the southwestern United States, thin cloud cover over Utah, Arizona, New Mexico, and Colorado is difficult to detect since the satellite sensor sees through the cloud and averages the ground's reflectance with the cloud's. In figure 6.13, an IR image taken at the same time, the cloud cover appears more extensive than it actually is. This is because the cloud is "smeared" as a result of the lower resolution of the IR sensor.

Figure 6.12. Thin cloud detection (GOES VIS, 2001 UTC, October 22, 1993)

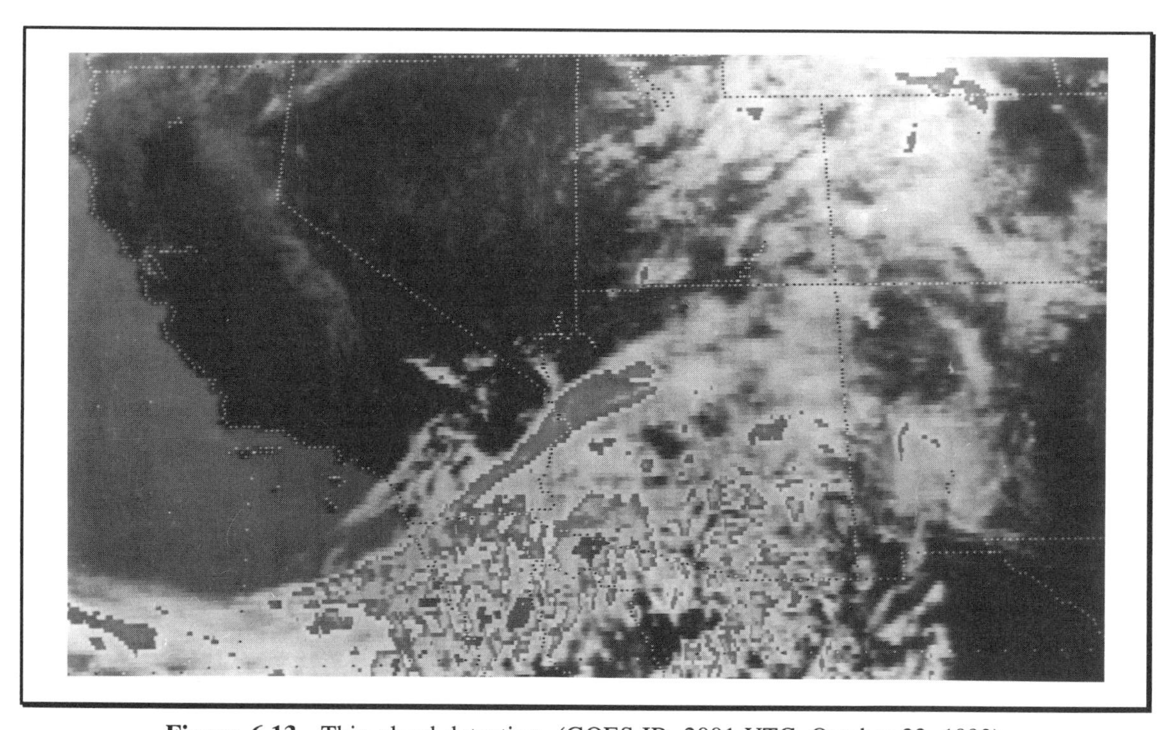

Figure 6.13. Thin cloud detection (GOES IR, 2001 UTC, October 22, 1993)

CHAPTER 7

Determining Wind Direction from Satellite Imagery

Introduction to winds

Wind is a very important phenomenon in our atmosphere. Strong winds can be a very destructive and dangerous force. They can destroy buildings, cause airplanes to crash, and create dangerous wave conditions at sea. Winds can also be beneficial by carrying rainfall-producing clouds to drought-ridden areas or by carrying dangerous smog and pollution away from a heavily populated area. Because of their importance to us, an understanding of winds is essential. One of the many advantages of weather satellites is that they can provide information about low-level wind direction and speed. Although wind itself is not visible in satellite images, many indicators exist in the cloud patterns that offer clues about wind speed and direction. Once the patterns are known, determining surface wind direction is relatively easy.

Winds blow because of differences in heating across the surface of the Earth and because of pressure differences. Air over warmer regions becomes heated and rises; air over cooler regions sinks. This leads to differences in surface air pressure. Pressure tends to be lower in regions of rising air and higher in regions where air is sinking. Horizontal differences in air pressure are referred to as **pressure gradients**. At the surface, air will flow from areas of high pressure to areas of low pressure. This horizontal flow of air is felt as wind. Generally, the stronger the pressure gradient, the stronger the wind. Pressure gradients develop on both global and local scales. **Global wind movement** consists of the movement of large air masses and jet stream flow, while **local wind flow** involves a smaller quantity of air that affects a smaller area. This chapter will focus mainly on determining the direction of winds on a local scale.

Since wind cannot be directly observed in a satellite image, the direction and strength of the winds must be inferred from patterns in the clouds. To understand what causes these patterns to appear in satellite imagery, you need to understand the concept of **converging** and **diverging** airflow. Air is a fluid and flows across the surface of the Earth in many directions. When two streams of flowing air come together, it is known as **convergence**. Areas of low-level convergence usually exhibit deeper convective cloud formation than surrounding areas. **Divergence** occurs when airflow spreads apart. Very little cloud development is associated with low-level diverging air. In general, low-level air converges near lows and weather fronts; it usually spreads apart in highs. At upper levels, air tends to diverge over lows and converge over highs, the reverse of what happens below, at lower levels.

Convective features indicating low-level wind direction

Convection typically occurs when the air above the ground is unequally heated. This causes warmer air to become less dense and to rise, while cooler air nearby sinks. The rising air cools and forms convective-type clouds and possibly precipitation. By studying these convective areas from space, we can often determine the wind direction from the shapes and patterns created by the convective clouds.

Cloud streets and lines

During the winter months, when the land is cooler than the oceans, cold air may move off the eastern and southern coasts of continents and over the warmer ocean water, where it is heated and moistened from below. This air rapidly rises, and as a result, cumulus clouds form (figure 7.1, top). These clouds are usually found in rows that are parallel to the wind direction. These rows, known as cloud streets, consolidate downwind into solid lines. If an inversion is present, vertical development is limited and the clouds will spread into stratocumulus lines. From space, these cumulus and stratocumulus lines—which, like cloud streets, tend to form parallel to the low-level wind direction—are easy to recognize (figure 7.1, bottom). They are usually good indicators of low-level wind direction when they are young and near the coast. Far off the coast, other factors affect the orientation of the lines and they are not reliable wind direction indicators.

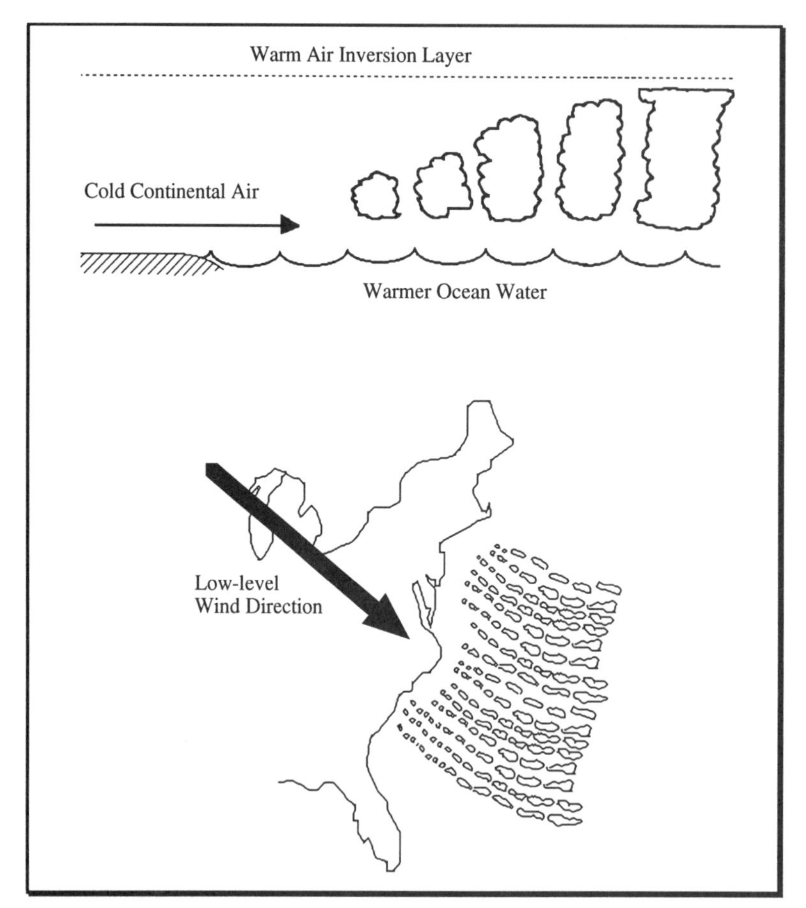

Figure 7.1. Formation of cloud streets. View from side (top) and from above (bottom).

Figure 7.2 is a VIS HRPT image of the east coast of the United States, and figure 7.3 is an IR image of the same scene. Cold air flowing from the northwest is moving across warmer ocean waters. As a result, convective cloud lines have formed parallel to the lower-level wind direction. The clouds are thick, so they appear bright in the VIS wavelengths. However, these clouds are relatively low and warm, and they do not appear bright in the IR image. Farther out in the ocean, the lines spread out and their orientation is influenced more heavily by larger-scale weather patterns than by low-level wind direction.

Figure 7.4 is a VIS HRPT image of Greenland. North is to the upper left, and Greenland is in the upper center portion of the image. Cold air from the northwest is blowing off the coast of Newfoundland and moving over the warmer waters of the Labrador Sea. Here the air is heated and numerous cloud streets have formed. Close to shore, these lines are parallel to the wind direction. Farther off shore, the orientation of these cloud lines is affected by the northeast winds blowing off Greenland and by the circulation around three polar low-pressure systems that are located off the southern tip of Greenland. In these low-pressure areas, the orientation of the cloud lines is not a reliable indicator of wind direction.

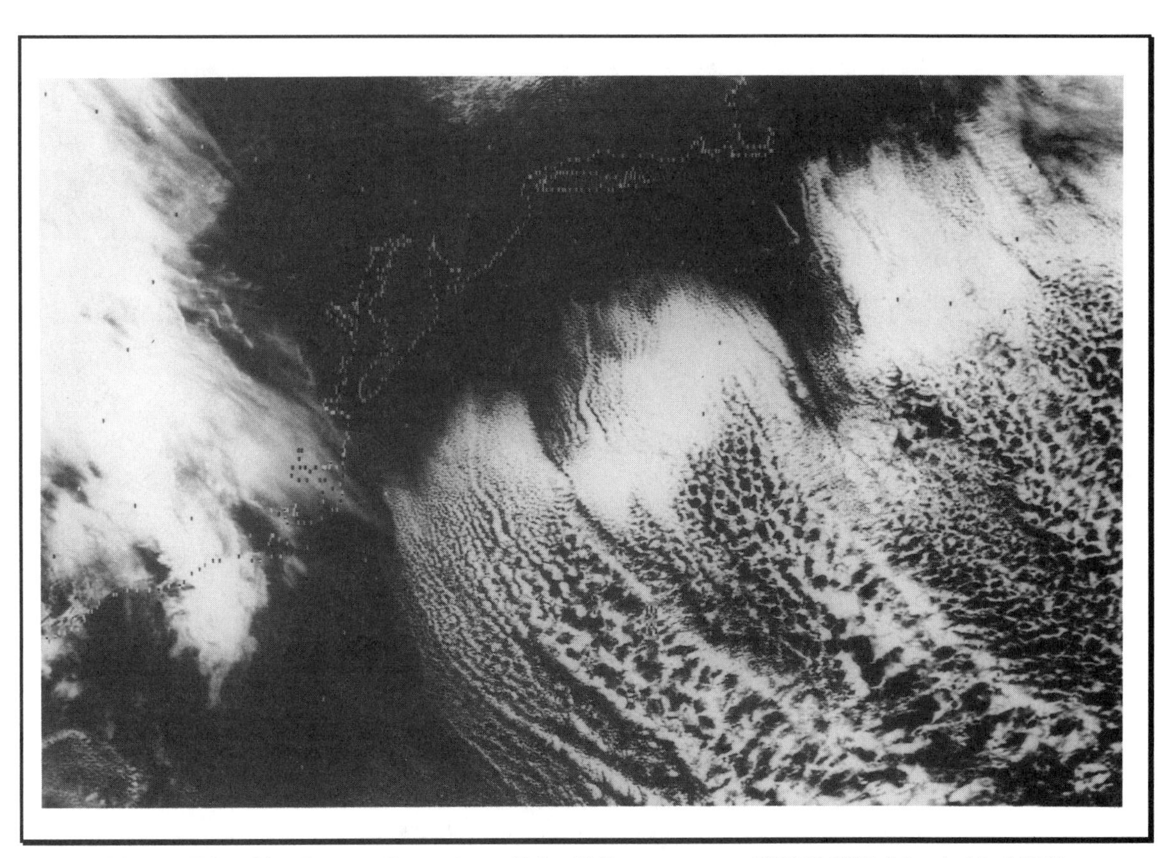

Figure 7.2. Cloud street formation off the U.S. east coast (HRPT VIS, March 12, 1991)

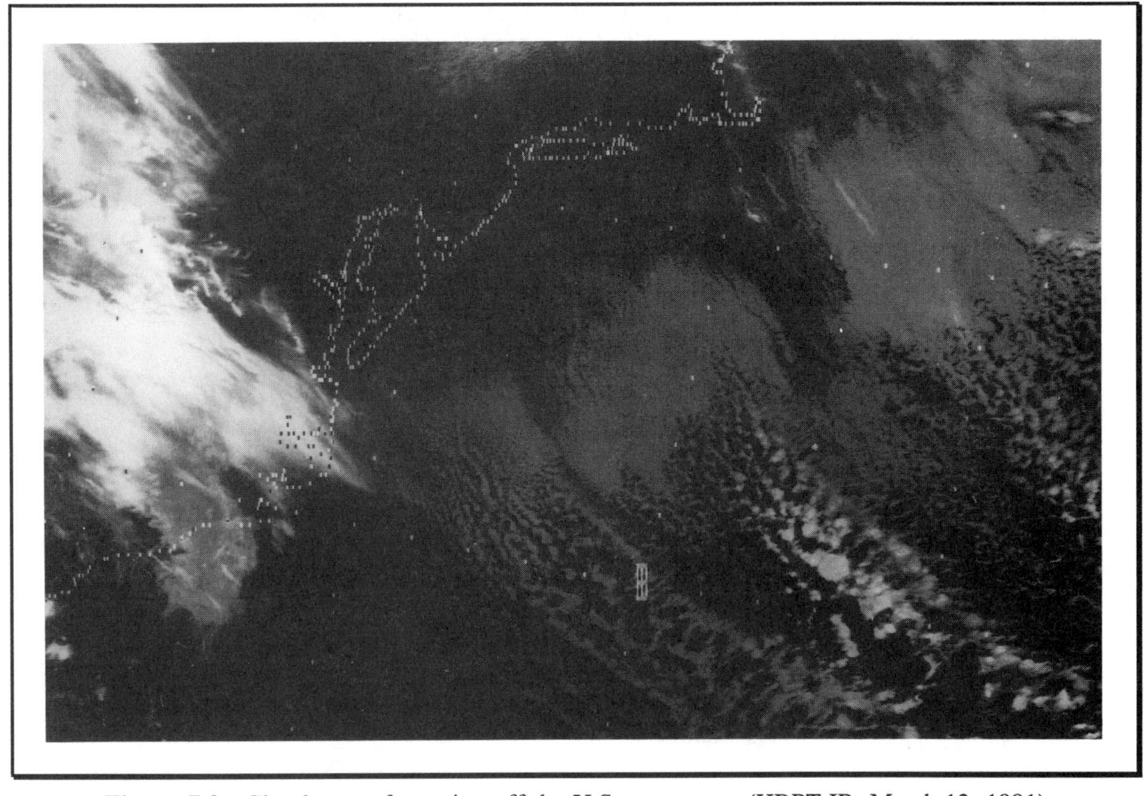

Figure 7.3. Cloud street formation off the U.S. east coast (HRPT IR, March 12, 1991)

Figure 7.4. Cloud streets forming in polar regions (HRPT VIS, January 30, 1991)

An ocean is not the only place cloud streets form. Cold winds blowing across the warmer waters of a large inland body of water will often form cloud streets over the lake, extending to the lee shores. In the Great Lakes region these clouds often bring localized heavy snow to parts of western New York, Michigan, and northern Ohio. These **"Great Lakes effect"** snows and their convective bands can often be seen in satellite imagery.

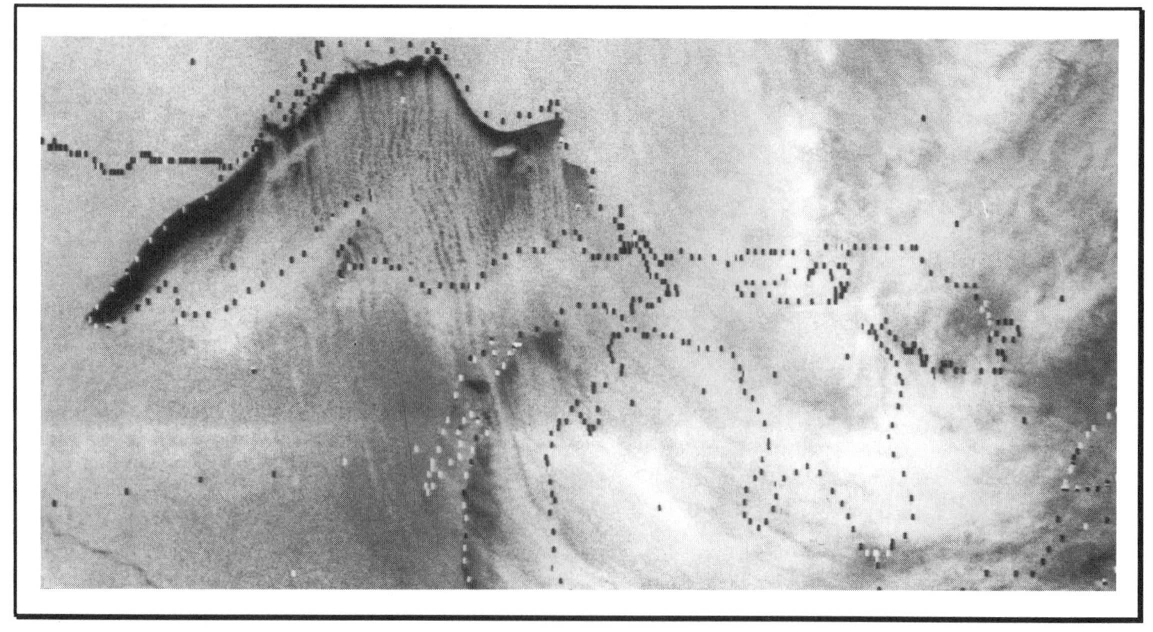

Figure 7.5. Cloud streets over Lake Superior (HRPT VIS, February 15, 1991)

Figure 7.5 shows what happens when cold Arctic air from the north is heated as it crosses over the warmer waters of the Great Lakes. On this day, cloud lines have formed over Lake Superior that extend over the south shore in Minnesota, bringing snow to the region. These "lake effect" cloud streets can extend for hundreds of miles, bringing snow to areas as far away as the east coast of the United States.

Sea and lake breezes

During the day, land surfaces heat up much faster than adjacent water surfaces. As the air in contact with the land is heated, it rises and is replaced by the cooler air over the water. The local onshore wind that is formed as a result of this convection is called a **sea breeze** (figure 7.6, top). Where the cooler and warmer air come together, air is often forced to rise due to differences in the air's density. Along this boundary, often called a **sea breeze front**, convective clouds and thunderstorms may develop. This is often a daily occurrence along coastlines in tropical regions. Evidence of a sea breeze front can be detected in figure 7.7. Clear skies over a coastline and adjacent offshore waters and an inland area of cumulus clouds are good indicators of an onshore, or sea, breeze.

Lake breezes also develop similarly around the larger inland bodies of water. Lake breeze fronts along the shores of the Great Lakes often appear in satellite imagery. In figure 7.8, an HRPT image of the Great Lakes region, the air over Lake Superior and nearby smaller lakes remains cloud free, while an area of cumulus clouds is apparent inland, indicating a lake breeze. For both lake and sea breezes the wind is onshore, generally perpendicular to the coastline.

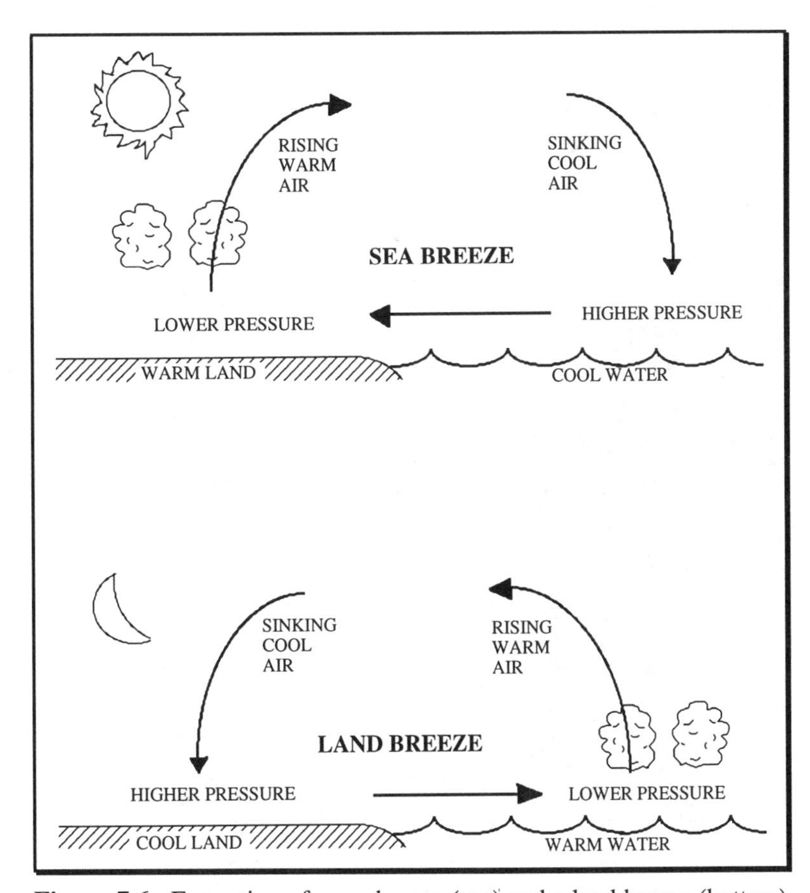

Figure 7.6. Formation of a sea breeze (top) and a land breeze (bottom)

Figure 7.7. Sea breeze (GOES VIS, May 25, 1993)

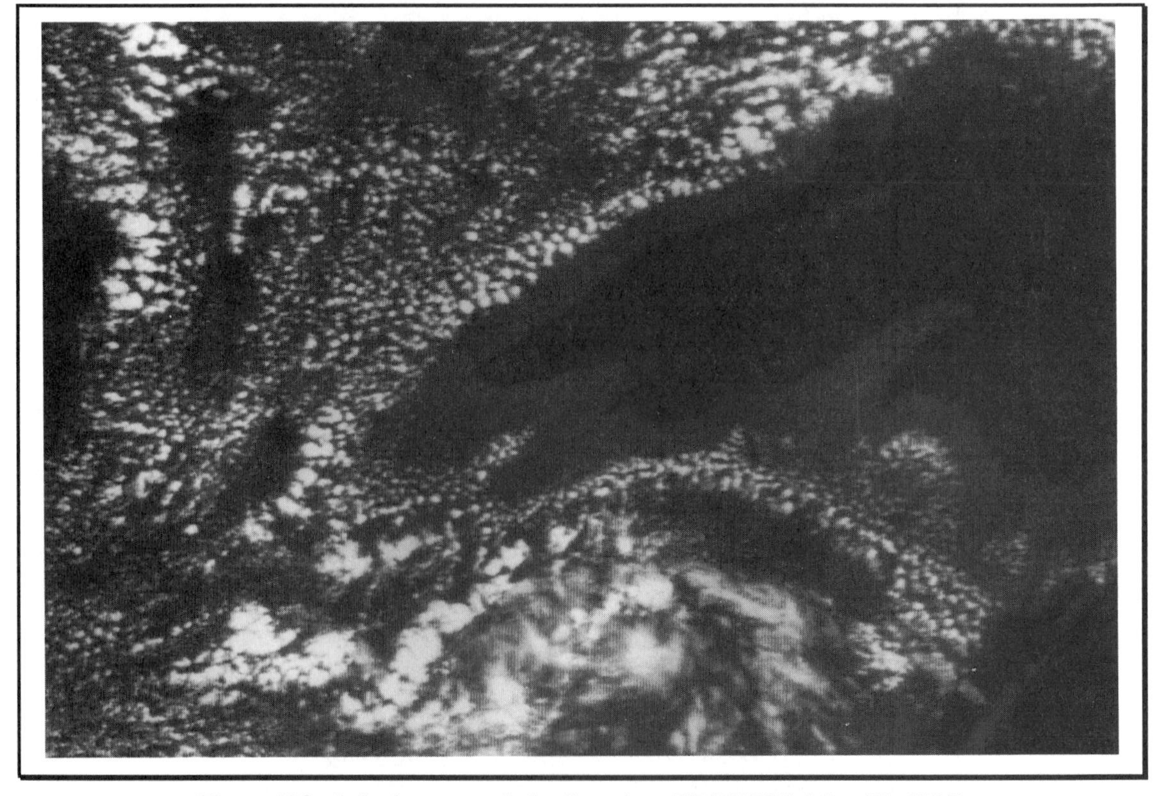

Figure 7.8. Lake breeze on Lake Superior (HRPT VIS, May 27, 1990)

Land breezes

At night, land surfaces cool more rapidly than the surrounding water surfaces and can become cooler than the water. When this occurs, cooler air from the land will flow offshore as a **land breeze** (figure 7.6, bottom). A line of cumulus clouds will often form along the leading edge of the breeze front just off the coastline. Localized surface winds will generally be perpendicular to the line of clouds. Figure 7.9 is a morning image of Florida. Cooler air flowing from the land is being heated by warmer ocean waters, resulting in a line of convective clouds and thunderstorms offshore.

Figure 7.9. Land breeze off the Florida coast (GOES VIS, July 15, 1993)

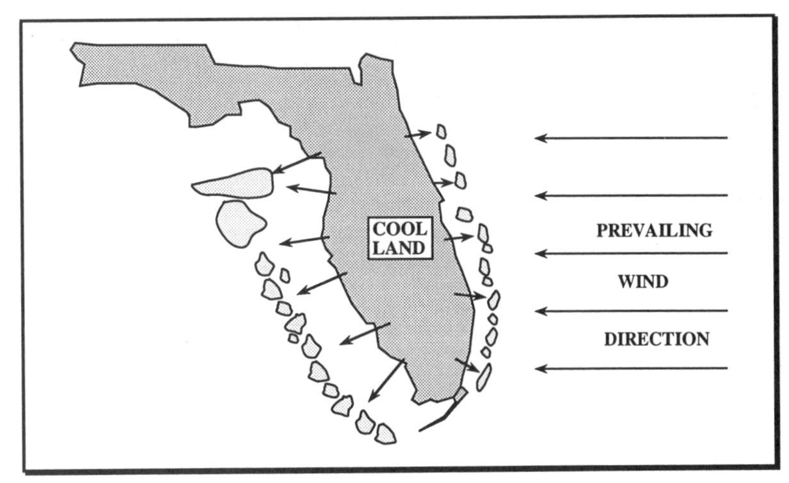

Figure 7.10. Influence of prevailing winds on land breeze front development

Prevailing winds can also influence the development of a land breeze front. Figure 7.10 illustrates the impact of prevailing winds on a land breeze front. In the diagram, prevailing (warmer) winds from the east have prevented a strong land breeze front from forming off the east shoreline of Florida; meanwhile, the stronger push of cooler air off the western shore has helped create a strong line of convective clouds west of the peninsula. The cloud line off the west coast of Florida is also more advanced downstream with respect to the prevailing winds.

The lake effect

The distribution of clouds around a lake can help you determine wind direction. If a cumulus cloud pattern exists, the clouds will extend to the **upwind (windward)** side of the lake and dissipate over the cooler water surface. The **downwind (leeward)** side of the lakeshore will be clear, and the clouds will re-form some distance away from the lee shore. Figure 7.11 illustrates this pattern. Cumulus clouds extend to the water line on the northern and western shores of Lake Erie and Lake Huron, while the southern and eastern shores remain cloud free. This indicates a prevailing wind direction from the northwest.

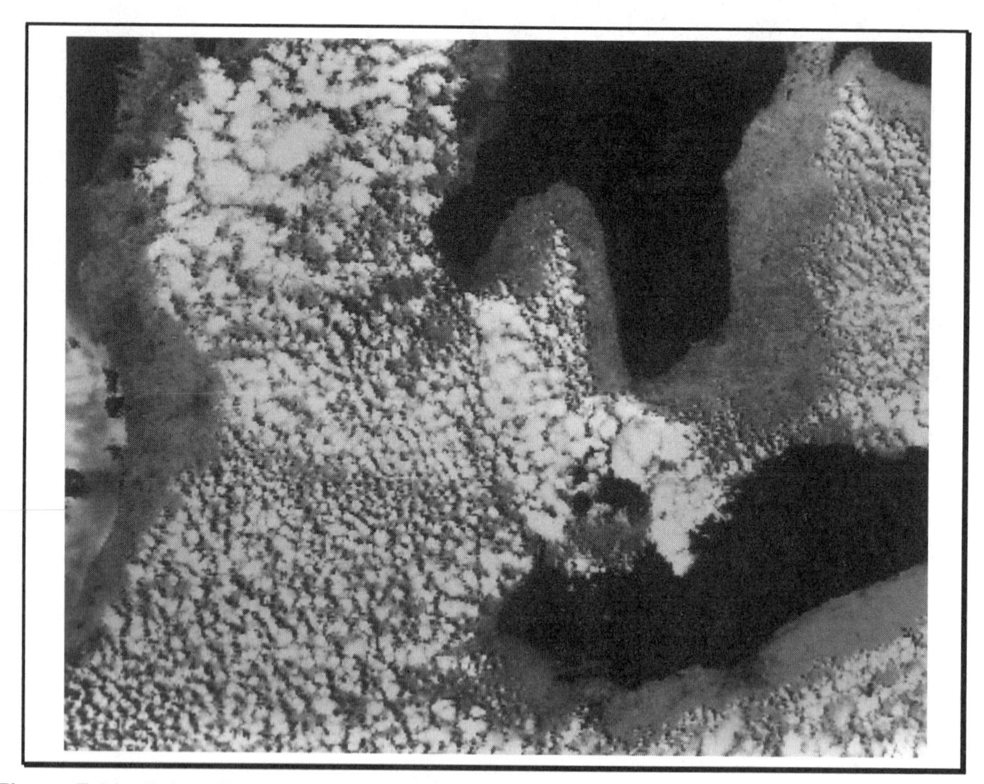

Figure 7.11. Lake effect on cloud patterns over the Great Lakes (HRPT VIS, May 28, 1992)

Mountain upslope winds

Local winds can also form along mountains and highlands as air flows up and down terrain slopes. As the sides of a mountain heat up during the day, the air adjacent to the mountain slope is warmed, causing it to rise. This is known as a **mountain upslope wind**. If the air cools below its dew point, clouds will form over the mountain. These areas are often easy to locate in satellite imagery when the areas surrounding the mountains are clear and convective clouds are only seen over mountainous regions. When the heating

Figure 7.12. Mountain convection (GOES VIS, August 20, 1992)

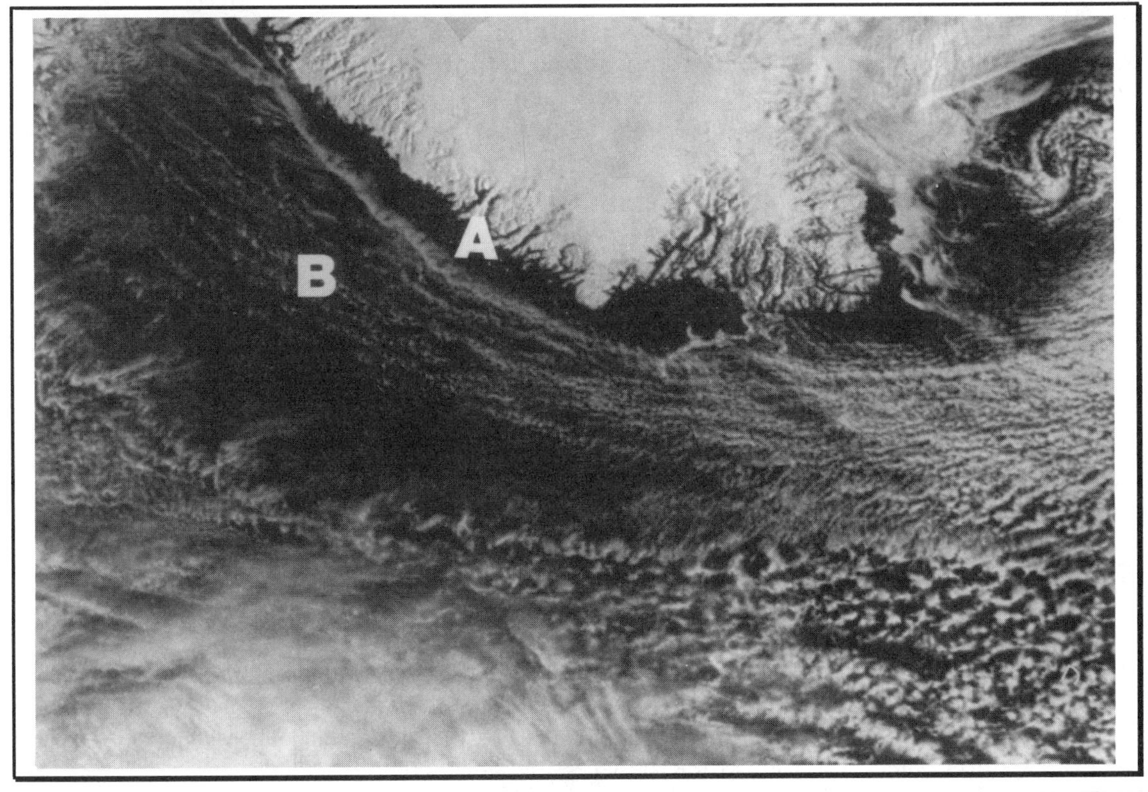

Figure 7.13. Katabatic flow off the coast of Greenland (HRPT VIS, April 2, 1981)

is very intense, thunderstorms can develop over the mountains. In figure 7.12, many of the thunderstorms are a result of mountain upslope winds. A long line of mountain thunderstorms is seen in Mexico along the Sierra Madre mountain range. Several large storms are also visible over the Rocky Mountains and the higher elevations in the southwestern United States. Smaller storms formed by mountain winds are visible along the highlands of Baja California.

Katabatic (mountain downslope) winds

Since cold air is denser than warm air, it tends to flow downhill, much as a liquid would. These downslope winds are called **katabatic winds**. When katabatic flow meets with other streams of wind, convergence occurs and a thick line of cumulus clouds may form. The influence of katabatic flow on cloud patterns can be seen in figure 7.13, a NOAA VIS image of southwestern Greenland and the Labrador Sea (north is to the upper left of this image). Cloud streets, caused by cold northerly winds coming into contact with the warmer ocean water, can be seen at point *B* in the Labrador Sea. Along the west coast of Greenland (at point *A*), very cold, dense air is flowing down the steep valleys and over the Labrador Sea. This katabatic flow converges with the northerly winds almost at a right angle. Along this zone of convergence, the band of clouds appears much thicker and exhibits more pronounced convection than the surrounding cloud streets.

Open-cell convection

When a cold unstable air mass moves over warmer ocean water, **open-cell convection** is likely to occur. Small, localized rain or snow showers occur, and the precipitation cools the air. As the showers quickly dissipate, a ring of cumulus clouds is left that marks the outer edge of the precipitation-cooled air. These clouds, known as **open-cell cumulus**, are different in appearance from the traditional **closed-cell convection** (described in chapter 6). Open-cell cumulus clouds will often appear as a nearly circular ring of clouds surrounding a cloud-free center. Clouds with this characteristic shape usually form in relatively calm conditions. Moderately strong winds will tend to give these rings an elongated shape, with the longer axis usually oriented parallel to the wind direction. In very windy conditions, the trailing windward edges of the rings in a field of cells are often cloud free; therefore the open end is usually pointing upwind.

Figure 7.14 is an HRPT IR image taken over the Pacific Ocean. In this image, many ring-shaped cloud features can be seen which are a result of open-celled convection. The rings are elongated, and in many cases the western portion of the ring is missing and the eastern portion of the ring is enhanced. This is an indication that the wind is flowing from the west.

Thunderstorm outflow boundaries

Wind direction in the vicinity of thunderstorms can often be determined by studying cloud organization and shape. In some thunderstorms, rain-cooled air sinks rapidly to the ground and moves out from under the thunderstorm cloud. As this air flows radially out of a storm, a line of clouds (called an **arc-cloud line**) forms along the leading edge of the localized, rain-cooled air mass. Often, new thunderstorms develop along this arc-cloud line, especially in an area where two different lines converge. Figure 7.15 is a GOES image that shows a line of thunderstorms extending from the Gulf of Mexico across the southern tip of Florida. An arc-cloud line can be seen along the southern edge of these storms. The local winds from these thunderstorms are radiating away from the storm centers and are perpendicular to the arc-cloud line.

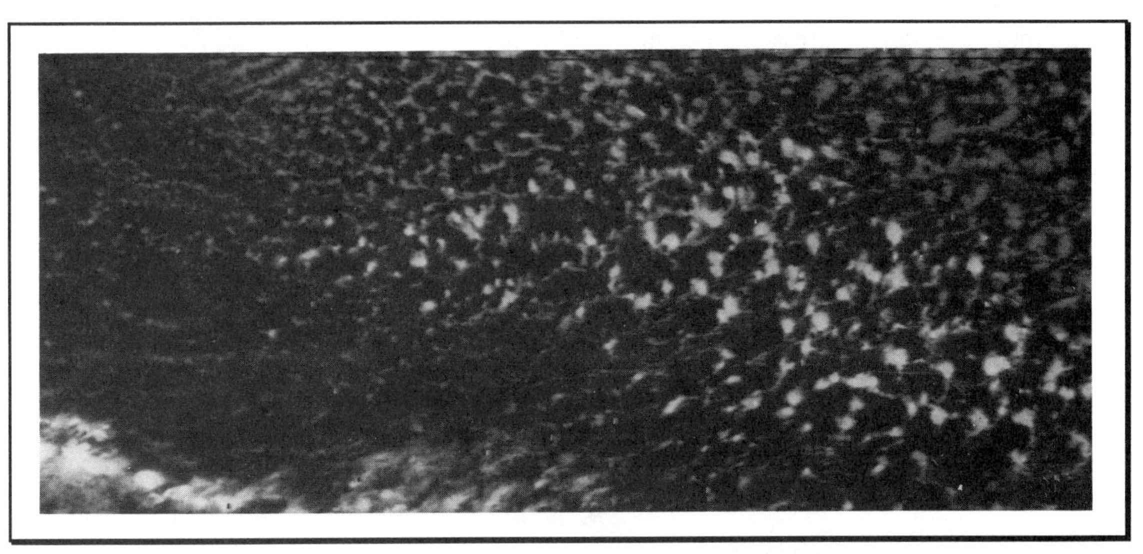

Figure 7.14. Open-cell convection over the Pacific Ocean (HRPT IR, March 11, 1988)

Figure 7.15. Thunderstorm outflow boundary (GOES VIS, February 18, 1986)

Flow around and over mountains

Air flowing around or over mountains often leaves distinct cloud patterns that offer valuable insight on the state of the atmosphere in that region. High mountains often block the flow of winds. On the windward side of a mountain range, large **cloud banks** may build up as the air is essentially dammed up by the mountains; at the same time, the lee side of the mountain range may remain clear.

A very distinct **mountain wave pattern** may appear on the lee side as winds flow over the mountain. This pattern consists of a series of low- to middle-level wavelike clouds that are oriented perpendicular to the wind direction. This pattern can be seen in figure 7.16, a GOES VIS image of the central Appalachian Mountains. As the westerly wind flows over the mountains, it becomes turbulent, rising and sinking in a wavelike manner. Where the air rises, a wave cloud forms. Where the sky is clear, the air is sinking.

The air within the mountain waves can be very turbulent and can become a hazard to aviation; therefore these features are studied extensively by aviation forecasters. The velocity of the winds within a mountain wave pattern can be estimated from satellite imagery using the formula

$$V = 6w + 12$$

[where V = wind velocity (mph) and w = wavelength (miles) of the waves]

To determine the wavelength (w) of the wave pattern in the clouds, you will need to measure the average distance from one wave to the next. To accomplish this, an

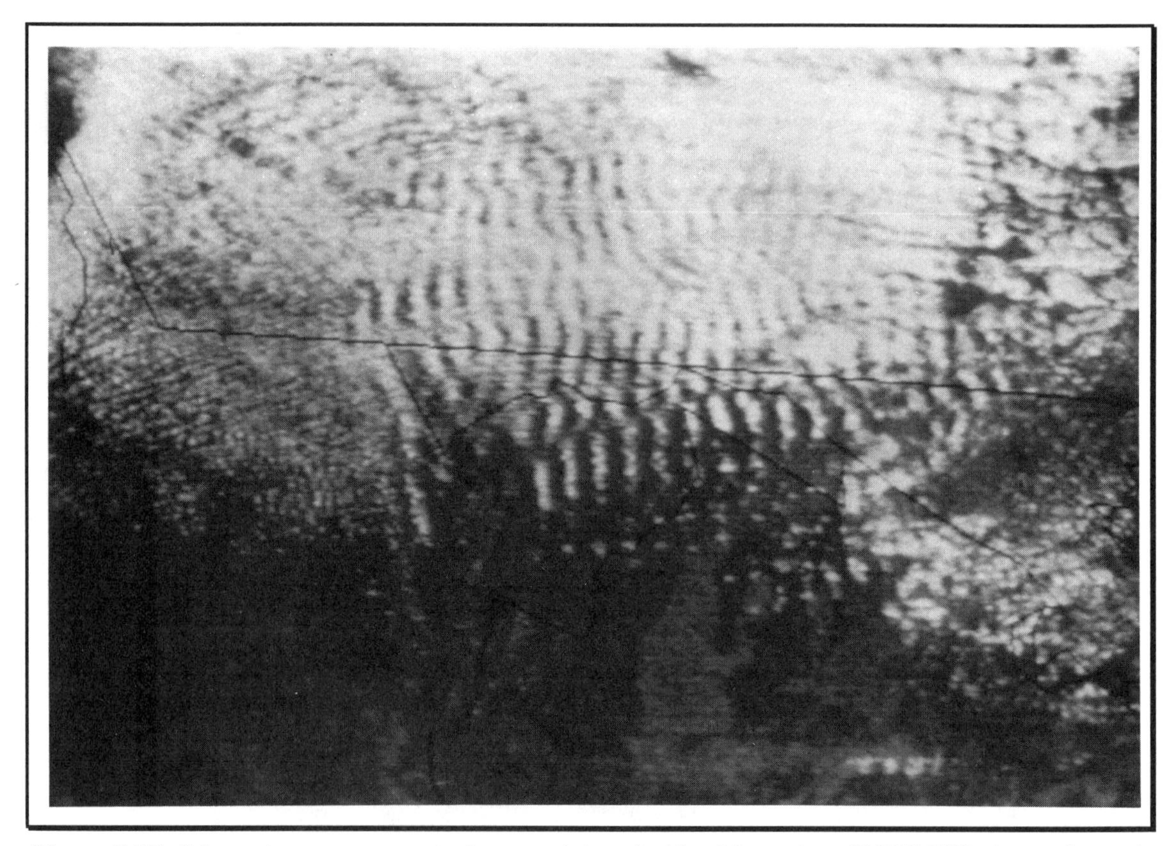

Figure 7.16. Mountain wave pattern in the central Appalachian Mountains (GOES VIS, date unknown)

approximate scale for the satellite image should be developed using two familiar geographical reference points and an atlas or map that contains the reference points. Then, the distance between two mountain waves can be measured on the satellite image and converted to an actual distance using the scale. The resulting distance can then be substituted into the formula.

Air traveling at a high velocity may also be lifted rapidly over a mountain range. Immediately downwind of the mountains, the air sinks, resulting in a narrow, cloud-free zone. Further downwind, the air rises again and may form a broad shield of cirrus clouds, a formation known as **lee high cirrus**. Figure 7.17 is a GOES VIS image of the Rocky Mountains in the United States and southern Canada. Westerly winds are forced to rise rapidly as they encounter the mountains. A cloud-free zone can be seen along the U.S.-Canadian border where the air is sinking on the lee side of the mountains. This zone is approximately at a right angle to the wind direction. A broad shield of lee high cirrus extends downwind in an eastward direction from this cloud-free zone.

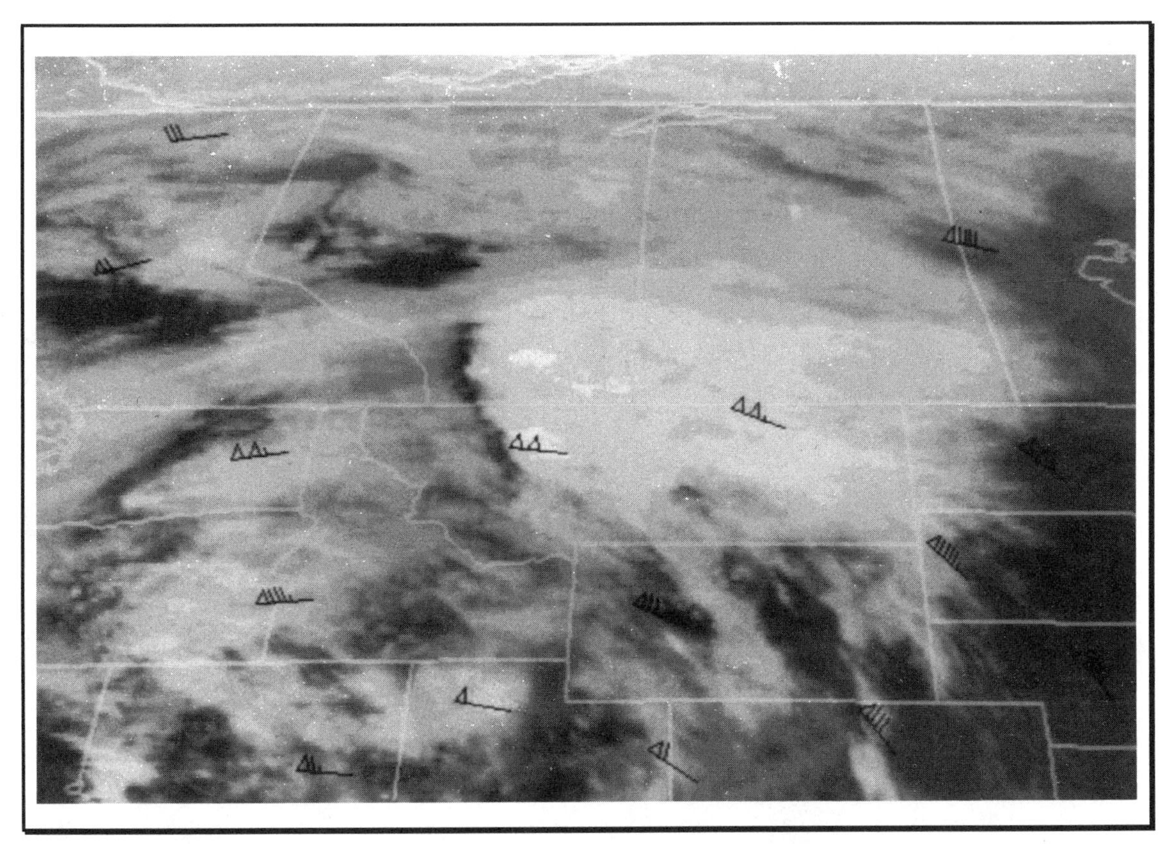

Figure 7.17. Lee high cirrus over the Rocky Mountains (GOES VIS, June 25, 1990) (see Appendix A for a description of the wind symbols)

Island barrier effects

Since islands are not usually very extensive features on the Earth, they do not block the flow of air as well as mountains do. Instead, airflow tends to be diverted around them. The patterns created often can be used as reliable wind direction indicators. In cloudy conditions, a clear "wake" may appear on the lee side of the island. A rounded arch may also appear on the upwind side of the island. Both of these features indicate that airflow is being disrupted by the island. This wake is often wedge shaped, with the point of the wedge pointing downwind. Air traveling over an island can also form an **island wave pattern** that is very similar to the mountain wave pattern, with the waves oriented perpendicular to the wind direction. The island wake and wave patterns often occur together and are reliable indicators of wind direction and turbulence.

Figure 7.18 is a VIS HRPT image of the northern Pacific Ocean. The Aleutian Islands extend across the center of this image (north is to the upper left corner of the image). It is normal for this area to be covered with clouds throughout much of the year, since cold polar winds often come into contact with warmer ocean water, initiating cloud formation. These clouds often exhibit complex patterns that form as a result of the influence of the islands on the air flowing around and over them. In this image, the air is flowing from the north. Where it comes into contact with the islands, it is forced to flow around them. On the right side of the image, along the Alaska Peninsula and the larger Aleutian Islands, many clouds are dammed up on the windward (north) side of the islands, while on the lee (southern) side of the islands there are several cloud-free wakes.

Figure 7.18. Island wake pattern downwind of the Aleutian Islands (HRPT VIS, June 28, 1991)

Island eddies

The wind field can also become deformed downwind of islands, and **island eddies** in the clouds may develop. These eddies appear as a swirling pattern in the clouds. While the individual cloud elements in an eddy are not useful indicators, observing the orientation of a series of eddies is a very reliable method to determine wind direction.

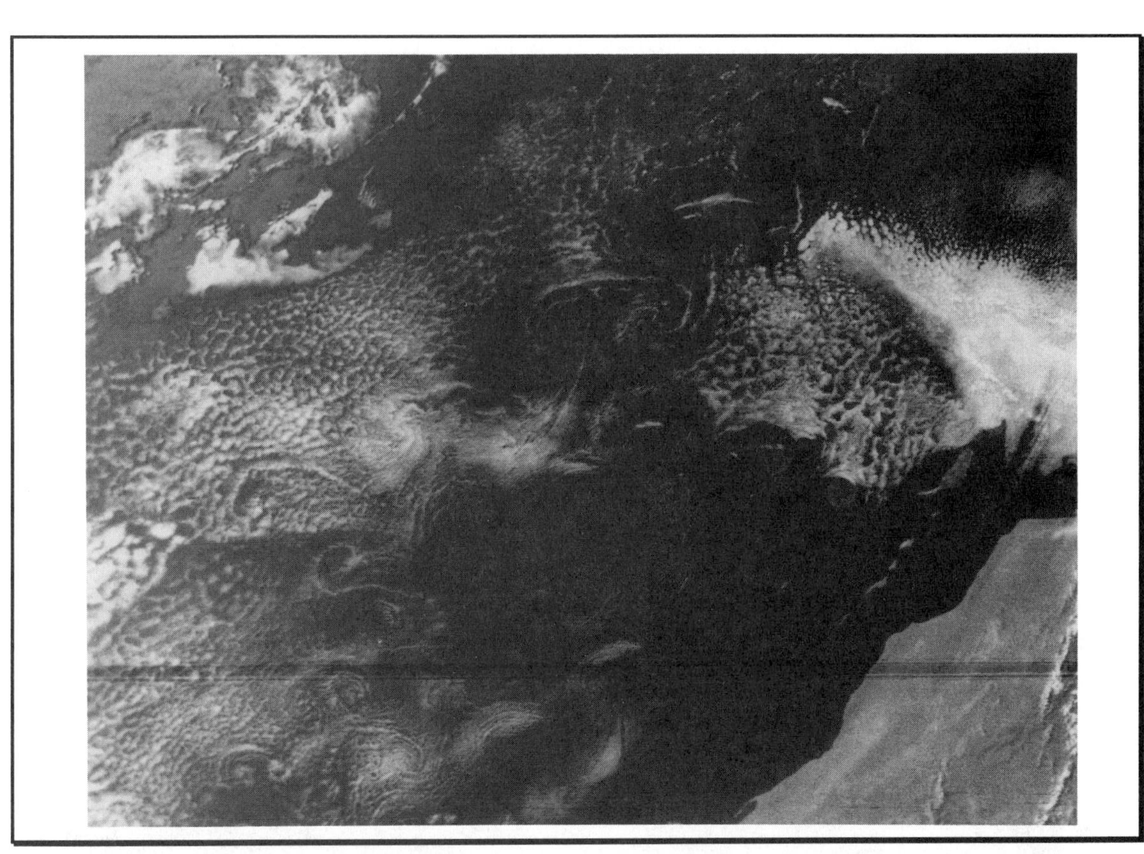

Figure 7.19. Island eddies downwind of the Canary Islands (HRPT VIS, July 26, 1988)

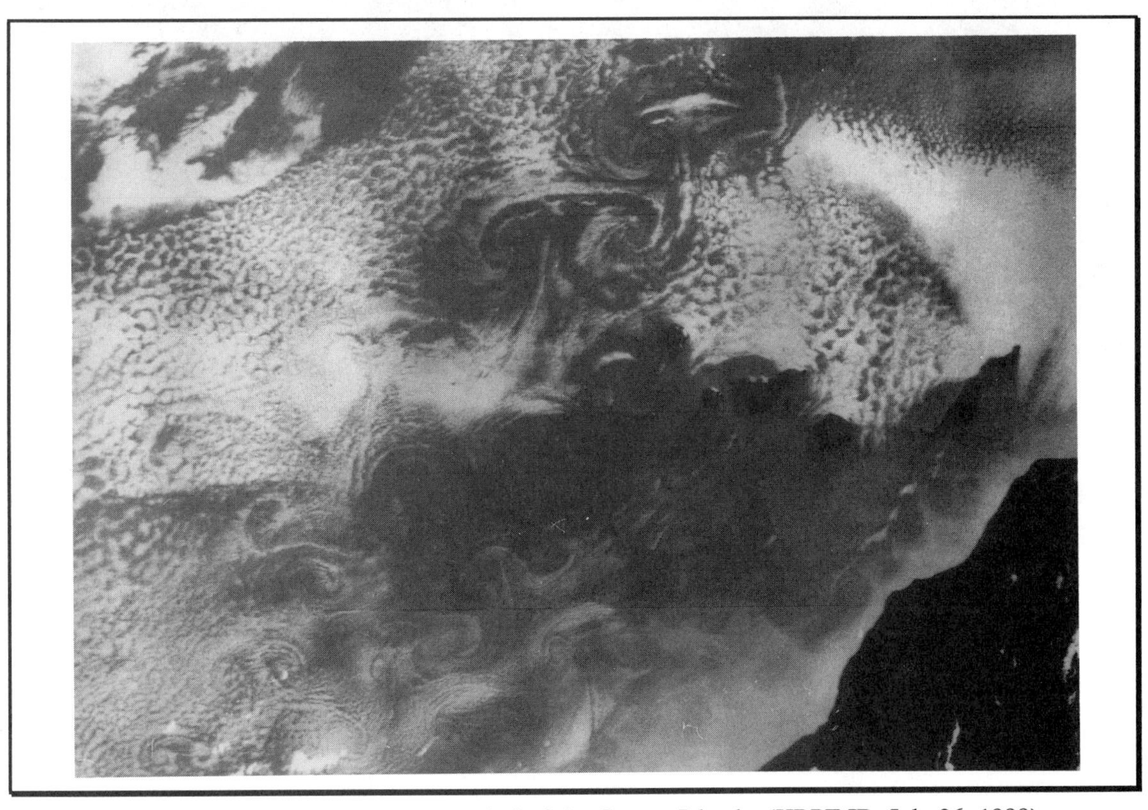

Figure 7.20. Island eddies downwind of the Canary Islands (HRPT IR, July 26, 1988)

Figures 7.19 and 7.20, a VIS and IR pair, were taken over the northwestern coast of Africa. The small islands immediately off the coast of Morocco are the Canary Islands. To the north of the Canary Islands are the Madeira Islands (not visible in these images). In this set of images, trade winds from the northeast have been disrupted by the islands. On the windward side of the Canaries, stratocumulus clouds have banked up behind the islands. There are no clouds immediately on the lee (south) side of the islands; however, farther downwind there are large formations of stratocumulus clouds. Within the stratocumulus formations there are numerous spiral eddies that are all oriented in a northeast-to-southwest direction, indicating a northeasterly flow. The high degree of spiraling and the large number of eddies downwind of each island indicate moderately strong winds.

Island lee lines

Island lee lines occur in tropical ocean regions when wind flows around an island and converges in a line downwind from the island. A small line of convective clouds will form along the convergence line that extends downwind from the island. Figure 7.21 is an HRPT image of Cuba and the surrounding tropical islands. Winds blowing from the east are diverted around the islands. Downwind of the islands the air converges and a narrow line of convective clouds has formed. These tropical island lee lines have formed off most of the major islands in the Bahamas and off the northwestern tip of Cuba.

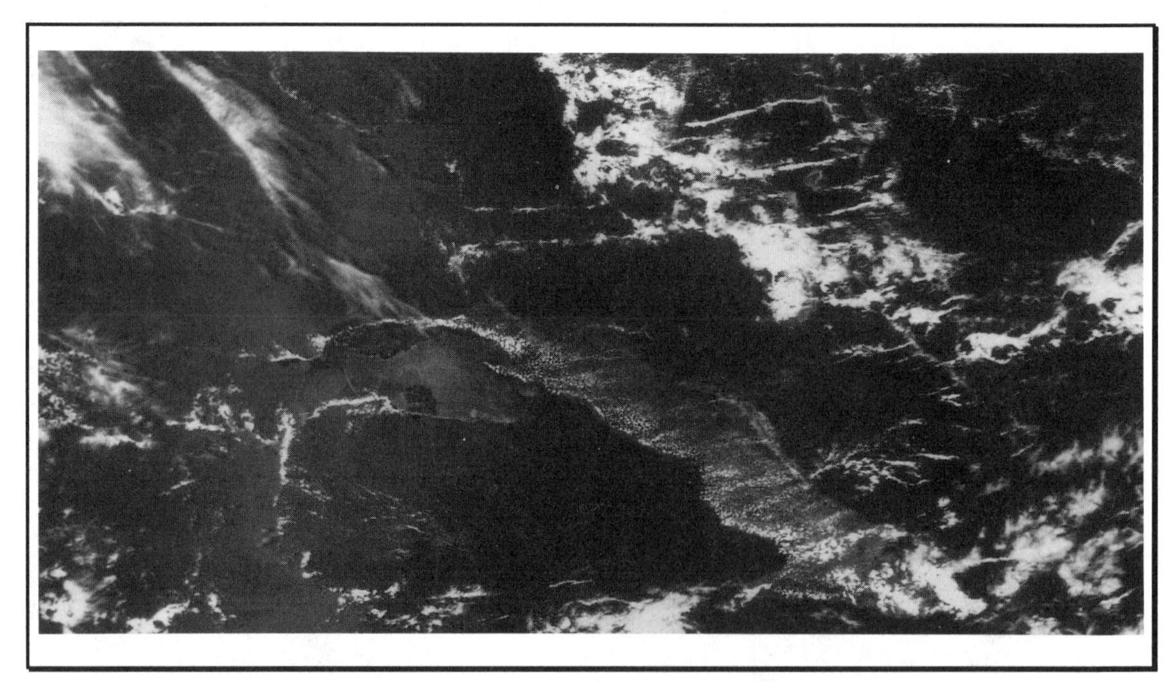

Figure 7.21. Tropical island lee lines (HRPT VIS, April 24, 1990)

CHAPTER 8

Global Circulation

Simple thermal circulation in the atmosphere
The Coriolis effect
A dynamic model of atmospheric circulation
The Intertropical Convergence Zone (ITCZ)

Simple thermal circulation in the atmosphere

Thermal circulation patterns in the atmosphere are responsible for carrying excess heat away from the equator and distributing it to higher latitudes. Thermal circulation is also responsible for movements of air on a global scale. These large-scale wind patterns play an integral part in determining the climate of any given region on Earth. Global winds carry large air masses, moisture, and storm systems across the Earth, causing the weather we experience each day. This chapter will discuss the global circulation patterns and serve as a foundation for later chapters that discuss large-scale weather patterns.

In a simple thermal circulation model of the atmosphere, warmer, less dense air over the equator rises and moves directly toward the poles, while cold, dense air sinks at the poles and moves across the surface toward the equator (figure 8.1). This model assumes that the Earth is not rotating and that its surface is uniform. Although this model illustrates the basic concept of what drives atmospheric circulation, it does not accurately describe the atmosphere. To do so, one must take into account the Earth's rotation.

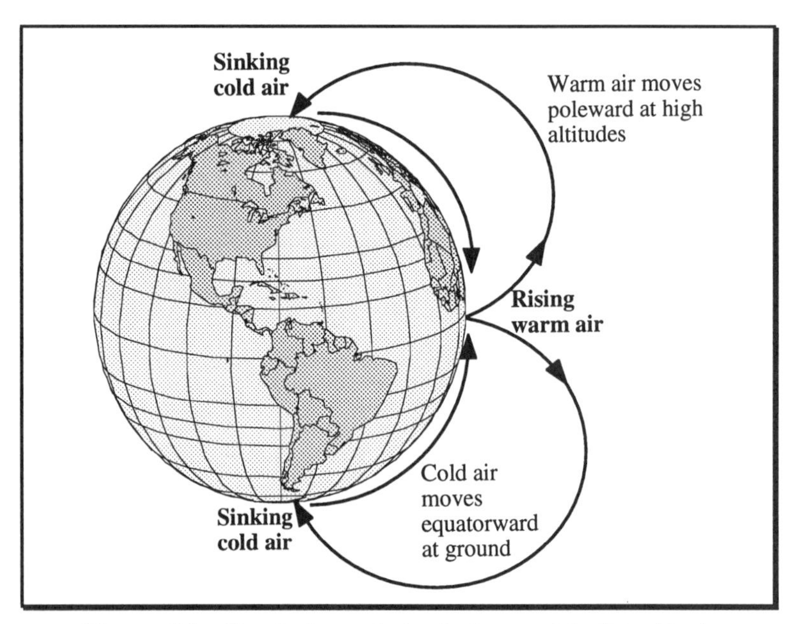

Figure 8.1. Simple thermal circulation model of the Earth

The Coriolis effect

To accurately describe the movement of winds on a global scale, it is necessary to understand the **Coriolis effect**. Picture an object, such as a missile, traveling over the planet in a straight line from north to south. While the missile is in the air, it is no longer connected to the Earth. The Earth rotates slightly under the missile during its flight, and when the missile lands it will be slightly to the west of its original target because the planet moved underneath it. To an observer on the Earth, it will appear as if an unseen force caused the missile to become deflected to the right of its path of travel. This apparent deflection of an object moving over the Earth's surface is known as the Coriolis effect.

Along the same lines, as winds blow across the surface of the Earth the planet rotates under the moving air. To an observer on the Earth, the winds in the Northern Hemisphere appear to have been deflected to the right of their original path. In the Southern Hemisphere the path of the winds appears to be deflected to the left. This deflection of the winds is caused by the Coriolis effect. A similar phenomenon occurs in the oceans as water flows across the surface of the Earth (see chapter 14).

As mentioned in the previous chapter, air will always flow from a region of high pressure to a region of low pressure. The path of the air becomes curved and flows around the high- and low-pressure systems as a result of the Coriolis effect. Air flowing around a high **(anticyclonically)** will circulate in a clockwise direction in the Northern Hemisphere and in a counterclockwise direction in the Southern Hemisphere. This is known as **anticyclonic** flow. Air flowing around a low **(cyclonically)** will circulate in a counterclockwise direction in the Northern Hemisphere and clockwise in the Southern Hemisphere. This is known as **cyclonic** flow. The winds around highs and lows are not exactly circular, but flow away from the center of a high and into the center of a low.

A dynamic model of atmospheric circulation

When one takes the Coriolis effect into account, the thermal circulation model of the Earth becomes more complex (figure 8.2). As air over the equator is heated, it becomes less dense and it rises, moving away from the equator toward both poles. This creates a low-pressure zone around the equator known as the **equatorial low**. As the air moves toward the poles it cools, and **subsidence** (sinking) occurs at about 30° on both sides of the equator, creating an area of high pressure known as the **subtropical high (STH)**. In the Northern Hemisphere the subtropical high over the Atlantic Ocean is known as the **Bermuda high**. The subtropical high over the Pacific Ocean is known as the **prevailing Pacific high**. Winds circulate clockwise around the STH in the Northern Hemisphere and counterclockwise in the Southern Hemisphere because of the Coriolis effect. These

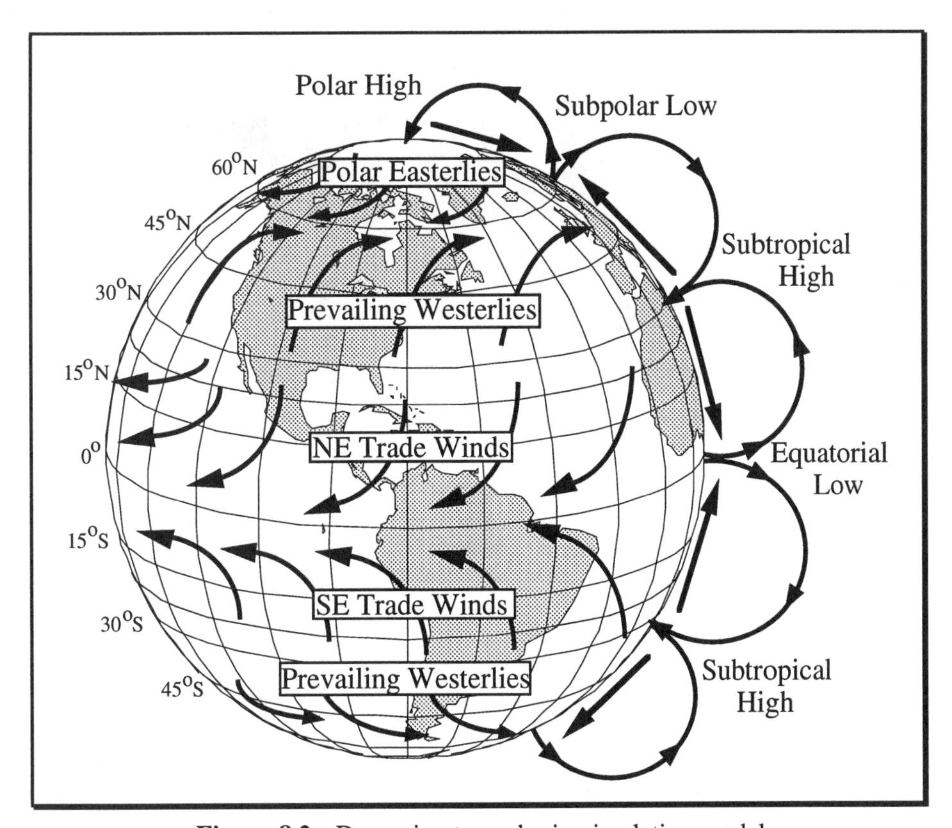

Figure 8.2. Dynamic atmospheric circulation model

oceanic regions are usually free of precipitation and are often covered by stratocumulus clouds. Over land, these regions within the subtropical high are generally characterized by dry weather.

The equatorial low and subtropical high can be seen in figure 8.3. In this image, the equatorial low can be seen as a continuous band of clouds along the equator that has formed as a result of rising warm air. On either side of the equatorial low are regions with little or no significant cloud cover. Here, air is sinking and little cloud formation occurs. These are the Pacific subtropical high regions. Figure 8.4 is an image of the Atlantic Ocean that shows the presence of a Bermuda high. This feature is seen as a relatively cloud-free region over the subtropical Atlantic Ocean. The clockwise circulation that characterizes a high-pressure region (in the Northern Hemisphere) is evident in the orientation of the many cumulus cloud lines along the shores of the southeastern United States.

Figure 8.3. Equatorial low and Pacific high (GOES IR, April 30, 1991)

The subsiding air associated with the subtropical high moves across the Earth as a surface wind. Poleward of the high, winds flow away from the equator. On the equatorward side of the high, the winds flow toward the equator. The surface winds moving away from the equator are deflected by the Coriolis force. In the Northern Hemisphere they come from the southwest, and in the Southern Hemisphere they come from the northwest. These winds are known as the mid-latitude **prevailing westerlies**. Across the continental United States, the majority of our weather comes from the west as a result of these winds. The winds that move toward the equator are deflected by the Coriolis force and come from the northeast (in the Northern Hemisphere) or the southeast (in the Southern Hemisphere). These easterlies are known as the **trade winds**, since early European sea merchants used them to sail their ships across the Atlantic Ocean on their way to the New World.

At the poles, cold dense air sinks toward the Earth, creating an area of high pressure known as the **polar high**. Surface winds blowing away from this high tend to move toward the equator. These winds are deflected so that they come from the northeast (in the Northern Hemisphere) or the southeast (in the Southern Hemisphere), creating a wind belt known as the **polar easterlies**. The air from the polar easterlies meets the prevailing westerly winds. At this point of contact, air rises, and a low-pressure region known as the **subpolar low** forms around 60° N or 60° S latitude.

It is important to understand that the preceding description is only a simplified model of the atmosphere. Other factors influence circulation in the atmosphere as well. Unequal surface heating, mountains, coastlines, ocean currents, and other factors help to modify global wind patterns. The model can be useful, however, in describing various phenomena that are observed in the atmosphere.

Figure 8.4. A Bermuda high (GOES VIS, May 25, 1993)

The Intertropical Convergence Zone

In the tropics, the northeasterly trade winds of the Northern Hemisphere meet the southeasterly trade winds of the Southern Hemisphere. As these trade winds converge, convective cloudiness is often enhanced. The result is a more or less continuous band of clouds over the tropical regions that is known as the **Intertropical Convergence Zone (ITCZ)**. Daily and seasonal position changes of the ITCZ can be observed in satellite imagery. Its position is generally dependent on the location of the sun's most direct rays. When the sun's most direct rays are in the Northern Hemisphere, the ITCZ will be located in the Northern Hemisphere. As the sun's most direct rays migrate towards the south, the ITCZ will follow. When it is winter in the Northern Hemisphere, the position of the ITCZ will be at its southernmost point. The north-south movement of the ITCZ is greatest

Figure 8.5. Pacific Ocean ITCZ (GOES IR, August 1, 1991)

Figure 8.6. Pacific Ocean ITCZ (GOES IR, January 31, 1991)

between Asia and Australia, where it can range from 20° S to 30° N. In the central and eastern Pacific Ocean the movement is very limited, as the ITCZ stays near the equator. In the Atlantic, it is always north of the equator. The average position of the Atlantic ITCZ is around 5° N.

This migration of the ITCZ often causes extreme differences in the weather of many tropical locations. Dry weather may be replaced by heavy rains, often referred to as **monsoons**. But it is not the rainy pattern that is the monsoon, but the seasonal shift of the ITCZ and its winds. These "monsoon" rains can be the primary source of rain in some locations, particularly India and the Sahel region of tropical Africa. As the ITCZ shifts southward later in the year, dryer, offshore winds bring an end to the rainy period. It should be noted that the ITCZ exhibits different characteristics in the western Pacific than in other places in the world. In the western Pacific the ITCZ is often referred to as the **monsoon trough**, and it is associated with southwest winds to the south of the trough rather than with southeast trade winds.

Figure 8.5 is a GOES IR image of the eastern Pacific taken during the summer season in the Northern Hemisphere. Note the nearly continuous, well-developed ITCZ across the center of the image, centered at approximately 10° N latitude. The regions to the north and south of the ITCZ in the STH do not have any substantial storm systems or precipitation. Most of the clouds that appear in the STH are warmer, low-lying stratocumulus clouds that formed as warm air came into contact with cold ocean currents. Poleward from the STH, there are many storms and storm systems in the band of westerlies, especially over North America, where more direct solar radiation caused many thunderstorms to form over the landmasses.

Figure 8.5 can be compared to figure 8.6, a GOES IR image taken during the winter months in the Northern Hemisphere. Notice that the ITCZ is not as well developed as it is during the summer months. Also, it is centered closer to the equator, at approximately 5° N. The subtropical high-pressure zones on either side of the equator remain relatively cloud free, and the storm systems are located in the mid-latitude regions. Since the sun's direct rays are now concentrated in the Southern Hemisphere, many thunderstorms have formed over South America.

CHAPTER 9

Jet Streams

Characteristics of jet streams

Due to large-scale temperature differences between the tropics and the poles, bands of strong westerly winds develop at high altitudes (generally 7–15 km, or 4–10 miles, above the ground) in the middle latitudes. The pattern is not a uniformly west-to-east wind flow but rather exists as a constantly changing and meandering wind field. Within these bands of westerly winds, zones of localized wind speed maxima can be found. The overall wind bands are often referred to as jet streams.

Two bands of westerly winds in the upper atmosphere are usually present in each hemisphere between about 25° and 60° latitude. The poleward band is referred to as the **polar jet stream**, and the southern branch is called the **subtropical jet stream**. These jet streams are an important part of the Earth's heat exchange system, since they aid in the transfer of surplus energy from the tropics poleward and of excess cold from the polar regions equatorward. Jet streams are also a key ingredient in weather forecasting. The polar jet stream is often the boundary between cold polar air and warmer subtropical air. The subtropical jet stream usually separates subtropical air from even warmer tropical air. Each of these boundaries includes a significant temperature gradient and a great deal of dynamic energy; therefore the areas they affect often experience very dynamic weather patterns. Jet streams are also important to pilots because high winds can either speed up or slow air travel depending upon the direction a plane is flying. Turbulence associated with jet streams can also affect aircraft safety and passenger comfort.

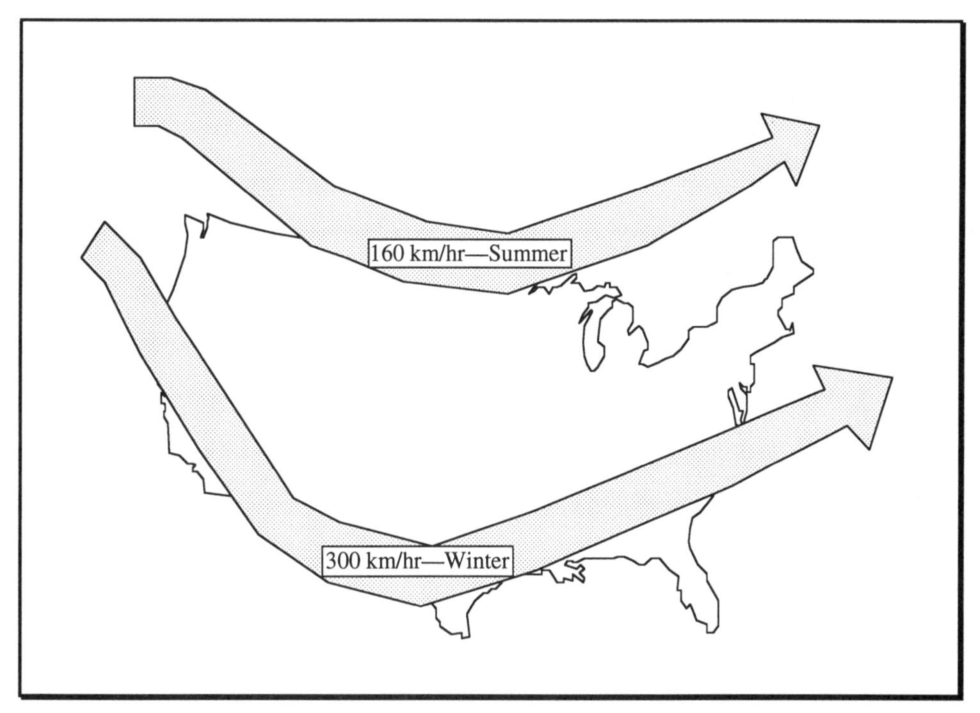

Figure 9.1. Seasonal migration of the polar jet stream over the United States

The characteristics of jet streams change with the seasons. In the winter, for example, the polar jet stream in the Northern Hemisphere is generally located between 30° N and 35° N, and its winds can reach 300 km/hr (200 mph). In the summer, however, the polar jet stream is much farther north (approximately 50° N), and wind speeds may only reach 160 km/hr (100 mph), as shown in figure 9.1. The subtropical jet stream goes through a similar seasonal variation. These changes are related to seasonal changes that correspond to the way in which the area of maximum heating on the Earth migrates throughout the year. The seasonal differences in wind speed within jet streams in the Northern Hemisphere occur because during winter, when the North Pole is in darkness, the temperature gradient between the pole and the equatorial region is at its greatest. This translates into a higher density gradient and a higher pressure gradient, which in turn help create stronger jet stream winds. During the summer months, the temperature difference between the pole and the equator is smaller and the winds are weaker. A similar scenario unfolds in the Southern Hemisphere.

Wave motion of jet streams

As the polar jet stream travels across the midlatitudes it often develops waves, or **meanders**, in its path of travel. These waves, like the waves shown in figure 1.1, can be said to have troughs and ridges. Troughs and ridges are important to understand, since each is associated with its own characteristic weather patterns.

In the Northern Hemisphere, a **trough** is the part of a jet stream wave where the wind direction shifts from northwesterly to southwesterly. In the Southern Hemisphere, a jet stream trough is a mirror image of a trough in the Northern Hemisphere. Therefore, in the Southern Hemisphere the wind direction in a trough shifts from southwesterly to northwesterly. In satellite imagery and on forecast maps, troughs are marked with a dashed line (figure 9.2). A **ridge** in the Northern Hemisphere is an area in which the winds change from a southwesterly direction to a northwesterly direction. A ridge in the Southern Hemisphere is a mirror image of a ridge in the Northern Hemisphere. Therefore, in the Southern Hemisphere, wind directions shift from northwesterly to southwesterly around a ridge. In satellite imagery or on forecast maps, forecasters use a saw-toothed line to mark the position of an upper-level ridge (figure 9.2).

As air flows through a trough in a jet stream, it tends to diverge at high altitudes, forming an upper-level low-pressure system. This causes air from the surface to flow

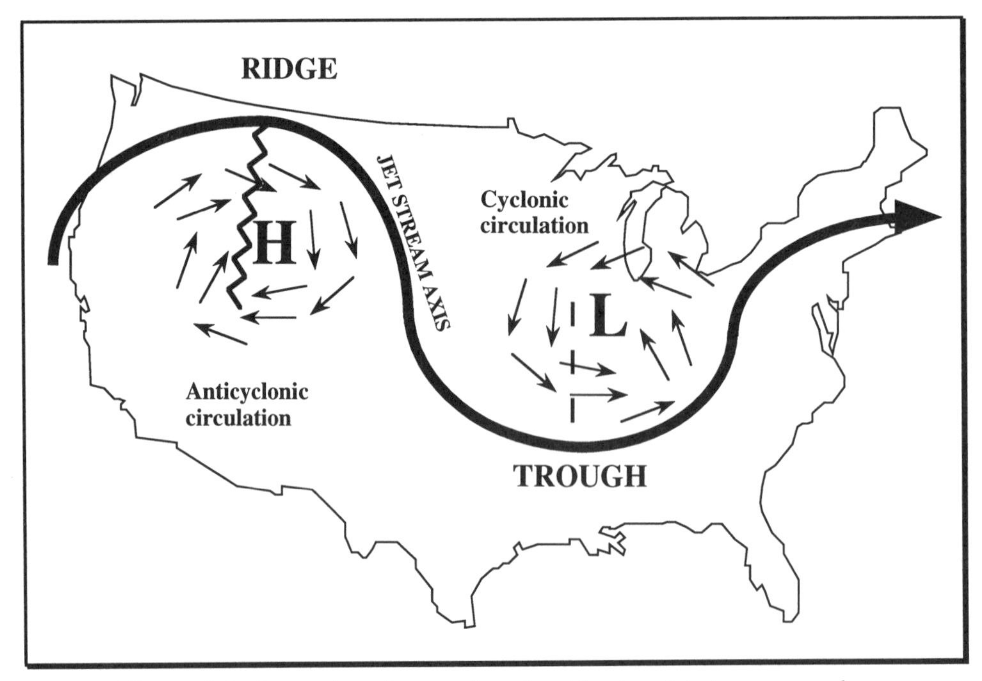

Figure 9.2. Formation of a trough and a ridge in a jet stream meander

upward, often forming a low-pressure system at the surface. As the air flows upward, it tends to spiral in a counterclockwise direction (in the Northern Hemisphere) or a clockwise direction (in the Southern Hemisphere). The trough is known as the **cyclonic region** of a jet stream (figure 9.2). In either hemisphere, the troughs and their associated **low-pressure** systems are characterized by rising air that cools and causes clouds to form.

Air flowing around the ridge of a jet stream wave tends to converge at high altitudes, creating an upper-level high-pressure area. Air flowing into an upper-level high is forced downward, since warmer air in the stratosphere prevents the air from rising. Air that is forced toward the ground often creates a high-pressure system at the surface. Air flowing out of this high tends to rotate clockwise (in the Northern Hemisphere) or counterclockwise (in the Southern Hemisphere). This area within a jet stream is known as the **anticyclonic region** (figure 9.2). A ridge and its associated **high-pressure** system are usually associated with sinking air that warms and produces fair weather.

These high- and low-pressure systems usually follow along the path of jet streams in the mid-latitudes, and they generally move from west to east, bringing frequent weather changes with them.

Zonal vs. meridional flow

Jet stream waves can also be said to have wavelength and amplitude. The wavelength of a typical jet stream wave is anywhere from 50° to 75° of longitude (or as much as 5000 km or 3000 miles). The wave amplitude is generally between 5° and 25° of latitude, and it can be very important in determining the weather. When the jet stream waves are low in amplitude this is called **zonal flow**, indicating that the air is flowing almost directly from west to east. With zonal flow, little mixing of warm and cold air occurs, and lows that develop are usually weak. Figure 9.3 is a GOES image that shows eastern North America under the influence of zonal jet stream flow.

High jet stream wave amplitude is called **meridional flow**. Meridional flow is characterized by troughs with low pressure and ridges with high pressure. These result in a greater transport of warm and cold air masses. With the jet stream supplying circulation and energy, meridional flow may contribute to the development of severe storms. Figure 9.4 is a GOES image in which North America is under the influence of meridional flow. Compare the more extensive north-south cloudiness that accompanies meridional flow with the stretched-out west-east cloud pattern associated with zonal flow.

Locating jet streams

Although locating the position of jet streams on a satellite image is not as easy as you might think, a variety of common patterns can be used to indicate where the maximum winds in the jet lie. The **jet stream axis** is the term used to describe the area of highest-velocity winds in a jet stream. One common indicator of the position of a jet stream axis is the cirrus clouds that tend to form on the southern or southeastern side of the axis. These cloud formations, known as **baroclinic zone cirrus**, have a well-defined border on their poleward edge. This border usually has a flattened S shape, and the jet stream flows about 1° of latitude above, and nearly parallel to, this border. Figure 9.5 illustrates the relationship between a jet stream and a baroclinic zone cirrus shield as seen from space. In figure 9.6, a cirrus shield associated with a jet stream can be seen across northeastern Texas extending northeastward to the Ohio Valley. The jet stream can be located just beyond the northern edge of this cloud shield.

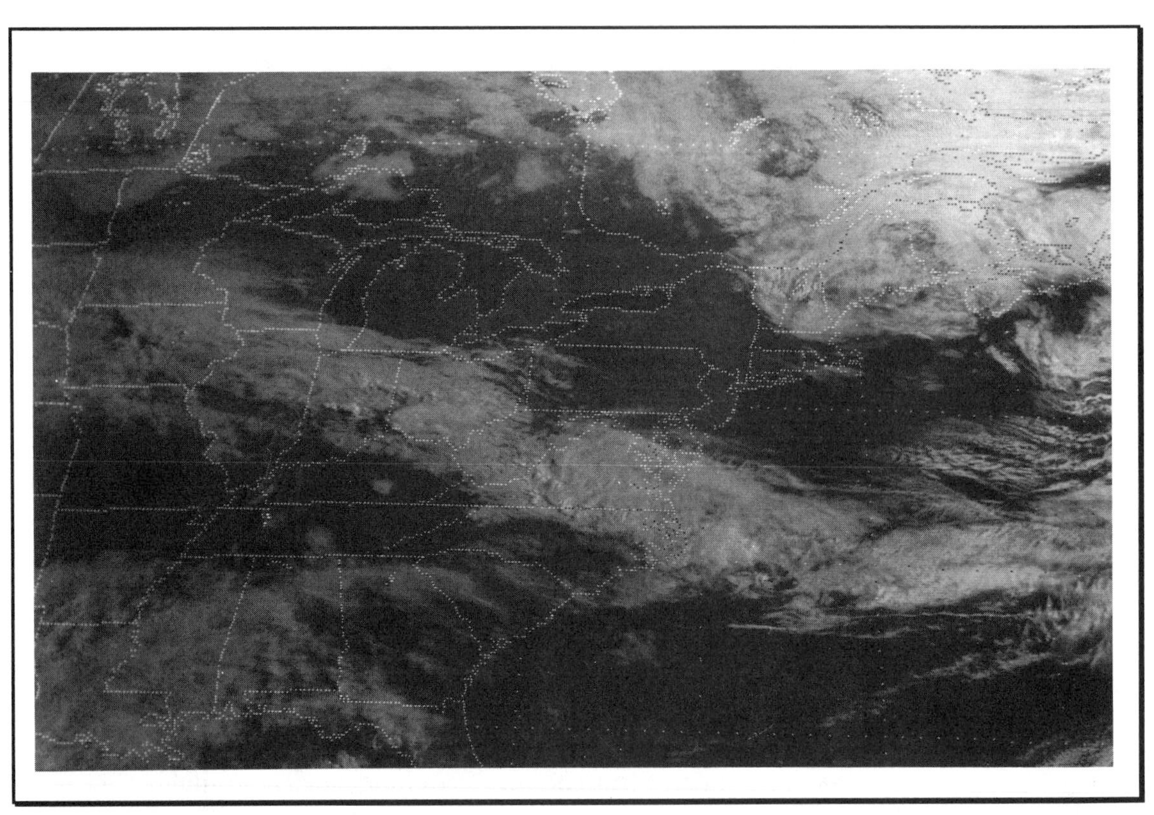

Figure 9.3. Cloud patterns indicating zonal jet stream flow (GOES VIS, April 22, 1988)

Figure 9.4. Cloud patterns indicating meridional jet stream flow (GOES VIS, April 7, 1991)

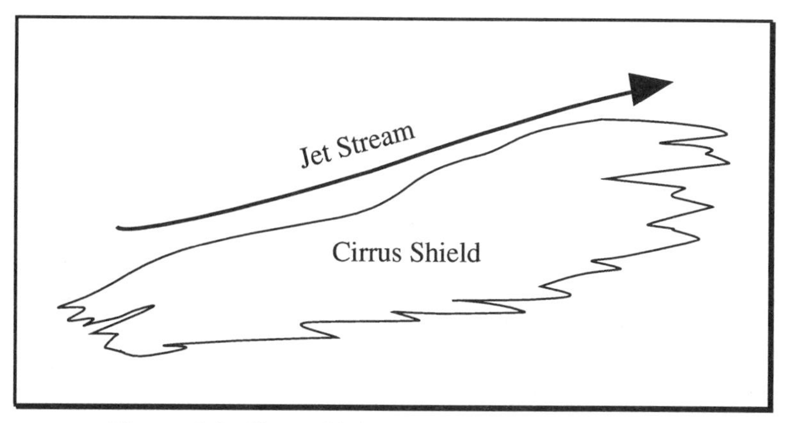

Figure 9.5. Cirrus shield associated with a jet stream

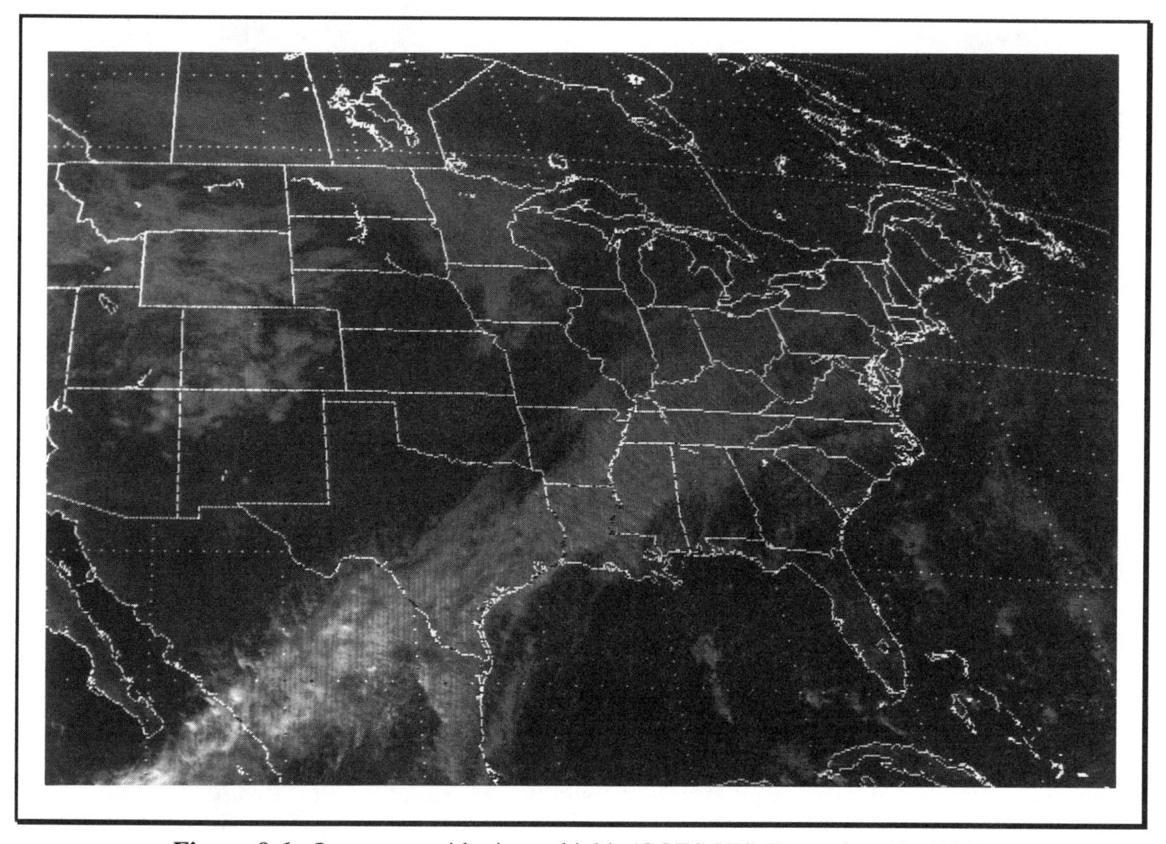

Figure 9.6. Jet stream with cirrus shield (GOES VIS, December 13, 1991)

Often a jet stream passes over a lower-level cloud feature and covers it with its associated **cirrus shield**. On a satellite image the texture of the clouds changes abruptly where the jet crosses the lower clouds. For example, comma clouds are comma-shaped cloud systems that often consist of a variety of lower- and mid-level cumulus clouds that have a very "lumpy" texture. Where the jet stream axis crosses the comma, the texture of the highest clouds (as seen by the satellite) changes. This can be seen just north of *A* in figure 9.7. When a jet with its cirrus shield moves over a lower cloud system, a long shadow that marks the edge of the cirrus shield may also be seen on lower-level clouds. Care must be taken, however, for this may or may not locate the position of the jet stream, since other upper-level cloud layers not associated with a jet stream can cast a shadow on lower clouds within a comma cloud system.

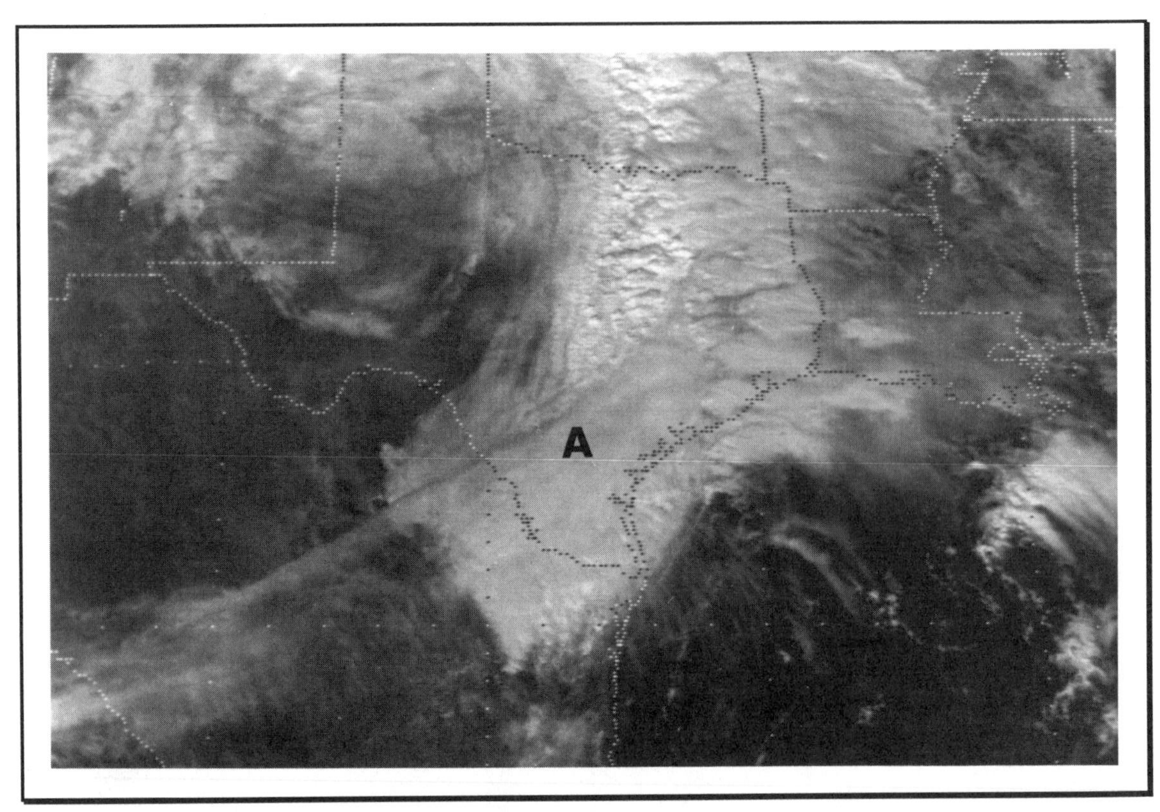

Figure 9.7. Jet stream with its associated cirrus shield crossing over lower-level clouds (GOES VIS, December 29, 1990)

When no baroclinic zone cirrus clouds are present but other high- and middle-level clouds exist, the jet stream can still be located. Where the jet crosses a cloud formation the cloud bands will be most advanced downstream. The upstream borders will also be more advanced than the surrounding clouds, giving these clouds a U or a V shape. Figure 9.8 shows the relationship between cloud movement and a jet stream. In figure 9.9, a GOES image of North America, the jet stream is crossing over a system of low- and mid-level clouds located over the eastern portion of the United States. Where the jet crosses these clouds (in Georgia), the clouds are notched, with the notch pointing downwind.

Abrupt changes in cloud type can also indicate the position of jet streams in satellite imagery. Often, over the oceans, open-cell cumulus clouds will exist right next to a field of closed-cell cumulus clouds. A jet stream is usually 1–3° of latitude poleward from this transition point (figure 9.10). This can be seen in figure 9.11, a VIS GOES image of the Pacific Ocean. A large area of open-cell cumulus clouds is located off the west coast of the United States at point *A*. To its southwest lies a region of closed-cell cumulus clouds, located at point *B*. A jet stream is typically located just north of this transition.

Often, small-scale cirrus lines appear along the main axis of a jet stream. These bands, known as **transverse bands**, resemble the mountain waves discussed in chapter 7. They are much more irregular, however, and they lie perpendicular to the upper-level wind direction. When transverse bands are identified, the jet is to the left of the lines and about 1° north of the shield. The winds in these bands may exceed 150 km/hr (90 mph) and tend to be very turbulent. In figure 9.12, a GOES VIS image, extensive transverse bands can be seen across Mexico and Texas. The jet stream can be located just north of these bands.

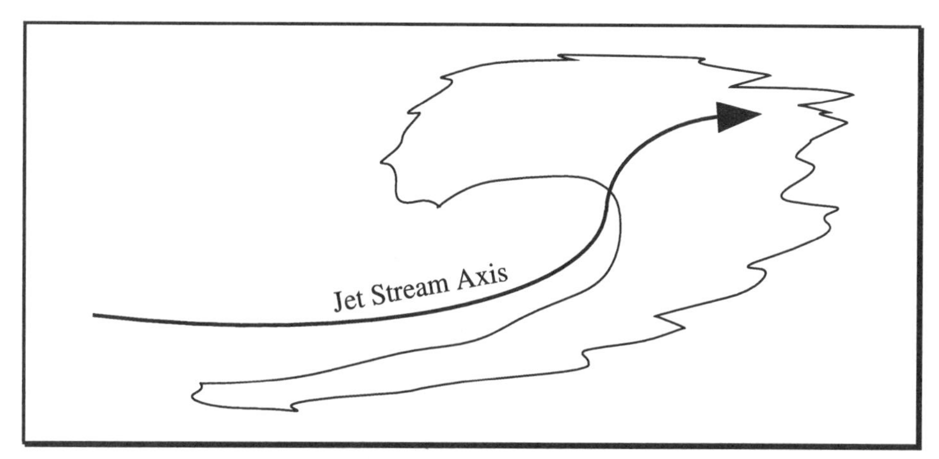

Figure 9.8. Advancement associated with a jet stream along a cloud band

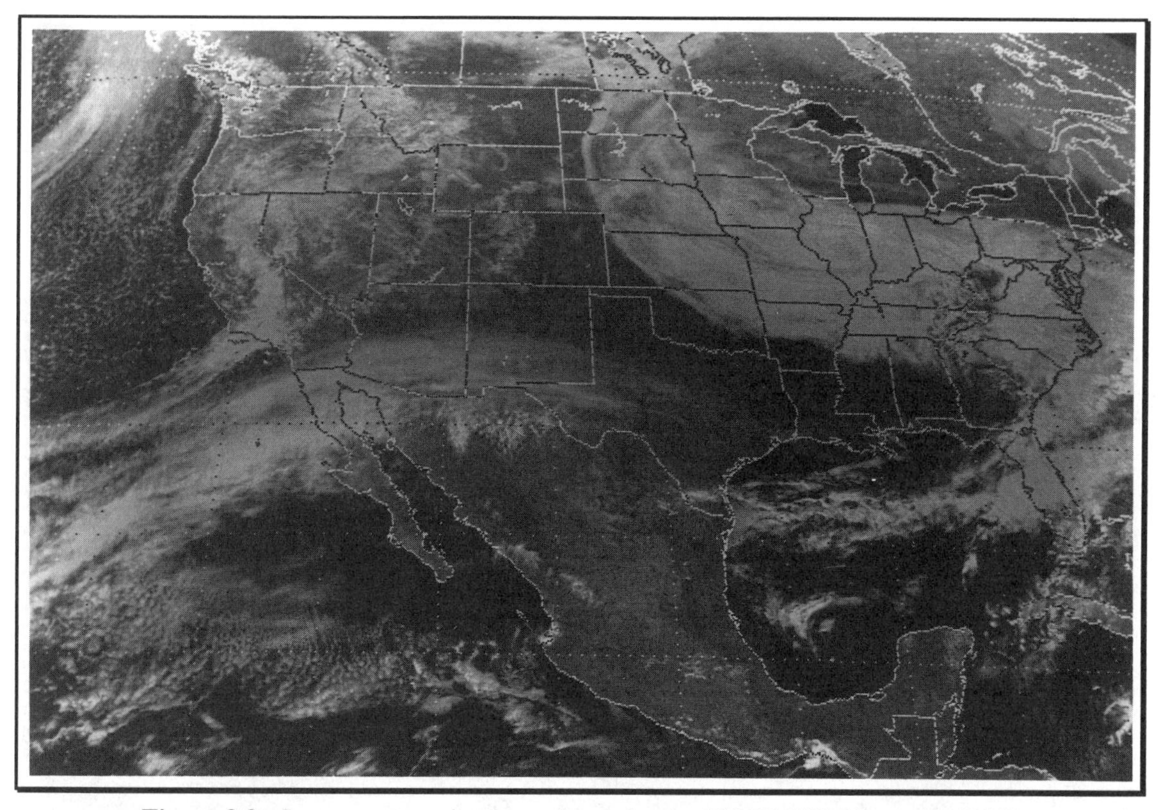

Figure 9.9. Jet stream crossing lower-level clouds (GOES VIS, March 13, 1991)

When there is not enough water vapor in the atmosphere to allow formation of a cirrus shield, **cirrus streaks** may appear. These small streaks are oriented parallel to the jet stream flow and are most often located on the west side of troughs. Winds in the cirrus streaks can be greater than 100 km/hr (60 mph). In figure 9.13, cirrus streaks can be seen across Mexico and southern Texas and into the southeastern United States (at *A* and *B*). A jet stream is usually located just north of these streaks.

Finally, when large thunderstorms form, their anvil tops may aid in determining upper-level wind direction. The anvil tops are very high clouds, and they can be swept downwind by the upper-level airflow. Although they can point in the direction of the upper-level winds, they are sometimes oriented in a manner that combines the effects of both low-level and upper-level winds.

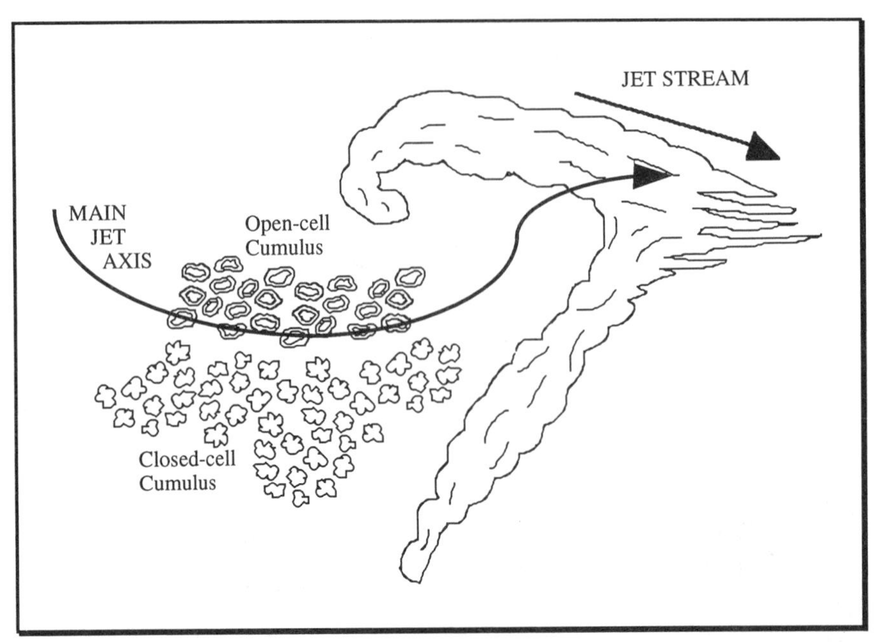

Figure 9.10. Locating a jet stream along an open cell–closed cell transition

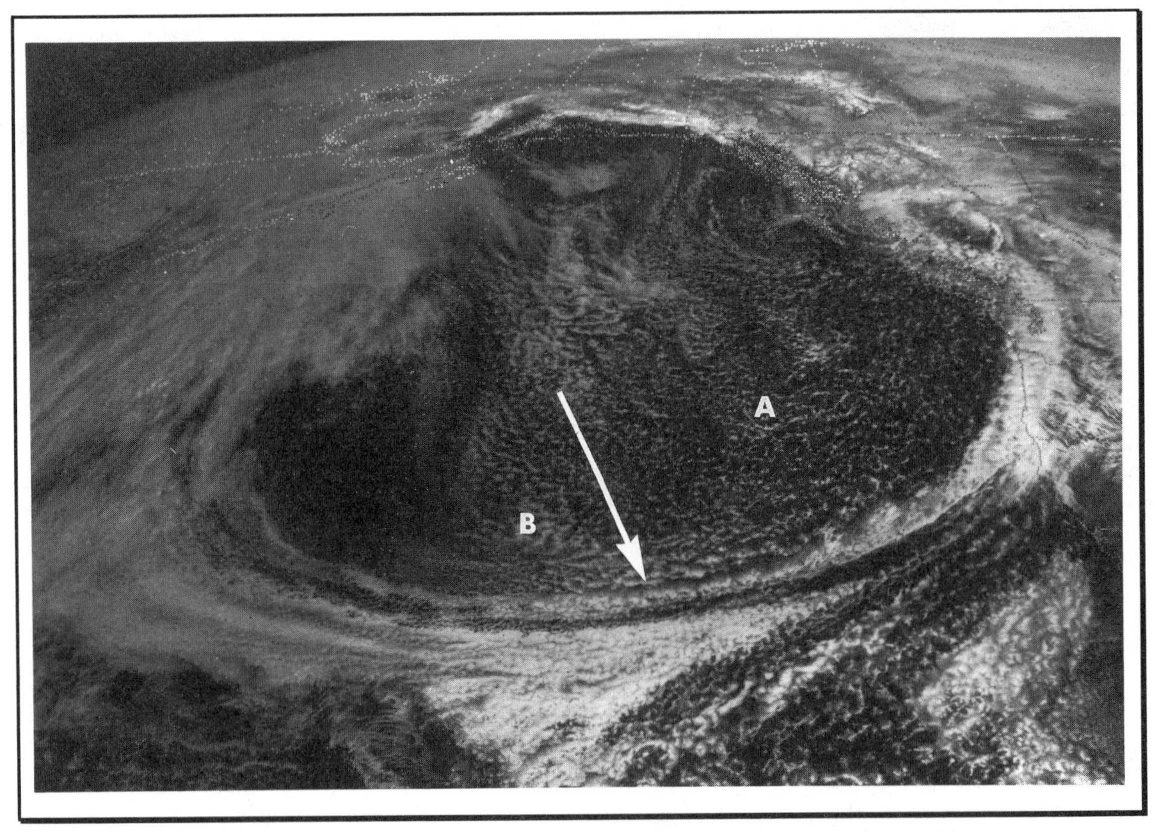

Figure 9.11. Locating a jet stream along an open cell–closed cell transition (GOES VIS, March 31, 1981)

Figure 9.12. Transverse cloud bands associated with a jet stream (GOES VIS, December 12, 1991)

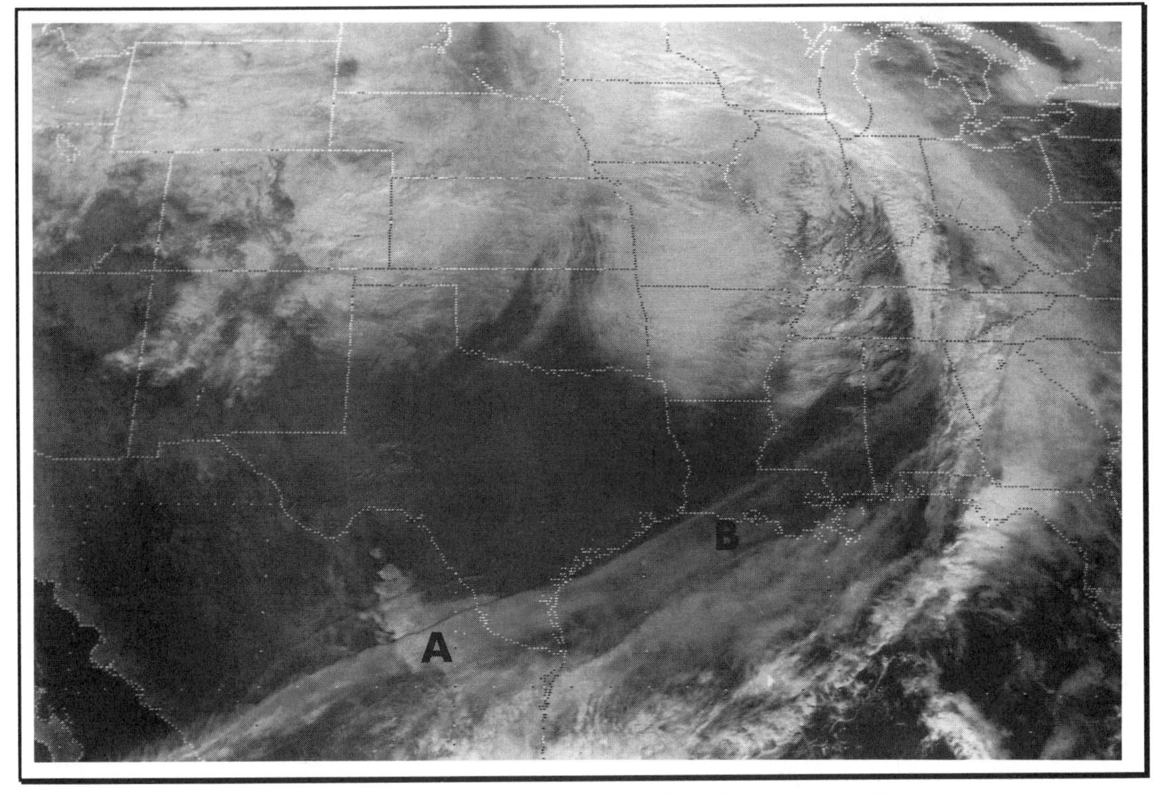

Figure 9.13. Cirrus streaks (GOES VIS, November 1, 1991)

A moderate ridge with less amplitude will exhibit a wider cloud band, possibly with a more southwest-northeast orientation. These bands will have a less distinct forward edge, and some high-level clouds will cross the ridge line (figure 9.14B). A very broad, low-amplitude ridge will develop a very wide cloud band that is almost east-west in orientation, and middle and high clouds will cross well over the ridge line (figure 9.14C).

In figure 9.15, a low-amplitude ridge can be seen over the south-central United States from New Mexico to Arkansas. The cloud band associated with the ridge is very wide, and the high cirrus clouds are well to the east of the ridgeline (point *A*). Off the east coast a very high-amplitude ridge can be located by the long, narrow cloud band with a north-south orientation (point *B*). The ridge is to the east of this cloud band. The clouds have a very sharp leading edge, and they do not extend over the ridgeline.

In figure 9.16, the MB enhancement curve makes the location of the higher-level clouds associated with the jet stream easier to identify. A moderate-amplitude ridge can be located over the Great Lakes region. The cloud band is broader than a high-amplitude ridge, and some clouds extend over the ridgeline, giving the cloud band a less distinct forward edge.

Locating troughs

Troughs can often be located in satellite imagery by locating the point at which the clouds in a frontal band start to become fragmented and the amount of cloudiness decreases dramatically (figure 9.17). The trough of a jet stream can often be found at this

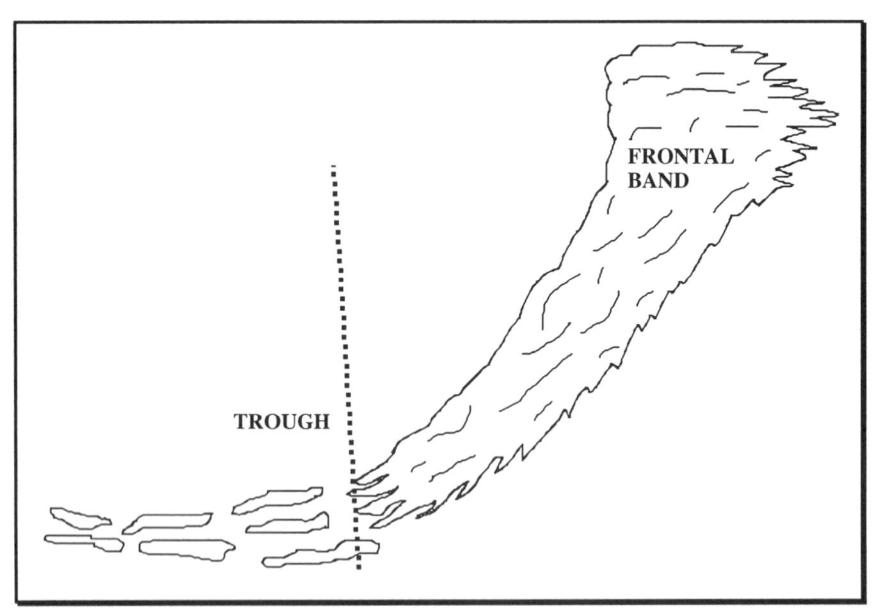

Figure 9.17. Locating a jet stream trough along a frontal cloud band

point. In figure 9.18, a jet stream trough can be located off the east coast of the United States. A cold front can be seen in the mid-Atlantic. The location of the trough can be identified where the continuous frontal cloud band becomes fragmented, just off the coast of North Carolina. Figure 9.19 is an enhanced IR image of the same scene. The MB enhancement used in this image highlights the coldest cloud tops in the frontal band. This makes it easier to locate the point at which the frontal band becomes fragmented.

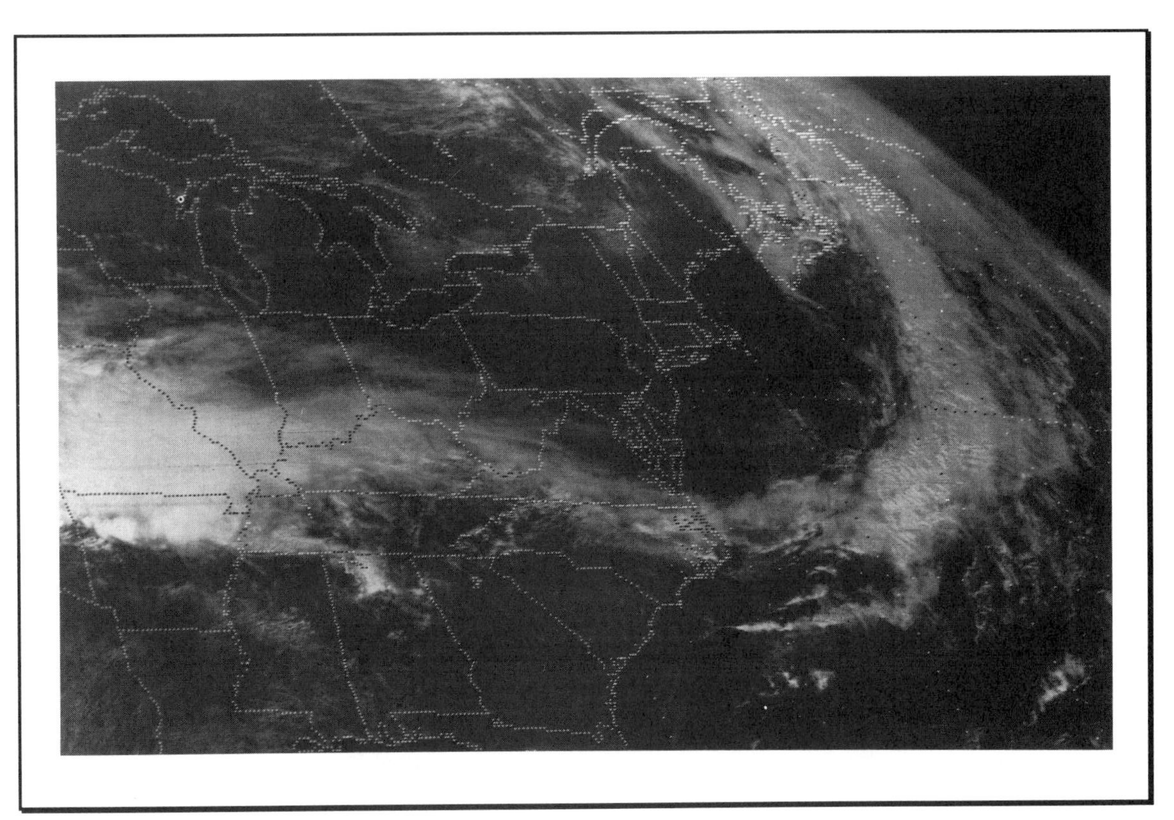

Figure 9.18. Locating a trough along a frontal band (GOES VIS, September 24, 1993)

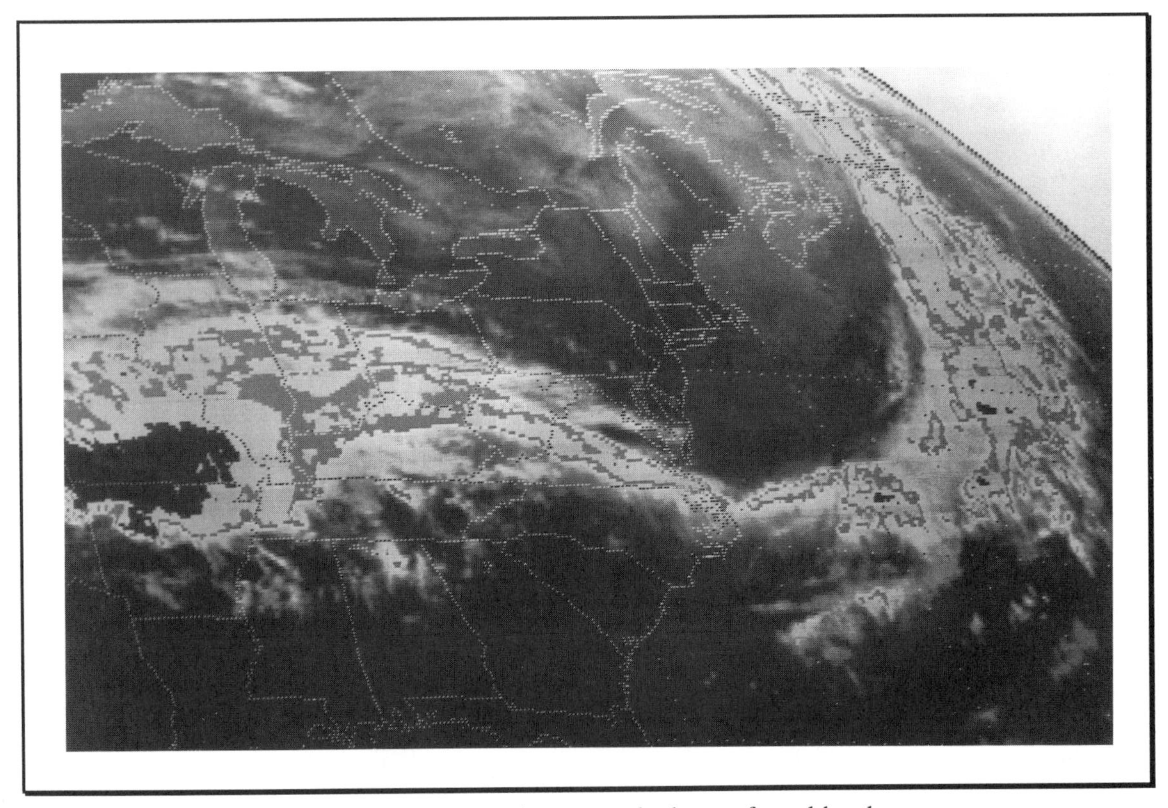

Figure 9.19. Locating a trough along a frontal band (GOES IR, MB enhancement, September 24, 1993)

CHAPTER 10

Synoptic-Scale Storm Development

Introduction to synoptic forecasting

Weather forecasting occurs on a variety of scales, ranging from global to local. **Synoptic-scale** forecasting is the branch of weather forecasting that interprets large-scale features in the atmosphere to determine what areas of the Earth may experience various types of weather. Synoptic forecasters study the position of the jet streams, the movements of air masses, and the changes in frontal systems in order to describe and predict the condition of large portions of the atmosphere during given time periods. This chapter explores the features in satellite imagery—including air masses, fronts, and developing storm systems—that meteorologists use to describe the state of the atmosphere on this scale.

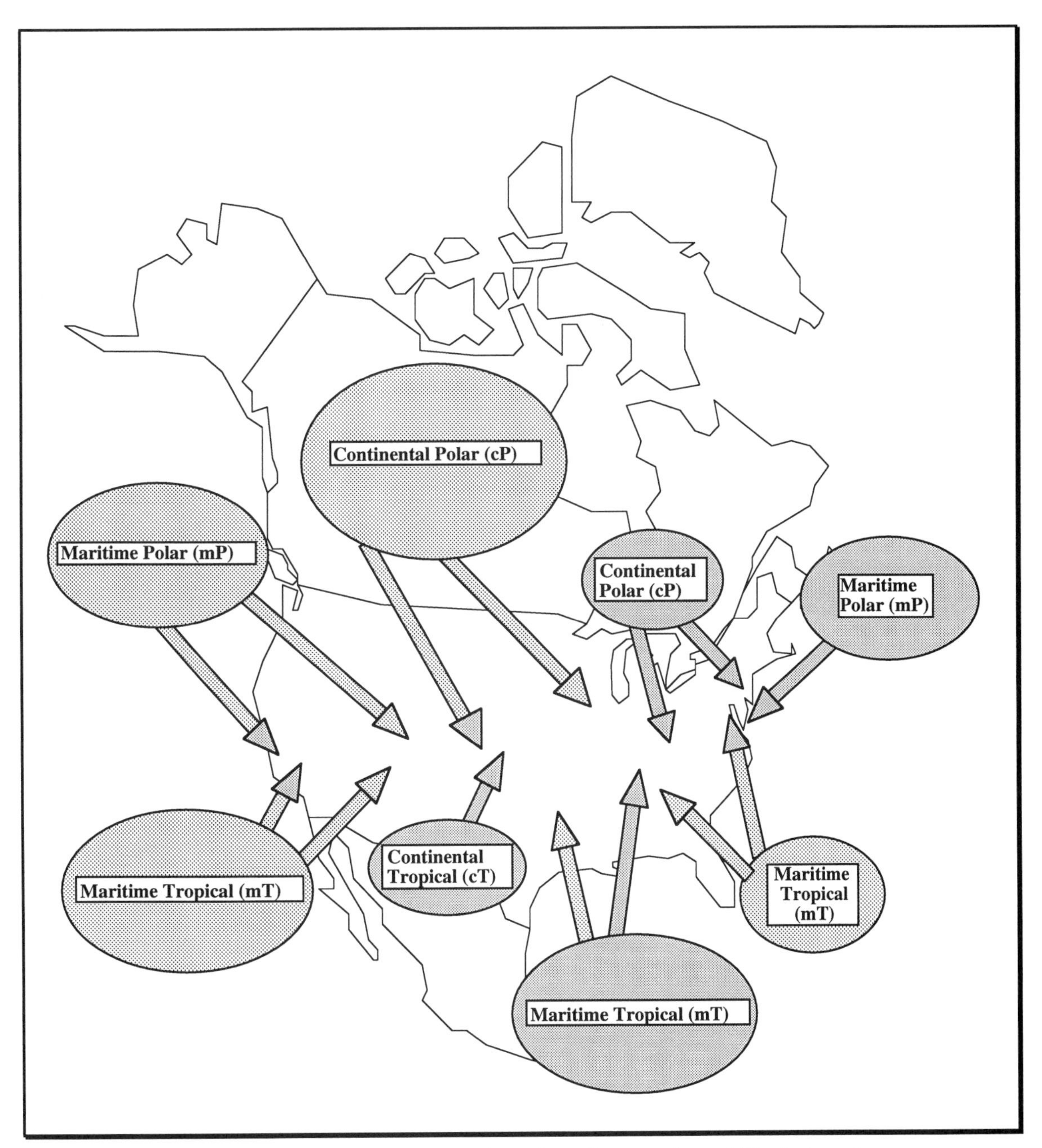

Figure 10.1. Air mass source regions that affect continental North America

In figure 10.4, a cold front can be seen just off the east coast of the United States. The front is identified by a very well-defined band of clouds with a northeast-southwest orientation. The lower clouds in the frontal band have a lumpy texture, and the higher, cirrus clouds along the front are smooth in texture. Fibrous cirrus clouds can be seen extending ahead of the frontal band as they are swept downwind by high-altitude winds. Notice how clear the air is in the cold air mass behind the front. The only clouds that can be seen are valley fog in the Appalachian Mountain region.

As a cold front evolves, the upper-level support for the front often weakens. The steep frontal slope flattens and the cold air sinks, spreading horizontally across the Earth's surface. Often a thin line of convective clouds, known as a **rope cloud**, will form along the forward edge of this cold air. Typically there is a clear area behind the rope cloud where the air is sinking. Rope clouds are most common along oceanic frontal systems where the relatively flat ocean surface provides very little disruption to the airflow. A rope cloud can be seen in the righthand portion of figure 10.4 at the leading edge of the cold front, indicating that this front is weakening.

Warm fronts

A warm front occurs when a warm air mass tries to displace a colder air mass (figure 10.5). Five hundred to a thousand kilometers ahead of a warm front, temperatures are cool and the skies are relatively clear. As a warm front approaches, warmer air comes into contact with denser cold air and slowly rises in a nearly horizontal manner. Far ahead of the front, cirrus clouds appear. As the front moves closer, the clouds become lower and thicker. Since the lifting along a warm front is nearly horizontal, the clouds that form are usually stratified in nature, and there are many vertically stacked layers of clouds (cirrostratus, altostratus, and stratus). As the front passes a location, it is usually accompanied by steady light rain, snow, and/or drizzle. After the front has passed, temperatures rise and the skies begin to clear.

On a satellite image, a warm front generally appears as a wide band of flat clouds with little, if any, shape or organization. A warm frontal zone covers a much wider area than a cold front, and the weather associated with a warm front often lasts longer. Warm fronts also travel relatively slowly (generally less than 25 km/hr, or 15 mph) when compared to cold fronts.

Figures 10.6 and 10.7 are a VIS and enhanced IR pair depicting a frontal system located over the southeastern United States. The system is centered around an area of central low pressure in the upper left corner of the image. A warm front extends from the central low

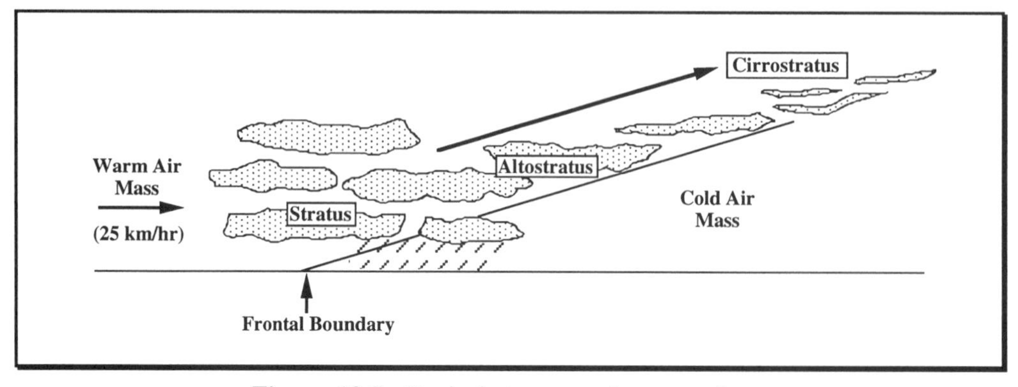

Figure 10.5. Typical structure of a warm front

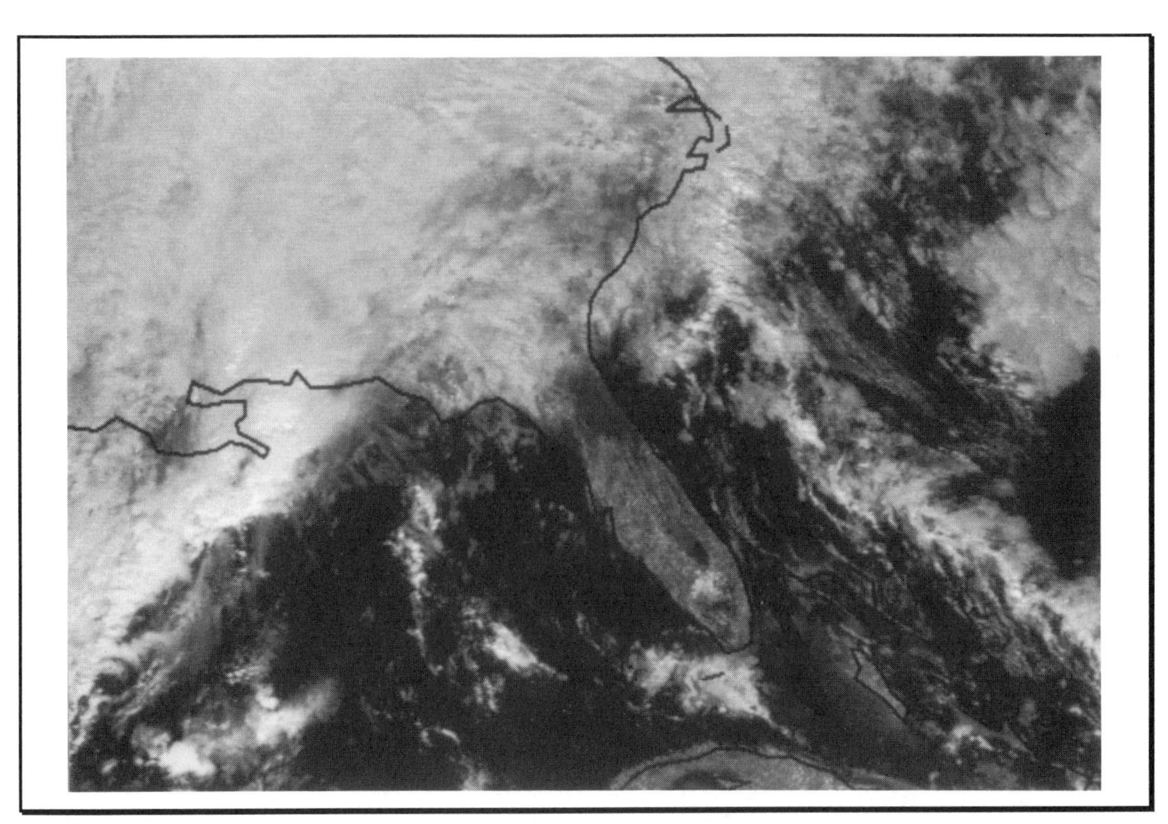

Figure 10.6. Frontal system centered over U.S. (GOES VIS, December 10, 1994)

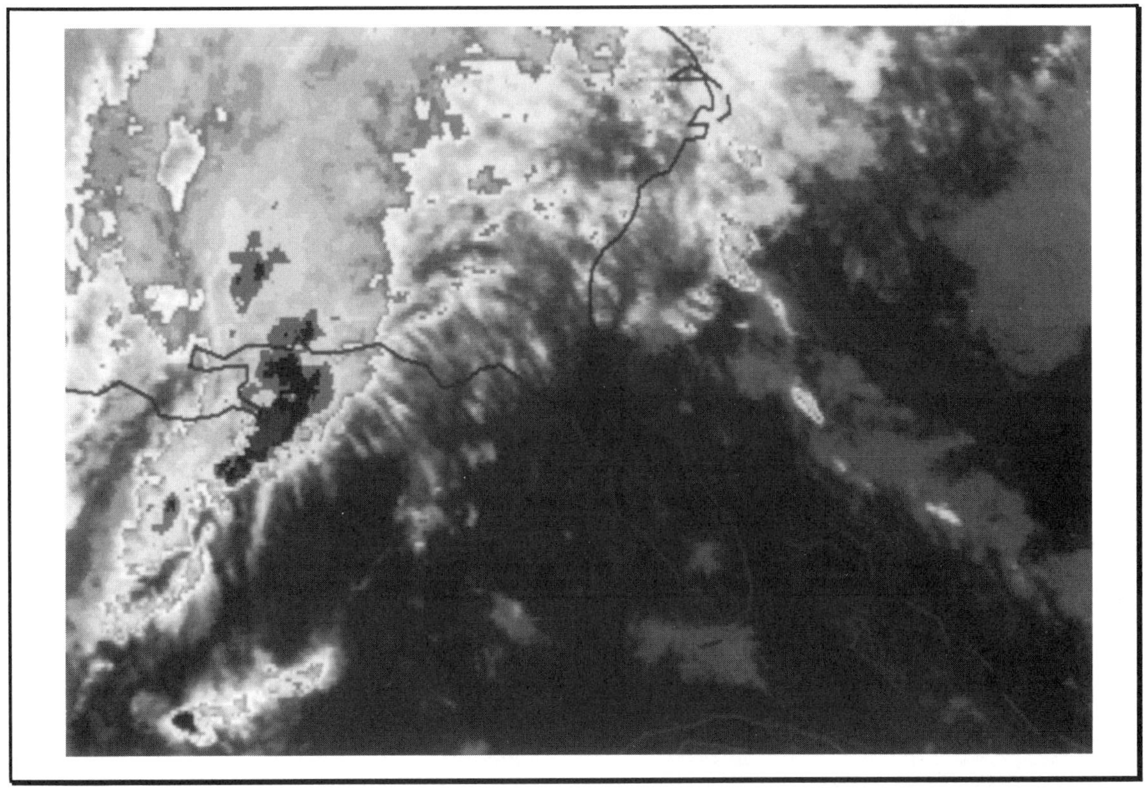

Figure 10.7. Frontal system centered over U.S. (GOES IR, MB enhancement, December 10, 1994)

Leaf stage

Storm systems often start out as a cloud formation known as a leaf cloud. This feature is usually found on the east side of an upper-level trough, and is often elongated and leaf-shaped. Leaf clouds have well-defined borders and contain vertically deep and thick clouds. They show up very clearly on both IR and VIS imagery.

The poleward side of a leaf cloud has a very distinct border that often forms a flat S shape. Leaf clouds can often be identified by a curved notch on the western or southwestern edge of the leaf. This pattern is caused by the jet stream pushing into the western edge of the cloud system. The first illustration in figure 10.8 depicts a typical leaf cloud formation. The highest cloud tops in a leaf cloud are often located over the eastern portion of the leaf. Cloud top height decreases westward, with middle-level clouds appearing over the westward portion of the notch. Low clouds are found along the top portion of the notch. Usually a cold front is located along the equatorward border of the leaf or within the leaf cloud structure. The leaf cloud is a significant region of clouds and precipitation, even if cyclogenesis does not occur. A leaf cloud that extends from the Gulf of Mexico to New England can be seen in figure 10.9. The western edge of the leaf has a well-defined border and associated notch. This leaf has formed on the eastern side of a relatively high-amplitude jet stream trough.

Open comma stage

Within a leaf cloud, the air is rotating or spinning around a point of **maximum vorticity** (the region of maximum spin). As the air circulates around this point, the cloud system becomes distorted. If the whole system remains stationary, the clouds will gradually develop into a comma shape. If the system moves eastward, the cloud system will undergo further distortion into a cloud pattern called a comma cloud, shown in the second and third illustrations in figure 10.8. Comma clouds appear in all sizes and shapes, depending on the development of the **vorticity** pattern. They range in size from small thunderstorms to large-scale mid-latitude cyclones, and they can change very rapidly.

Comma clouds have many characteristics that are important to identify when studying these storms. The back edge of a comma cloud, the part of the comma that is most easily identified, has a well-defined S shape. The point at which the back edge curvature changes from cyclonic (counterclockwise) to anticyclonic (clockwise) is called the **inflection point**. The front edge of a comma cloud is less clearly defined and tends to become very ragged as upper-level winds spread the clouds out. A comma cloud usually has the beginnings of a **dry slot**, also known as the **surge region**, where the jet stream crosses the system. This region usually forms as the notch in the leaf cloud expands. Here the jet causes the clouds to move more rapidly with respect to the other clouds, and they become more advanced downstream. The comma head generally lies to the west of the maximum winds. It tends to lag behind and to show the greatest tendency to rotate. The **tail** of the comma extends from the surge region southward. It generally lies more parallel to the axis of the maximum winds. A cold front is usually located along the tail.

As the comma develops, pressures in the storm system usually fall. The surface low migrates toward the western edge of the cloud mass, near where the inflection point and the jet stream are located. A warm front (often hard to see in satellite imagery) extends eastward from the inflection point.

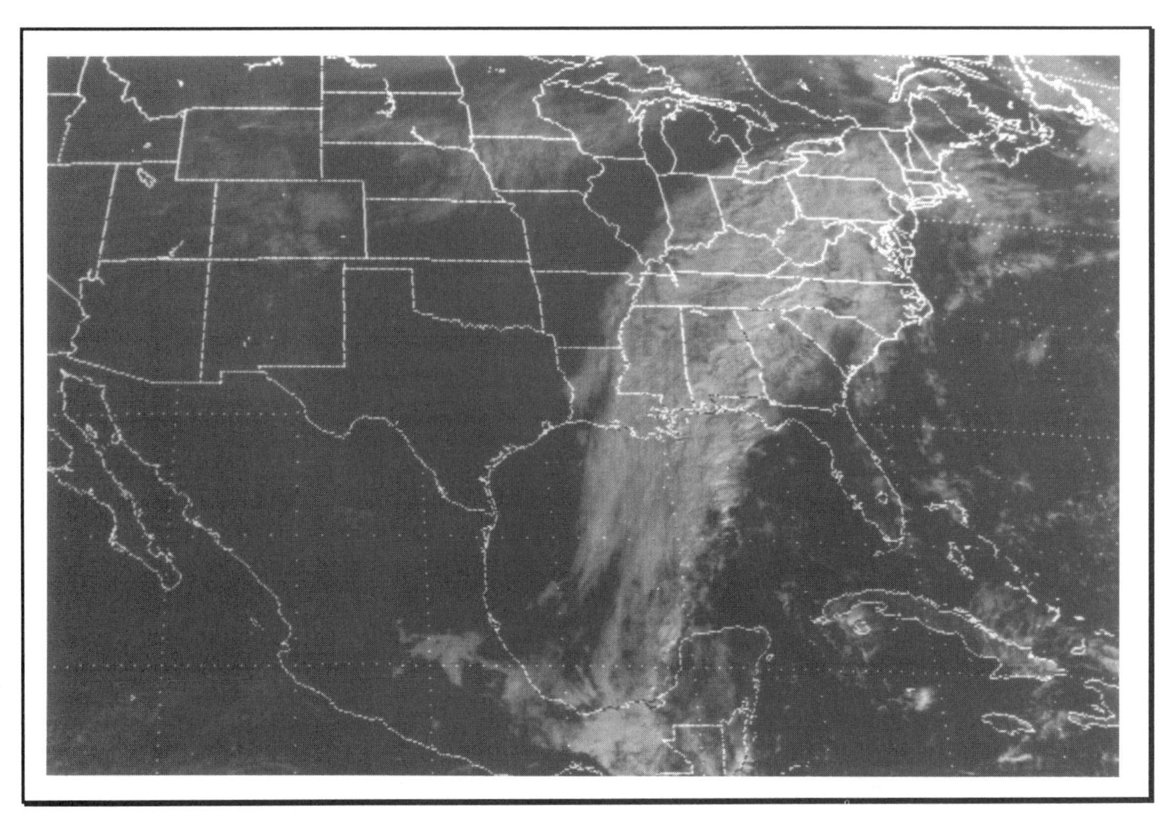

Figure 10.9. Leaf cloud over the eastern United States (GOES VIS, November 21, 1991)

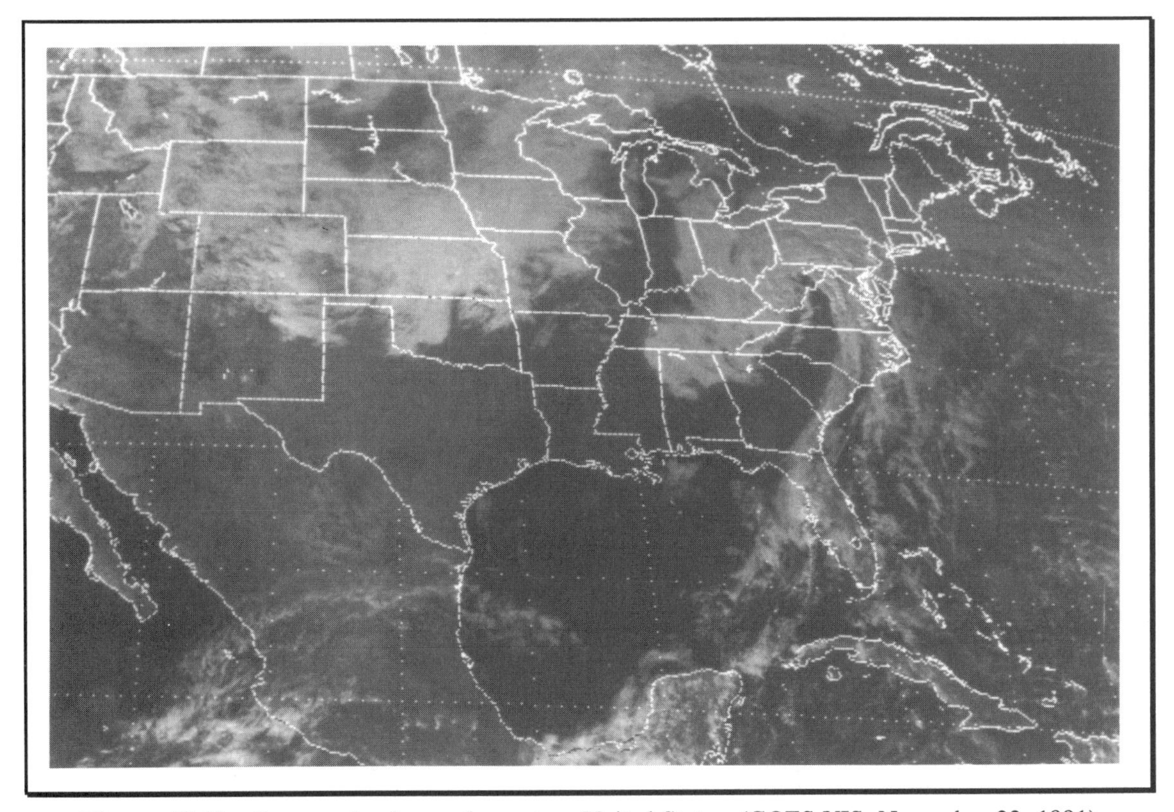

Figure 10.10. Comma cloud over the eastern United States (GOES VIS, November 22, 1991)

In figure 10.12, four mid-latitude cyclones in the Pacific Ocean can be seen in different stages of cyclogenesis. The storm at point *A* is a young storm that is in the comma cloud stage. The head of the comma is just beginning to spiral around the center of circulation. Storm *B* is a mature, occluded comma cloud. Notice the increase in spiraling around the center of rotation. Generally, when the comma head spirals around itself, the storm has become occluded. A very well developed cold front can be seen along the tail of the comma cloud. The spiral clouds at point *C* are part of a cut-off low-pressure system. This was the head of a comma cloud from an earlier storm. Storm *C* is well occluded. At point *D*, a new storm is developing. *D* is in the initial stage of comma cloud formation.

Synoptic forecasting in Atlantic winter storm monitoring

Mid-latitude ocean storms are of special interest to forecasters since they often produce very high winds, heavy rains, and rough seas that pose a threat to air and sea travel. Monitoring storms over ocean areas is especially challenging to meteorologists, since relatively little traditional data is collected over the oceans except within well-traveled air and ocean shipping lanes. Recently, methods have been developed to better estimate the intensity of a mid-latitude cyclone using satellite imagery. There are two similar techniques, one for Atlantic storms and one for Pacific storms. This section describes the Atlantic technique.

This method can only be applied to storms located poleward of 30° latitude and moving toward the east and/or toward the pole. It is limited to Atlantic Ocean storms during the winter months. Pacific Ocean storms have slightly different characteristics; therefore, estimates using the Atlantic technique will be somewhat inaccurate. Storms occurring over land areas are affected by such factors as terrain features (such as mountains) and friction with the ground. At the present time, their central pressure cannot be accurately estimated from satellite imagery. Although this technique cannot be used to provide accurate central pressure estimates under all conditions, it does provide useful information about the general characteristics and changes in pressure within storms year-round. For example, in a large extratropical cyclones over the midwestern portion of the United States, the greater the degree of spiraling, in general, the lower the pressure in the storm.

The method described here assumes that a relationship exists between cloud patterns, stages of storm development, and associated central surface pressures. The cloud patterns are identified and matched with corresponding central pressure estimates. Central pressure is important in storm intensity monitoring, since lower central pressure is often associated with stronger winds. This translates into the potential for higher ocean waves.

The average atmospheric pressure at sea level is 1013 mb. As a storm develops, the pressure drops. Figure 10.13 shows the expected pressure changes in 12-hour intervals as an ocean storm and the associated cloud patterns develop. In the graph, the pressure would **deepen** (drop) steadily to a minimum level (942 mb); then it would begin to **fill** (weaken). Using this graph as a **conceptual model**, we can compare a storm to other winter storms to determine whether it is developing at a faster or slower rate. We can also use the cloud patterns (shown at the bottom of the graph) to determine the developmental stage and intensity of the storm. Note that in the early stages of storm development, the cloud band has a small-amplitude wave, or S shape. As the storm develops and intensifies, the amplitude increases and the cloud takes on a comma-like or spiral shape. As the storm continues to develop, the degree of spiraling increases around the center of rotation. The cloud pattern then begins to break apart as the storm weakens. Most storms follow this pattern of development, although some will progress through each stage more rapidly or more slowly than the conceptual model.

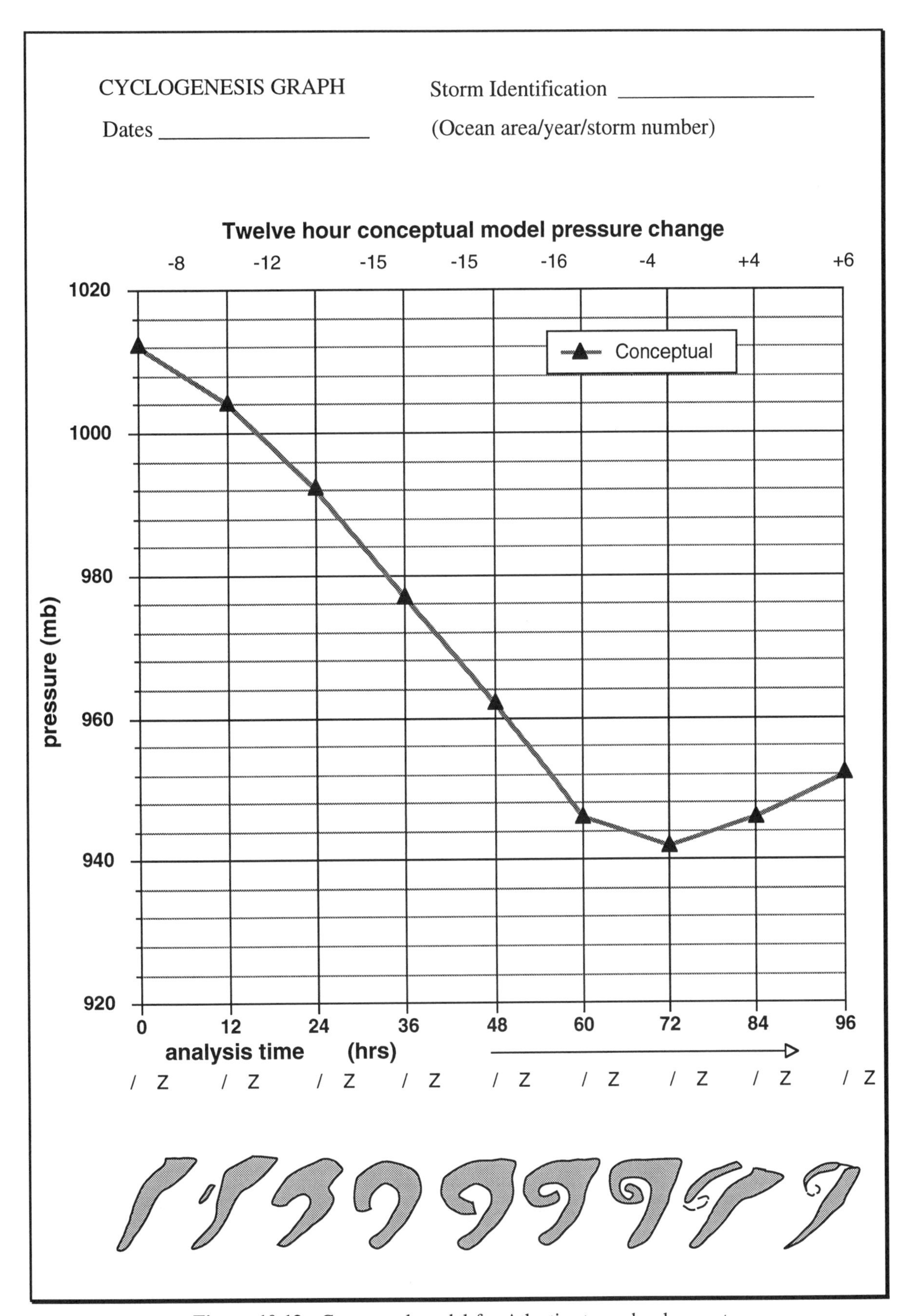

Figure 10.13. Conceptual model for Atlantic storm development

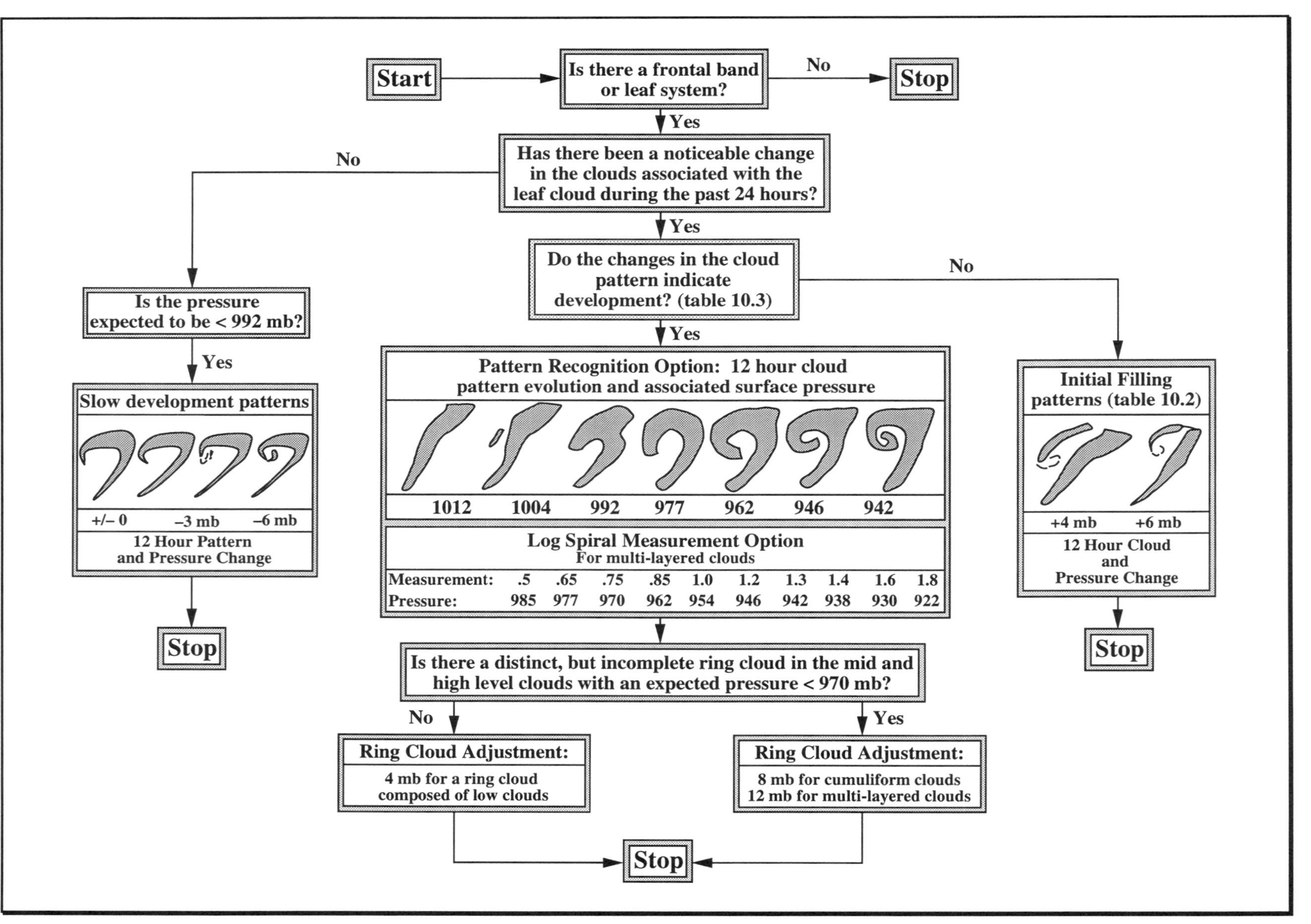

Figure 10.15. Flowchart to use for estimation of central pressure in Atlantic storms

Characteristics of a Filling System
* Comma cloud head becomes detached from the cloud band.
* Cloudiness in the comma cloud becomes thin and decreases.
* The frontal cloud becomes thin, or the cloud tops within it warm.
* A ring (composed of middle and high clouds) completely encircles the cyclone center; minimal central pressure is reached.

Table 10.2

Characteristics of a Developing System
* The cloud band increases in brightness.
* An S shape begins to develop and amplifies.
* An area of vorticity (spin) is located within 10° latitude west or poleward of the baroclinic leaf and approaches the leaf with time.
* A dry slot forms and expands behind the cloud band.
* An anticyclonically curved polar jet streak approaches the bottom portion of the cloud band from a west-northwest direction.

Table 10.3

The log spiral method is more quantitative, as it involves directly measuring the degree of spiraling in the cloud system and using this measurement to estimate central pressure. To make this measurement, copy the 10° log spiral in figure 10.16 onto a transparency. Place the overlay of the log spiral over the cloud system on the image, and rotate it so that it best fits the middle and high cloudiness associated with the developing storm system and lies within the cloud mass. The measurement should be made using only the high- and middle-level clouds, from the head of the comma to where the spiral curvature parallels the tail of the comma cloud. Then count the number of spiral spokes bounded by the top of the comma head and the tail of the comma. Divide this by 10 to get the log spiral measurement. Record this in the data chart in the appropriate column. The log spiral measurement can then be matched up with the corresponding central pressure estimate on the flowchart (figure 10.15). If the measurement falls between two values on the flowchart, estimate the pressure (to the nearest 0.05 units) between them.

Finally, an adjustment must be made if a **ring cloud** exists near the center of the comma. A ring cloud is a circular cloud feature that is associated with the head of a mature comma, especially when central pressure is lower than 970 mb. Winds are relatively strong in the ring itself and weaker in the central area inside the ring, as in the eye and eyewall of a hurricane. Different ring cloud features affect the final central pressure estimate of a storm. A storm with a high degree of spiral curvature and more complete rings will generally have a lower pressure. To account for the presence of a ring cloud, subtract 12 mb from the final pressure estimate if a distinct ring exists, is detached from the comma cloud, and is composed of middle and high (multi-layered) clouds. If the ring is composed of cumuliform clouds, subtract 8 mb, and if the ring is mainly low clouds (e.g., stratus clouds), subtract only 4 mb. This adjustment is also recorded in the data chart.

it at approximately 962 mb on the pattern recognition estimate. The log 10 spiral measurement is approximately 8.5, which would correspond to a central pressure of approximately 962 mb. Finally, a ring cloud composed of cumulus clouds can be clearly seen within the center of the system. This results in lowering the central pressure by 8 mb, giving a final estimate of 954 mb. Again, when plotted against the conceptual model graph, this storm seems to be developing more slowly than expected.

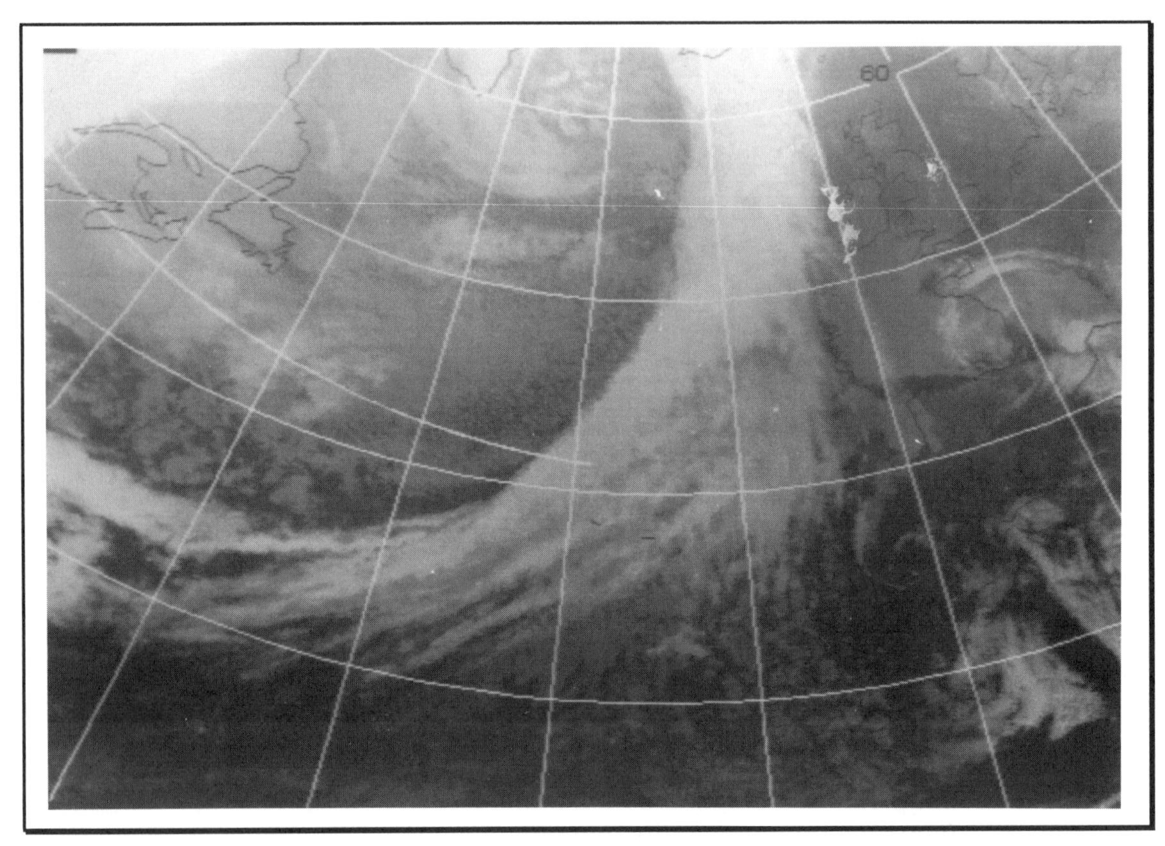

Figure 10.19. Atlantic Ocean winter storm: central pressure estimated at 954 mb (METEOSAT IR, 0013 UTC, January 15, 1991)

Chapter 11

Thunderstorm and Severe Weather Forecasting

winds are weak, the thunderstorm will grow nearly vertically and appear globular in a satellite image. Young, developing thunderstorms also have this shape, since the cloud tops have not yet grown into strong upper-level winds. In figure 11.2, several thunderstorms with a globular shape can be seen. When upper-level winds are very strong and the thunderstorm top has grown to that height, the top of the storm spreads downwind. This gives the storm top a triangular shape, which often points upwind in relation to the upper-level winds. The upwind portion of the storm is usually sharp and well defined while the downwind portion is more fuzzy in appearance. Since this portion of the thunderstorm cloud is usually composed of cirrus clouds, the texture is very smooth. Wedge- or carrot-shaped thunderstorms can be seen in figure 11.3 over Oklahoma and Texas, with the cirrus shields extending as far as Arkansas and Missouri.

In some cases, towering cumulus clouds will break through the cirrus top and give the cloud a small area with lumpy texture. This region of the cloud is referred to as an overshooting top, and it occurs in the strongest updraft region of the storm, where the updraft actually breaks through the cloud top. Locating these features can help you identify areas of intense convection and more severe weather within a larger thunderstorm area. Overshooting tops are best seen in VIS satellite imagery during early morning or late afternoon hours, when the sun angle is low and the shadows cast by the overshooting tops are easily seen. Several overshooting tops can be seen in figure 11.4, a VIS image of a thunderstorm complex located just south of the Mexico-Texas border. When it is suspected that thunderstorms are embedded in a larger mass of clouds, they can often be located by the overshooting tops breaking through higher clouds in the cloud formation. For example, in figure 11.3, overshooting tops can be located in the cloud tops by the shadows they cast on the storm's cirrus shield. Two overshooting tops are shown by black circles. There is also one in the thunderstorm in the Texas panhandle.

Figure 11.2. Globular-shaped thunderstorms (GOES VIS, 2230 UTC, July 20, 1986)

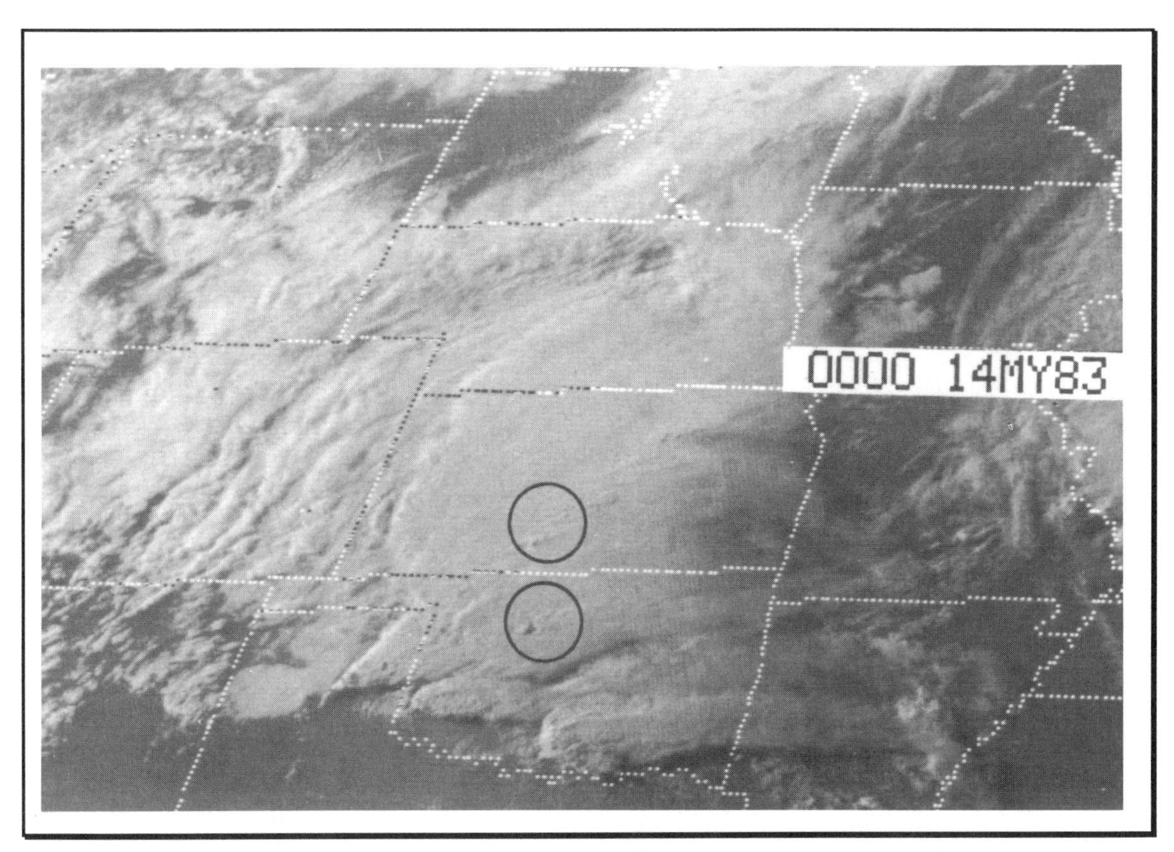

Figure 11.3. Wedge-shaped thunderstorms with overshooting tops (GOES VIS, 0000 UTC, May 14, 1983)

Figure 11.4. Overshooting tops along Mexico-Texas border (GOES VIS, October 8, 1994)

In IR imagery, overshooting tops may be located by small regions in the cloud tops where the temperatures are colder than in the surrounding clouds. It is important to understand that overshooting tops alone do not necessarily indicate that severe weather is occurring in a thunderstorm. Persistent overshooting tops (lasting one hour or more) are a more reliable indication of severe weather. In figures 11.5 and 11.6, the overshooting top associated with a series of thunderstorms in Oklahoma can be detected for several hours, indicating a strong thunderstorm that would likely contain severe weather.

Enhanced IR imagery

There are other reasons that thunderstorms are often studied in IR imagery. Some involve cold temperatures (which often signify strong updrafts); others involve shapes seen in the cloud top temperature field. The MB enhancement curve (refer to chapter 3 for further discussion of enhancement curves) is a commonly used enhancement technique for locating thunderstorms. When this enhancement curve is used, temperatures between –32° C (–26° F) and –58° C (–72° F) are highlighted as various shades of gray. Temperatures between –58° C and –62° C (–80° F) are highlighted in black, and temperatures of –62° C and colder are light gray to white. In the case of a thunderstorm, this produces a more or less oval-shaped contour pattern of gray to black to white in the cloud tops. Using this technique, very cold cloud top temperatures associated with thunderstorms can be easily identified. Figures 11.5 and 11.6 are IR images of the same storm shown in figure 11.3. These images have been enhanced using MB enhancement. White areas within the thunderstorm indicate areas where the cloud tops are the coldest and the convective activity is most intense. The colder overshooting tops associated with several Oklahoma storms can be located in these two images, since they appear as small white spots within the larger cloud system.

In enhanced IR imagery, a severe thunderstorm often exhibits a V-shaped notch of cold temperatures in the cloud tops. The narrow end of the V points upwind. Downwind of the notch, there are often warmer temperatures. This cloud top temperature pattern is frequently seen with thunderstorms that produce tornadoes, hail, strong winds, and intense up- and downdrafts. However, this pattern does not always indicate severe weather, and not all severe thunderstorms exhibit this pattern. This enhanced-V signature can be seen in several thunderstorms in figures 11.5 and 11.6.

Care must be taken when using cold cloud top temperature enhancement in thunderstorm detection. This is because temperature alone is not always an accurate indicator of thunderstorm activity. Not every thunderstorm reaches the tropopause, which varies in altitude across the globe. In the tropics, where there is a greater volume of warm air, the troposphere is at a higher altitude than in upper latitudes. Although cloud tops in the tropics may be very cold, there may not be any thunderstorm activity occurring. Thunderstorm cloud tops in the tropics will also tend to be much colder than those in middle and upper latitudes, but this does not mean they are very intense. It is simply a function of a higher tropopause. In middle and upper latitudes, where the tropopause is at a lower altitude, thunderstorm cloud tops are not as cold, yet the storms may be more intense. The altitude of the tropopause also varies with the seasons, being higher in the summer and lower in the winter. Therefore, summer thunderstorms tend to have colder cloud tops than winter thunderstorms. In all of these cases, the use of the MB enhancement curve, which locates convective storms with cloud tops colder than –58° C (–72° F), may not always pinpoint the most important thunderstorms. In some cases, other characteristics of the clouds, such as shape and movement, must be observed to accurately determine whether a storm is a severe thunderstorm.

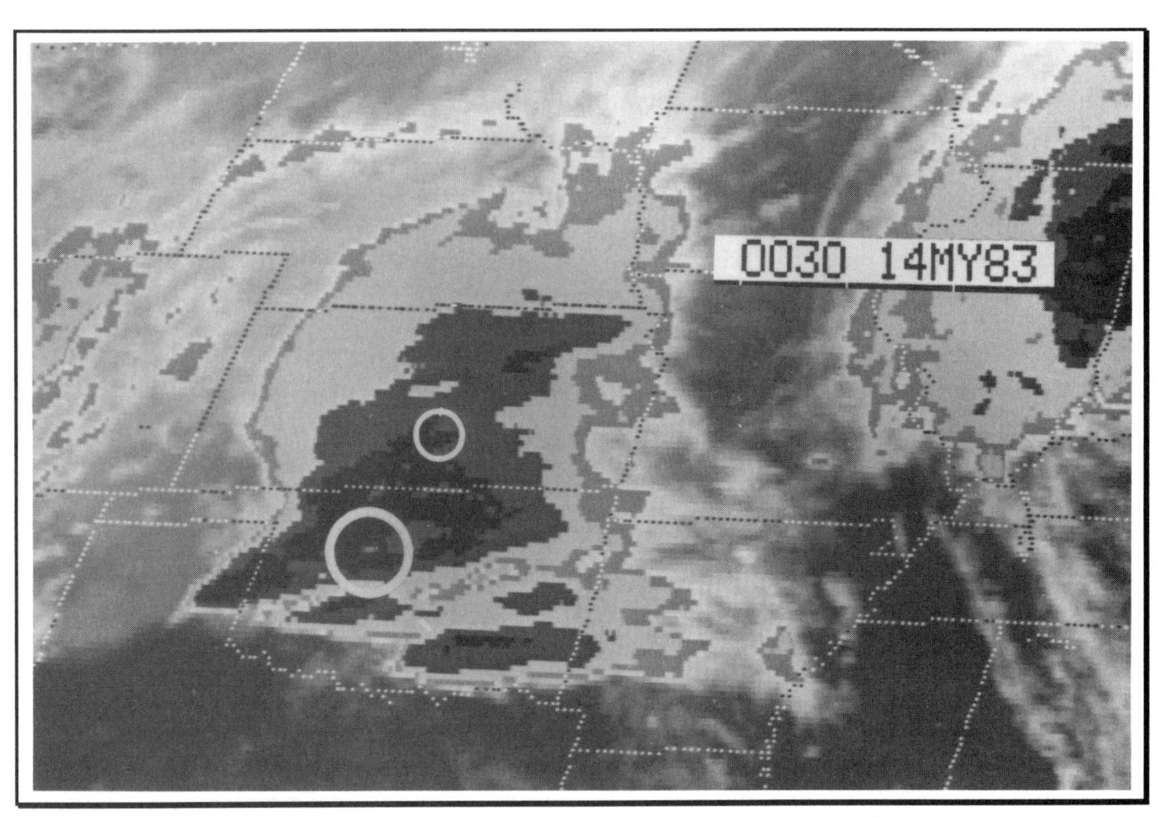

Figure 11.5. Thunderstorm enhancement showing the enhanced-V signature (GOES IR, MB enhancement, 0030 UTC, May 14, 1983)

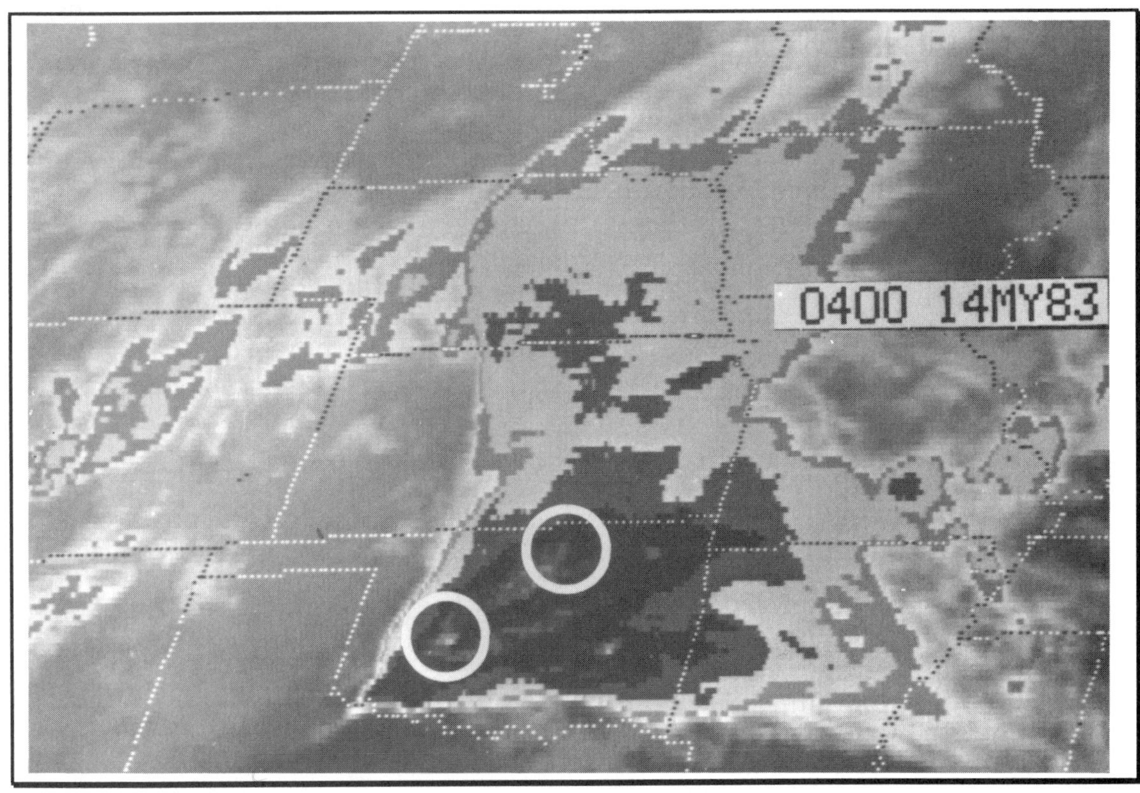

Figure 11.6. Thunderstorm enhancement showing the enhanced-V signature (GOES IR, MB enhancement, 0400 UTC, May 14, 1983)

Locating regions of possible thunderstorm development (synoptic scale)

Since thunderstorms develop very rapidly, it is important to identify regions where their development is most likely. Observations at the synoptic scale are useful for identifying regional conditions that are necessary to produce strong convective storms. The position of frontal boundaries, the jet stream, mid-latitude cyclones, and the intertropical convergence zone all offer clues that aid in the location of such areas. Careful monitoring of the **storm-scale** (or mesoscale) processes within these broad regions will permit more timely and accurate severe weather forecasts and warnings.

Frontal boundaries

Severe thunderstorms are often associated with organized convergence lines such as cold, warm, and occluded fronts. Thunderstorm development is especially common along cold fronts that are associated with mid-latitude cyclones. Converging surface winds often occur along these fronts. The winds ahead of the cold front are usually from the southwest, while winds behind the cold front come from the northwest. This results in a convergence of air along the cold front, forcing the air above the front to rise. This vertical acceleration of air often leads to the development of thunderstorms in a nearly continuous line, known as a **squall line**. A squall line can be seen over the High Plains in figure 11.7, an enhanced IR image. The squall line is seen as a continuous dark band (colder and higher cloud tops) that is located along the cold front in a mid-latitude cyclone.

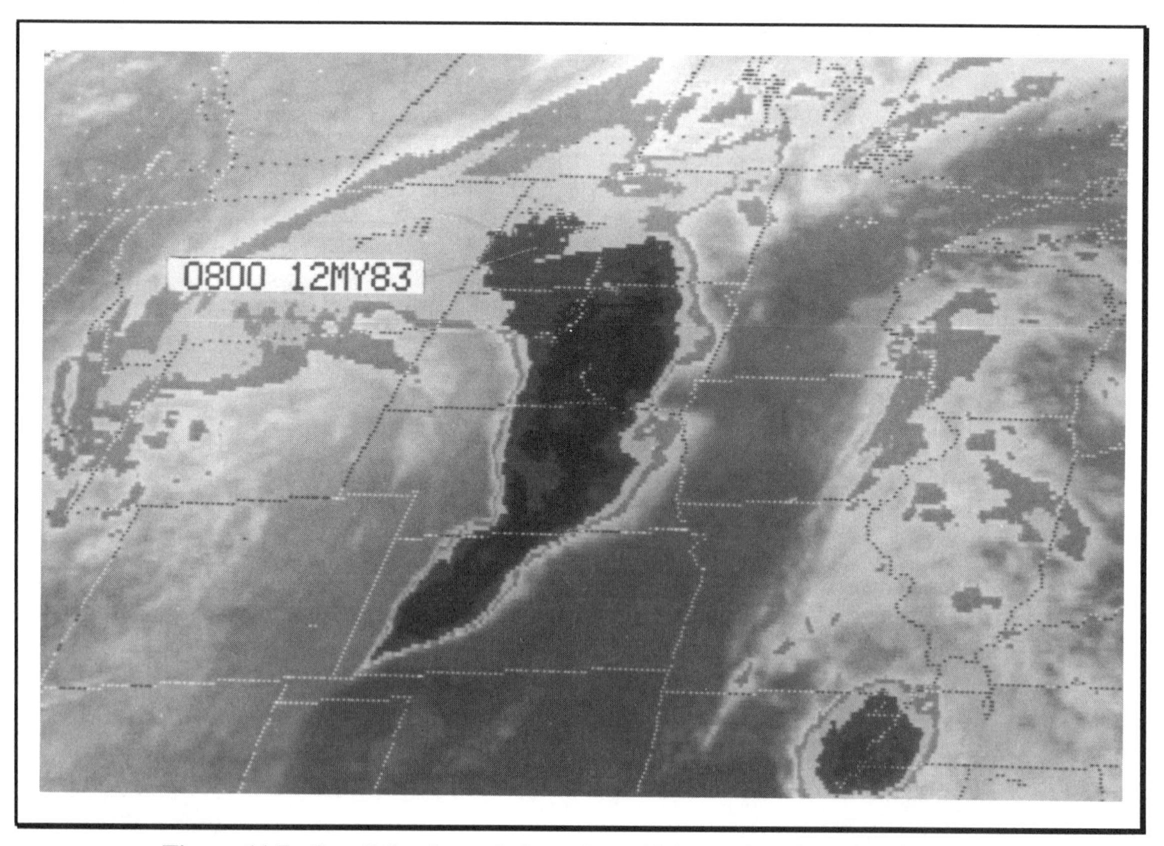

Figure 11.7. Squall line located along the cold front of a mid-latitude cyclone (GOES IR, MB enhancement, 0800 UTC, May 12, 1983)

Squall lines do not always form along fronts. Sometimes in the central United States there is a boundary between the dry cT (continental tropical) air from the desert southwest and the moist mT (maritime tropical) air from the Gulf of Mexico. If such a moisture boundary exists, it is called a **dry line**. The formation of a dry line is illustrated in figure 11.8.

Along the dry line, air is very unstable. The temperature in the cT desert air falls off rapidly with height. When this air pushes into the moist air from the Gulf of Mexico, potentially explosive unstable conditions can occur. It is along this dry line that many intense squall lines develop. At times the squall line associated with the dry line will interact with a frontal boundary, providing even stronger convergence and thunderstorm development. Figures 11.9 and 11.10 show a severe thunderstorm outbreak along the dry line. In figure 11.9, a line of thunderstorms can be seen forming along the dry line, which extends in a southwest-northeast line through Nebraska. To the west of the line lies a cloud-free zone, showing dry air from the southwest. To the east of this line, southerly winds are feeding moisture into the thunderstorms, forming low-level cloud lines with a north-south orientation. As the dry air moves further into the warm, moist air, the region becomes increasingly unstable, since moist air is less dense than dry air. In figure 11.10, an image made one hour after figure 11.9, the thunderstorms have grown considerably and produced severe weather in eastern Nebraska.

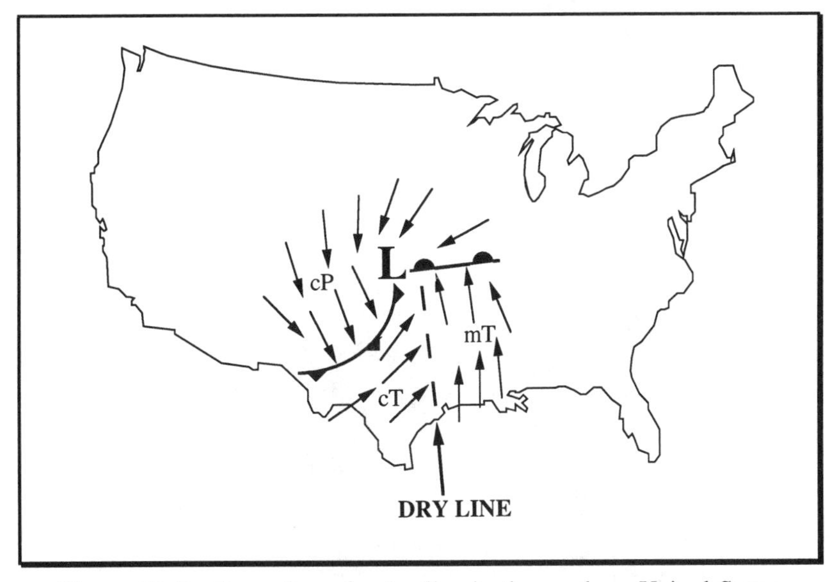

Figure 11.8. Formation of a dry line in the southern United States

Jet streams

Jet streams also have an influence on the development of convective storms. Most severe thunderstorms form in regions where there is a cold front and the polar and subtropical jet streams are **diffluent** (spreading apart from each other). Severe thunderstorms will usually form to the east of the cold front (in the warm moist air mass). Also, along the axis of the jet stream there are specific regions where air rises and sinks. The areas of rising air interact with locally unstable conditions to foster thunderstorm development. If a high-level jet stream can be located on a satellite image, thunderstorm development can often be seen along the left front and right rear quadrant of the maximum winds. In figure 11.11, a large thunderstorm can be seen in eastern Arkansas that has formed under conditions known as **split flow** aloft. This occurs when the jet stream

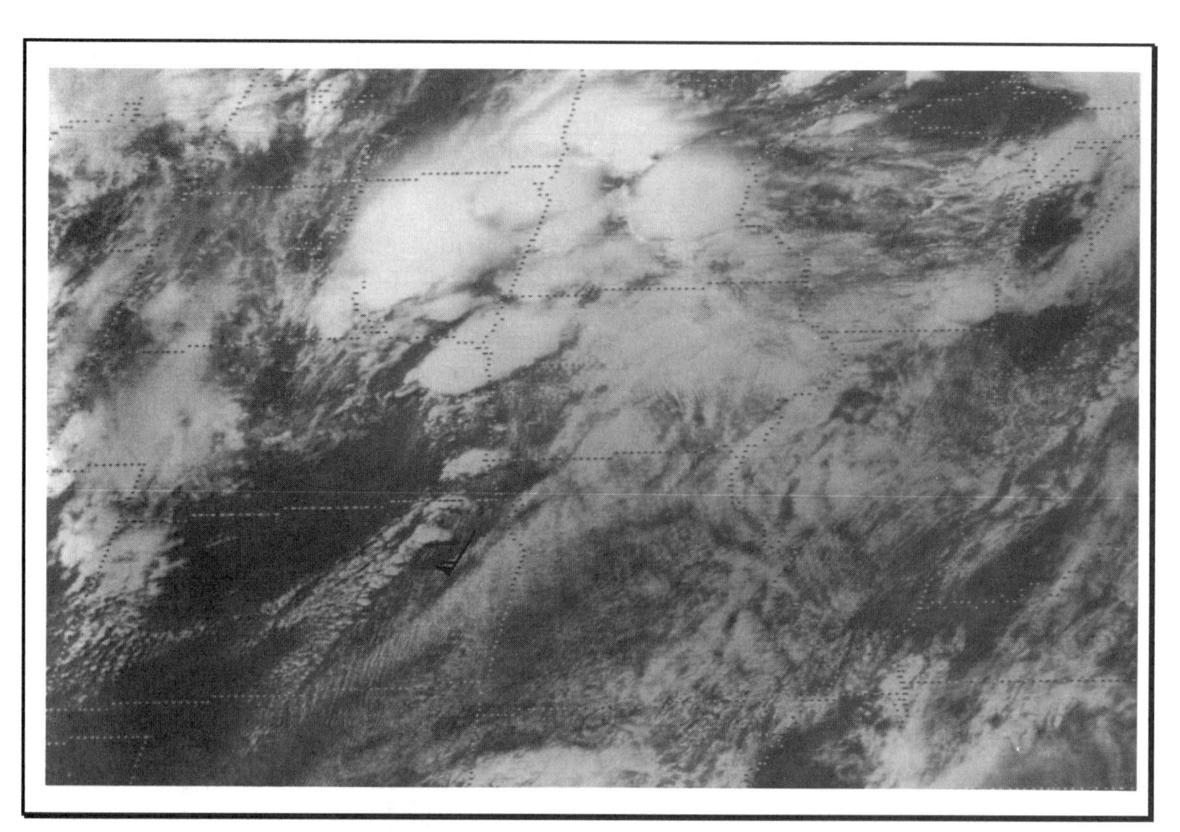

Figure 11.9. Severe thunderstorm outbreak along the dry line (GOES VIS, 2000 UTC, June 7, 1984)

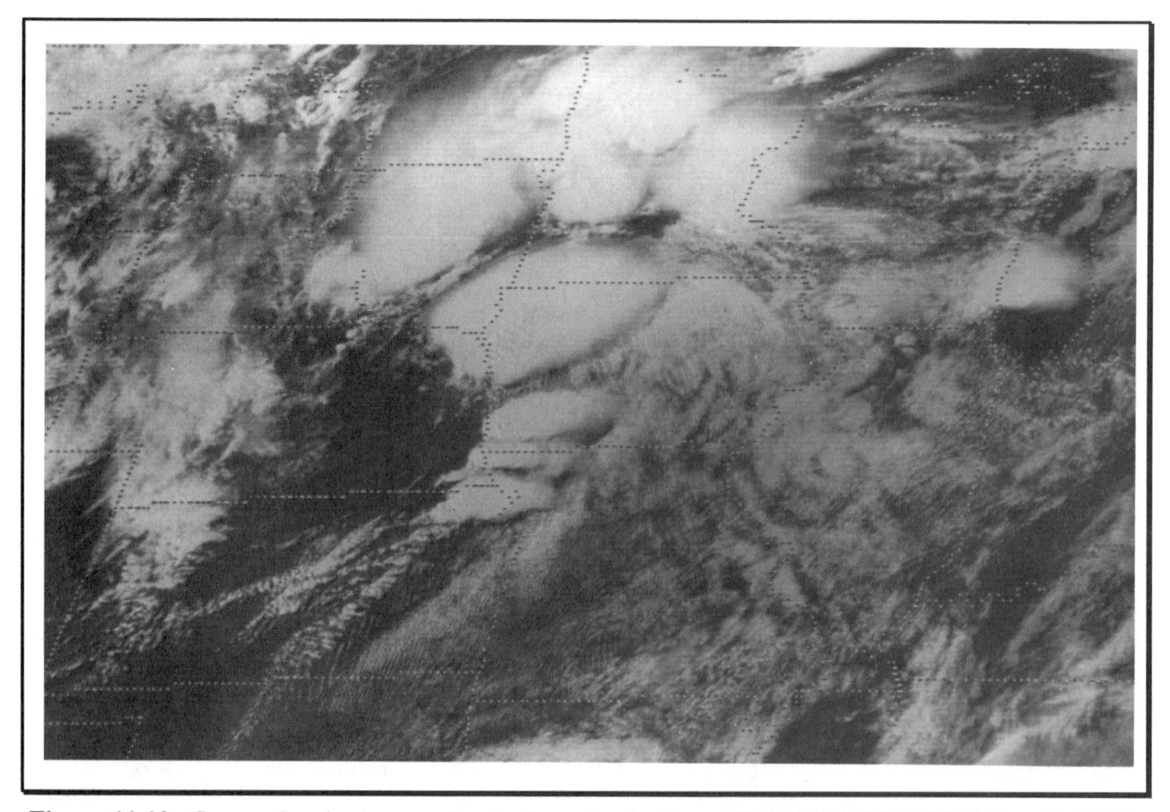

Figure 11.10. Severe thunderstorm outbreak along the dry line (GOES VIS, 2100 UTC, June 7, 1984)

Figure 11.11. Large thunderstorm complex in split jet stream flow (GOES IR, MB enhancement, 0200 UTC, May 15, 1983)

actually splits, or branches, creating a region where upper-air flow is diffluent. This enhances convection and often leads to intense storm development in these regions. In this image, the cirrus clouds associated with the jet can be seen splitting in eastern Texas and Louisiana. The storm has formed in the region just downwind of this split flow.

Mid-latitude cyclones

Thunderstorms may also form east or southeast of the center of circulation (maximum vorticity) in the dry slot of a mid-latitude cyclone. This region shows up best on water vapor imagery as a dark (therefore dry) region near the maximum vorticity. Figure 11.12 is a water vapor image that shows a mid-latitude cyclone over the midwestern United States. The dry slot shows up clearly in this image as a dark area. East of the dry slot, several severe thunderstorms with very high winds can be seen over Oklahoma, Kansas, and Nebraska. Four hours later these storms had moved eastward with the mid-latitude cyclone, as shown in figure 11.13. The individual storms are harder to identify in this water vapor image since the anvil cirrus shields of the individual storms have merged together.

The Intertropical Convergence Zone (ITCZ)

Finally, along the ITCZ, heating of the air above a warm-water surface, coupled with convergence of the trade winds, causes convection nearly on a daily basis. As a result, thunderstorms often develop in large masses along the ITCZ. In figure 11.14, a nearly continuous band of thunderstorms has formed along the ITCZ, which is centered at approximately 10° N latitude as a result of intense heating in the tropics during August.

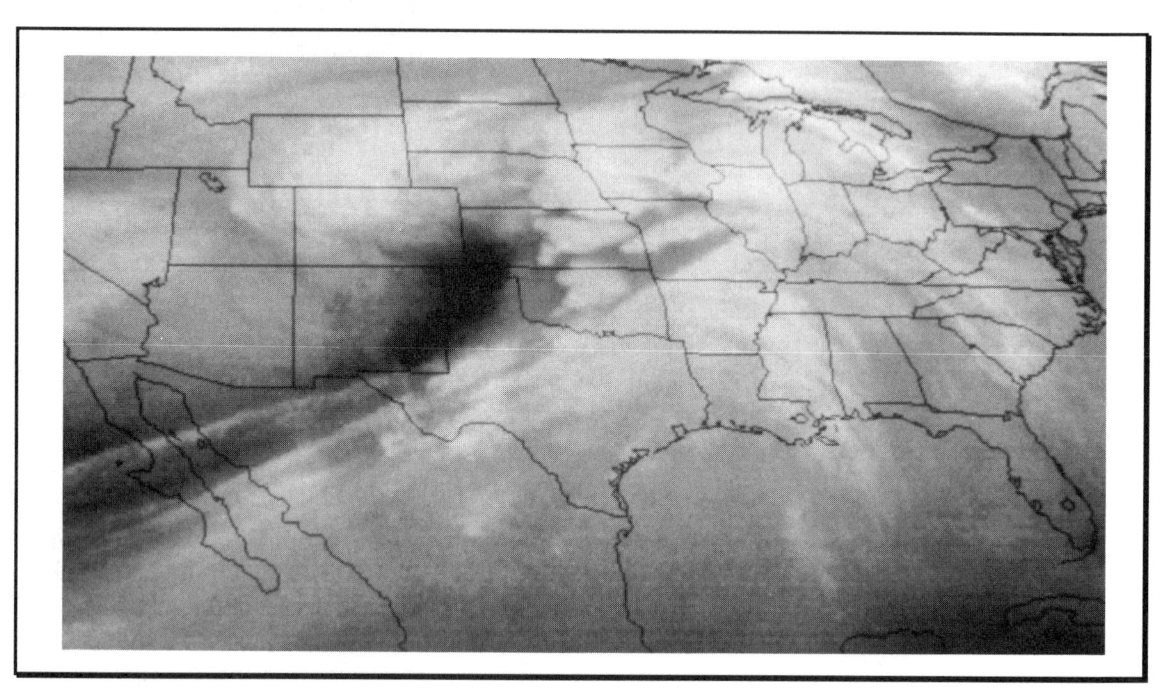

Figure 11.12. Thunderstorm formation in the dry slot of a mid-latitude cyclone (GOES WV, 0530 UTC, March 27, 1991)

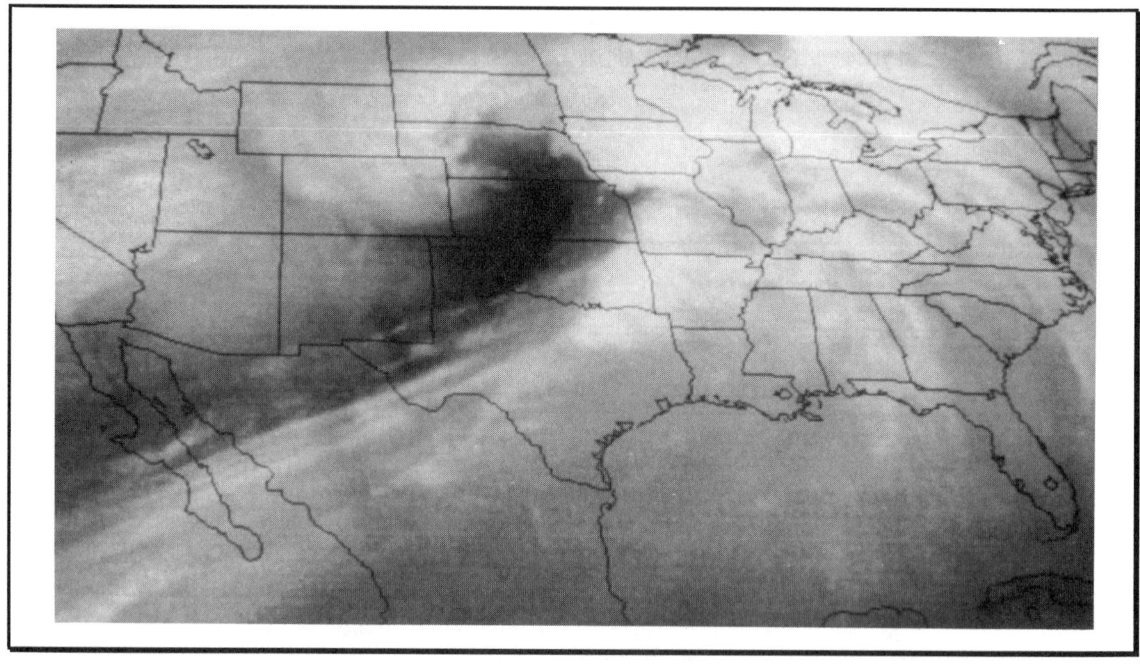

Figure 11.13. Thunderstorm formation in the dry slot of a mid-latitude cyclone (GOES WV, 0930 UTC, March 27, 1991)

Figure 11.14. Thunderstorm formation along the ITCZ (GOES IR, August 25, 1991)

Locating local areas of thunderstorm development (mesoscale)

Studying thunderstorm development at the mesoscale level is necessary to determine exactly where a storm is most likely to form within the more broadly defined synoptic-scale region. This allows you to examine local conditions and pinpoint the area where intense convective activity is most likely to occur. It is in this manner that severe weather watches and warnings can be issued for specific locations.

Arc-cloud lines

In satellite imagery, the leading edge of a gust front (or outflow boundary) appears as an arc-shaped line of convective clouds moving away from the center of a dissipating thunderstorm area. This feature, referred to as an arc-cloud line, is normally composed of cumulus or cumulonimbus clouds. Since these clouds are often small, they are best observed in 1 km GOES imagery. In figure 11.15, outflow from a dissipating thunderstorm in Arkansas and Mississippi has formed into a large arc-cloud line that extends across three states. Many smaller, new thunderstorms are beginning to develop along and behind the gust front.

Observations have shown that arc-cloud lines are very important in the formation and maintenance of strong convective storms. An arc-cloud line may retain its identity long after its parent storm has dissipated; therefore, it can continue to influence the weather in its area. Arc-cloud lines are especially important because new thunderstorms often form where two or more arc-cloud lines intersect. In this manner, a dissipating thunderstorm can "feed" a developing thunderstorm. Figure 11.16 is composed of three images that

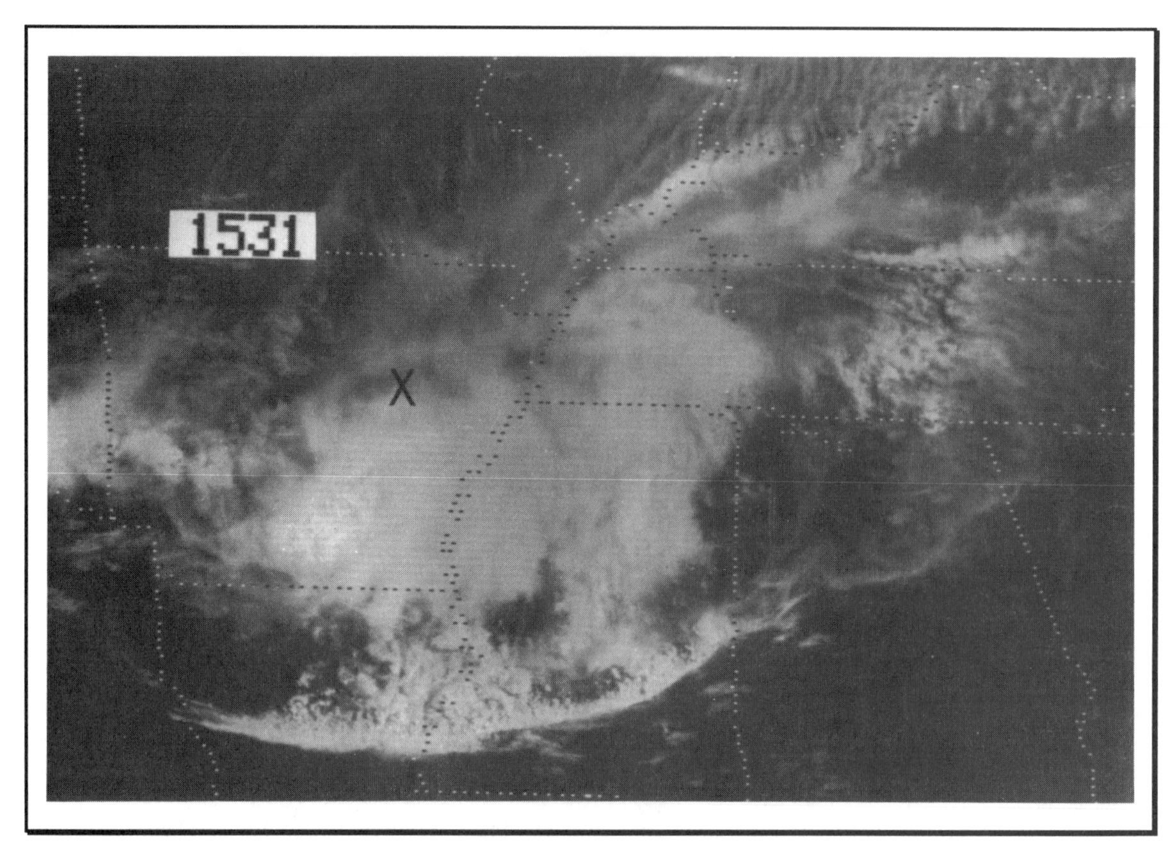

Figure 11.15. Arc-cloud line over the southeastern United States (GOES VIS, August 2, 1986)

show an arc-cloud line intersection. The top image shows a portion of two dissipating thunderstorms over South Dakota and Iowa. In the middle image, the arc-cloud lines from these older storms have intersected and a new thunderstorm can be seen forming at the point of intersection. In the third image in this sequence, the new storm has grown into a severe thunderstorm while the parent storms have weakened.

New convection can also occur in regions where an arc-cloud line interacts with a convectively unstable region such as a frontal boundary or a sea breeze front. Convectively unstable regions are often marked by cumulus clouds or cumulus lines. Where an arc-cloud line intersects such a region, thunderstorms are more likely to develop. In figure 11.17, a rope cloud associated with an oceanic cold front can be seen as it moves southeast and interacts with a convective cloud line over the Gulf of Mexico. In figure 11.17A, the two lines can be seen as they first converge. In figure 11.17B, small thunderstorms have developed in the region of intersection, and in figure 11.17C, the thunderstorms have matured and likely reached the tropopause.

Squall lines

Within a squall line, certain locations are especially favorable for severe thunderstorm growth. Along the line, many thunderstorms are occurring at one time, and they often influence each other. Outflow from one storm in the squall line can feed air into another storm, and the stronger convergence can help storms grow larger than they would as individual storms. Some storms in a squall line will often become severe when the squall line "bows" rapidly eastward. Where a squall line intersects a frontal boundary or a dry line, the chance of severe weather is higher. Finally, the southernmost storm in a squall

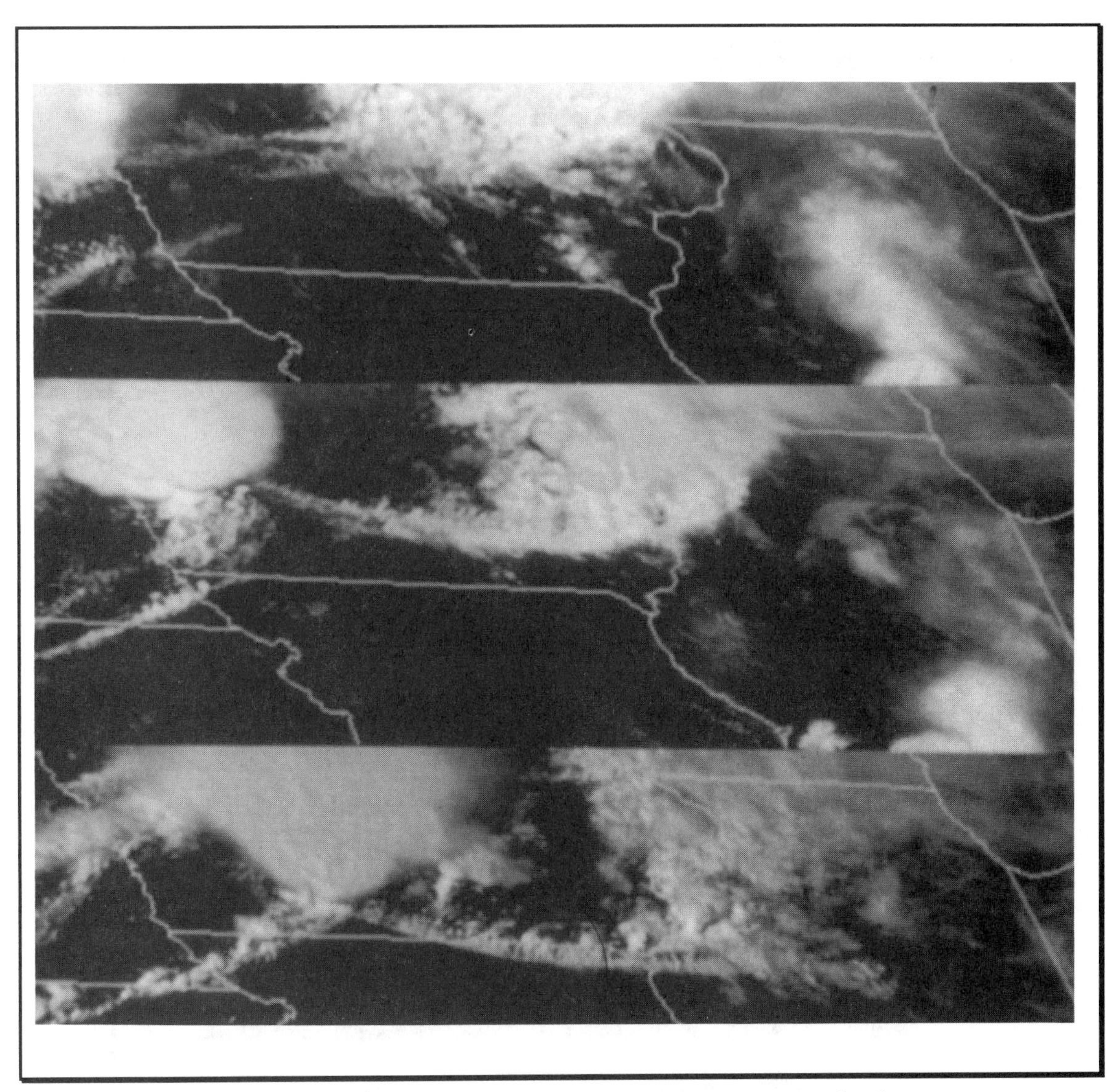

Figure 11.16. Converging arc-cloud lines leading to new storm development (GOES VIS, August 18, 1993)

line, or the storm nearest the center of low pressure in a mid-latitude cyclone is most likely to become a severe storm. This can be seen in figure 11.18, an enhanced IR image of a mid-latitude cyclone with an associated squall line. The many white regions in the cloud tops at the southern end of the squall lines indicate the region where the thunderstorm activity is concentrated.

Land and sea breezes

Land and sea breezes are a common phenomenon whose effects are routinely observed in satellite imagery. These breezes are a consequence of differential heating between land and adjacent water. A mesofront known as a **breeze front** forms along the boundary of warmer and cooler air. Cumulus and cumulonimbus clouds form along this front. Different curvatures of the coastline can cause areas where breeze fronts converge. This leads to strengthening of cumulus activity in these regions. Where breeze fronts intersect, thunderstorms often form. This is especially common on peninsulas where sea breezes

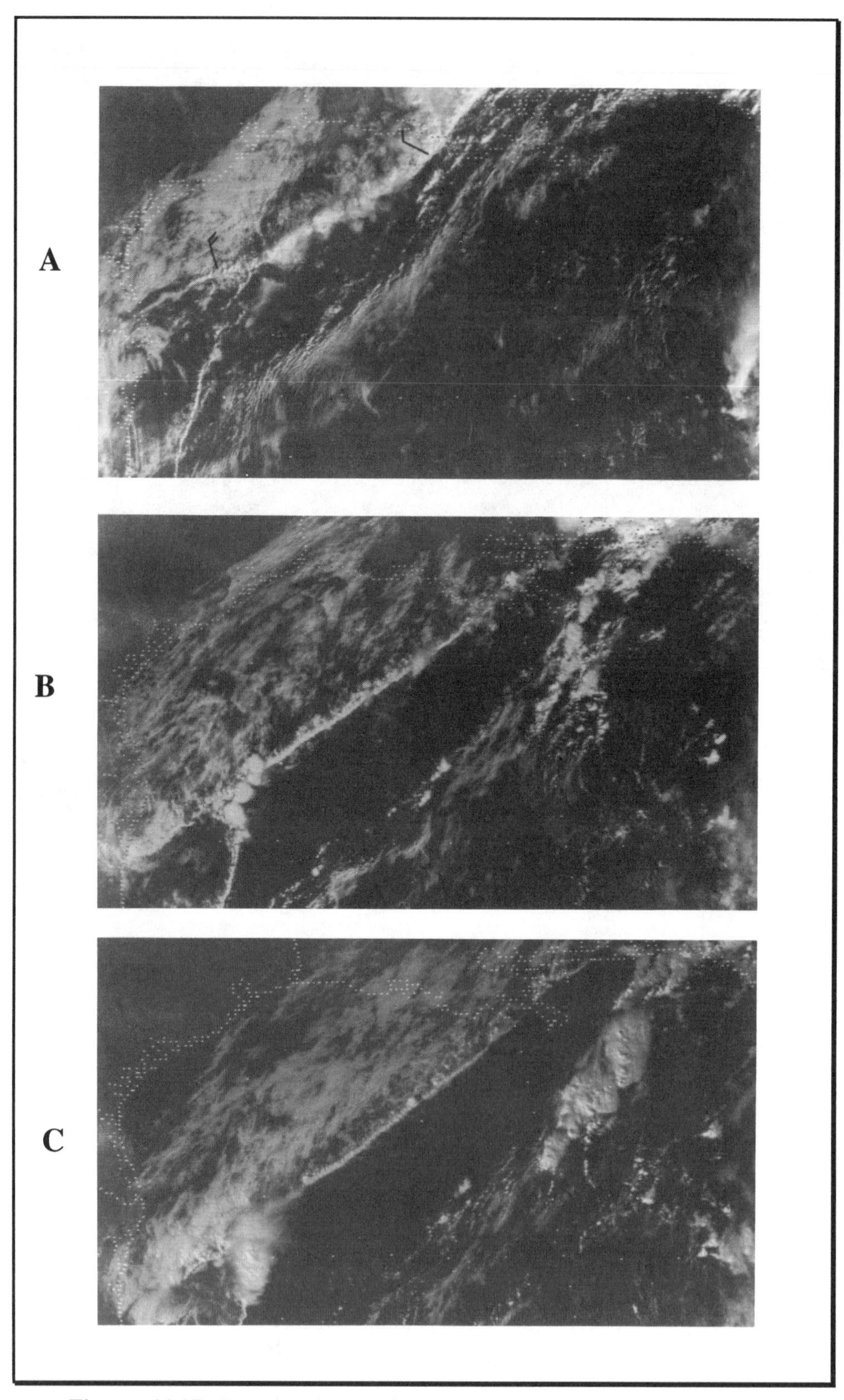

Figures 11.17. Boundary intersection leading to thunderstorm development (GOES VIS, [A] 1530 UTC, [B] 1900 UTC, [C] 2100 UTC, December 1, 1985)

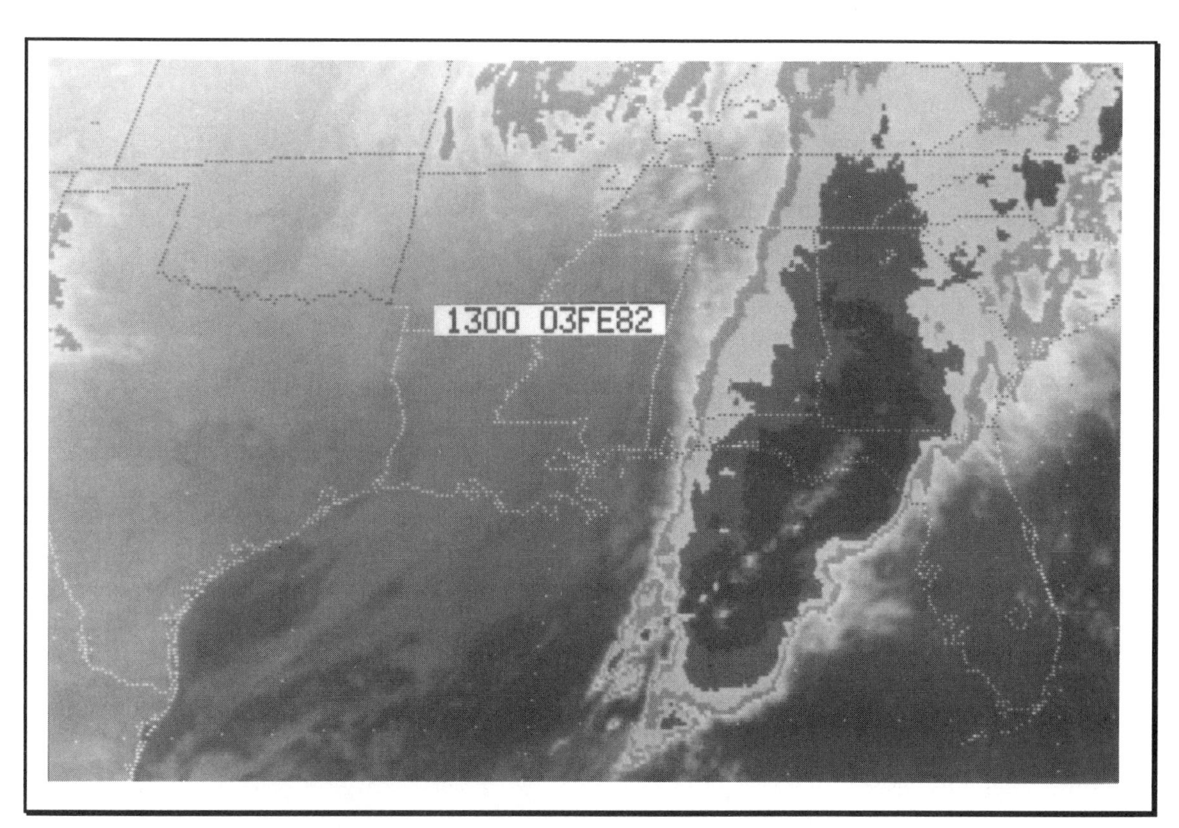

Figure 11.18. Thunderstorm intensification along the southern end of a squall line (GOES IR, MB enhancement, 1300 UTC, February 3, 1982)

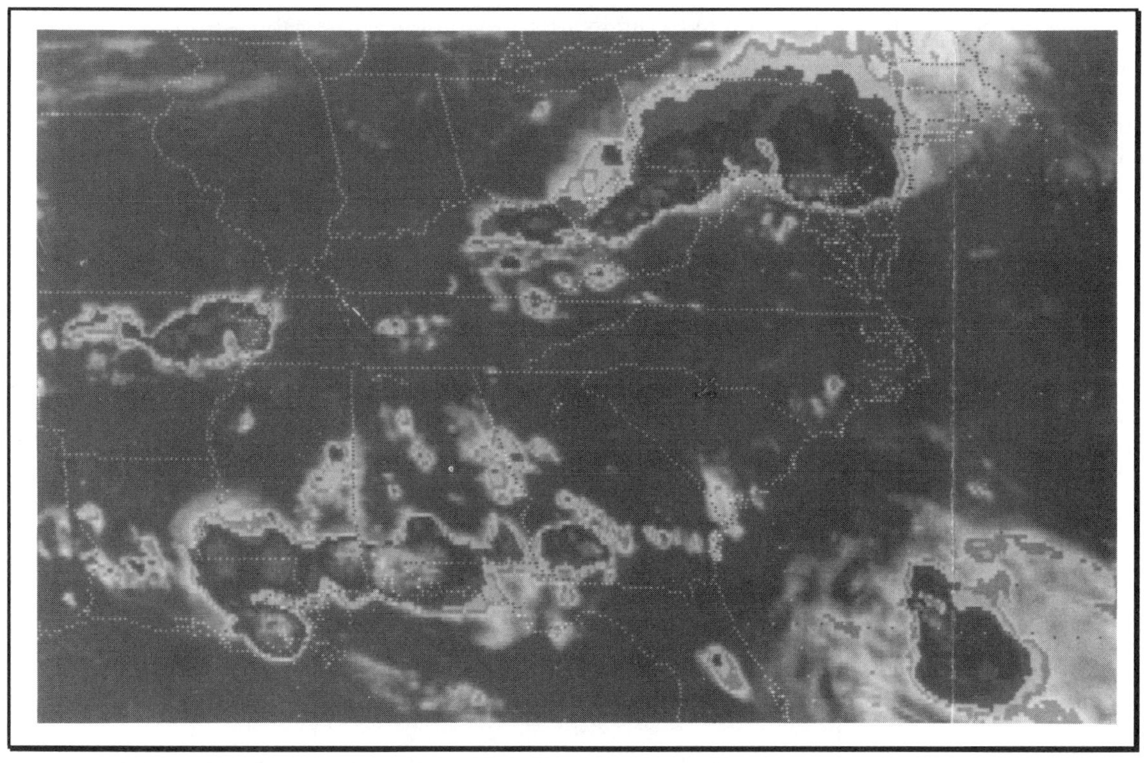

Figure 11.19. Thunderstorm formation along a sea breeze front over the Gulf Coast (GOES IR, MB enhancement, 2300 UTC, July 20, 1986)

from different directions converge. Although sea breezes most commonly lead to thunderstorm development, lake breezes can also do so. Figure 11.19 is an MB-enhanced IR image of the same thunderstorms in figure 11.2. The line of thunderstorms that has formed along the coastline along Gulf of Mexico is a result of convergence along a sea breeze front. Notice the shape of the thunderstorm in southern Louisiana. Here, a breeze from Lake Pontchartrain and the breeze from the Gulf of Mexico are interacting to produce the "hook-shaped" thunderstorm cloud pattern.

Early morning cloud cover

Early morning cloud cover or fog slows the heating of the land surface. Adjacent cloud-free areas heat up more rapidly, thus creating differential heating. If the clouds dissipate, a boundary still exists between cooler and heated air. This can be thought of as another kind of breeze front. Often, thunderstorms will form along this boundary. In figure 11.20, a ridge of high pressure is centered over the eastern United States. The convective cloud lines that have formed are oriented to the anticyclonic winds in this system. In Alabama and Louisiana there is a large cloud-free area. In this region, there was early morning cloud cover. The surface is cooler, since the clouds prevented heating; therefore, convective activity and cloud formation have not occurred as in the surrounding areas. Along the edges of this area, the convective clouds are enhanced, since the temperature gradient is greater. This boundary line is an area that could produce afternoon thunderstorms, especially if it interacts with a sea breeze front along the Gulf Coast or a similarly unstable region.

Ocean currents

Ocean currents also play a part in thunderstorm development over the oceans. Convective storms often develop along the edge of warm ocean currents, which heat the air from below. Air over adjacent cooler water is not heated, and a temperature gradient forms, along which thunderstorms may develop. In figure 11.21, a nearly continuous line of thunderstorms has developed along the Gulf Stream, a warm-water current that flows northward nearly parallel to the southeastern coast of the United States.

Mountain convection

Air flowing over mountain ranges is forced to rise very rapidly. In very unstable conditions, this rising air forms convective clouds. Air also tends to rise quickly over mountains due to solar heating of the mountain slope. This process, known as **mountain convection**, can lead to thunderstorm formation over the mountains. Storms created in this way are seen in satellite imagery as a cluster of storms located over mountain peaks. An example of mountain convection can be seen in figure 7.12.

Identifying areas of heavy precipitation

Meteorologists are also very concerned with locating areas where heavy rainfall may occur. Heavy rainfall from a thunderstorm can cause local streams and rivers to swell rapidly. As these streams overflow, the water may inundate croplands, highways, and buildings, causing widespread property damage and loss of life. This type of flooding, known as **flash flooding**, is one of the most threatening of all weather events.

Figure 11.20. Breeze front created due to differential heating of fog and fog-free regions over the southeastern United States (HRPT VIS, April 24, 1990)

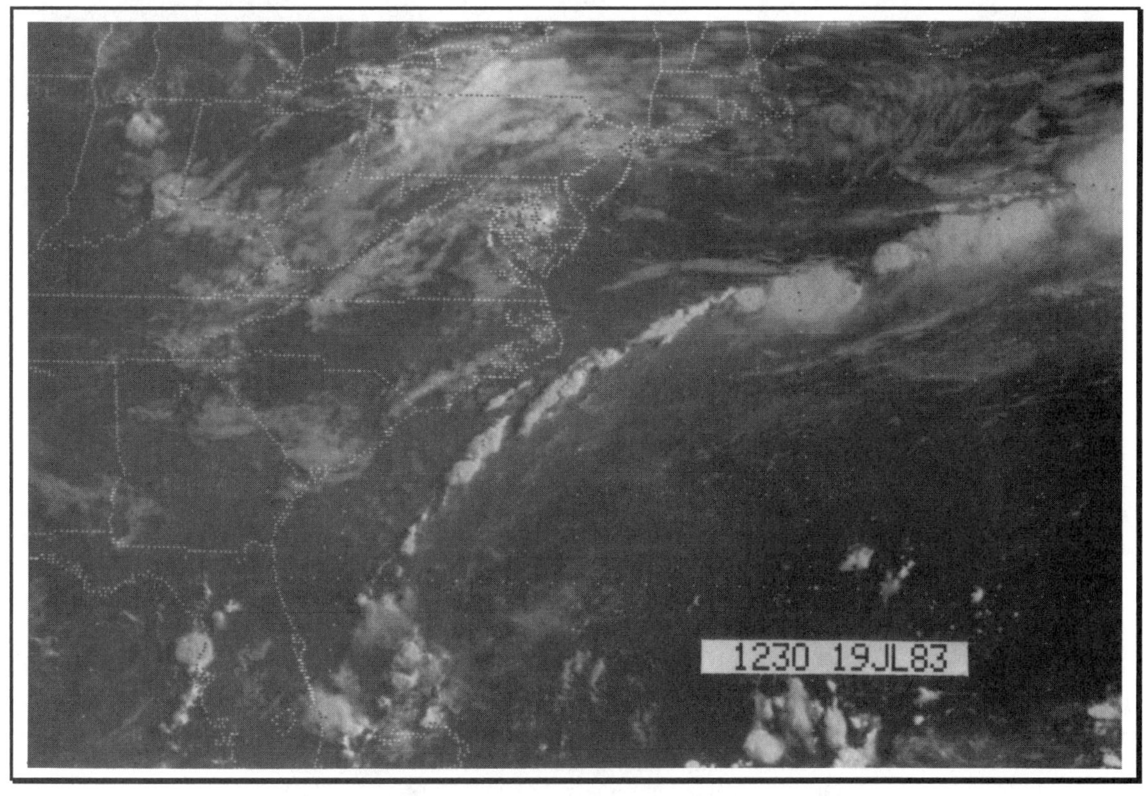

Figure 11.21. Thunderstorm development over the Gulf Stream (GOES VIS, 1230 UTC, July 19, 1983)

One common method used to identify areas of heavy precipitation in satellite imagery is examination of the growth and the temperature characteristics of thunderstorm clouds. Rapid storm growth and rapidly cooling cloud top temperatures indicate that heavy rain is likely. The shape of the storm cloud may also offer clues about rainfall potential. A wedge-shaped thunderstorm is likely to produce very heavy rains, with the heaviest rain near the apex (or upwind edge) of the cloud. Overshooting tops (or colder cloud tops) in a thunderstorm occur in the region of the strongest convection. Areas near these tops are usually associated with very heavy rainfall. In figure 11.22, a large thunderstorm cluster is located over the Gulf of Mexico. The coldest temperatures are near the apex of this thunderstorm; however, overshooting tops are evident in this storm due to the bright areas in the downwind portion of the storm. All of these are areas where heavy precipitation could occur.

Figures 11.23 and 11.24, a VIS and IR pair taken at the same time, show a large thunderstorm in central Texas. This storm exhibits a wedge shape, with the coldest cloud tops at the storm's apex. This is where the heaviest rainfall and the greatest flood potential are located.

Within a storm system there may be multiple wedges, indicating more than one location with heavy precipitation. In figure 11.25, a VIS image of Texas and the Gulf Coast, the high clouds associated with two thunderstorms are distinguished from lower-lying stratus clouds by shadows cast on the lower clouds and by the different characteristics of the cloud tops. Figure 11.26, an enhanced IR image of the same storm, shows two separate wedges, indicating the presence of two separate storms within the clouds.

Figure 11.22. Thunderstorm cluster over the Gulf of Mexico (GOES IR, MB enhancement, 1430 UTC, May 2, 1980)

Figure 11.23. Large wedge-shaped thunderstorm over central Texas (GOES VIS, 1731 UTC, October 29, 1991)

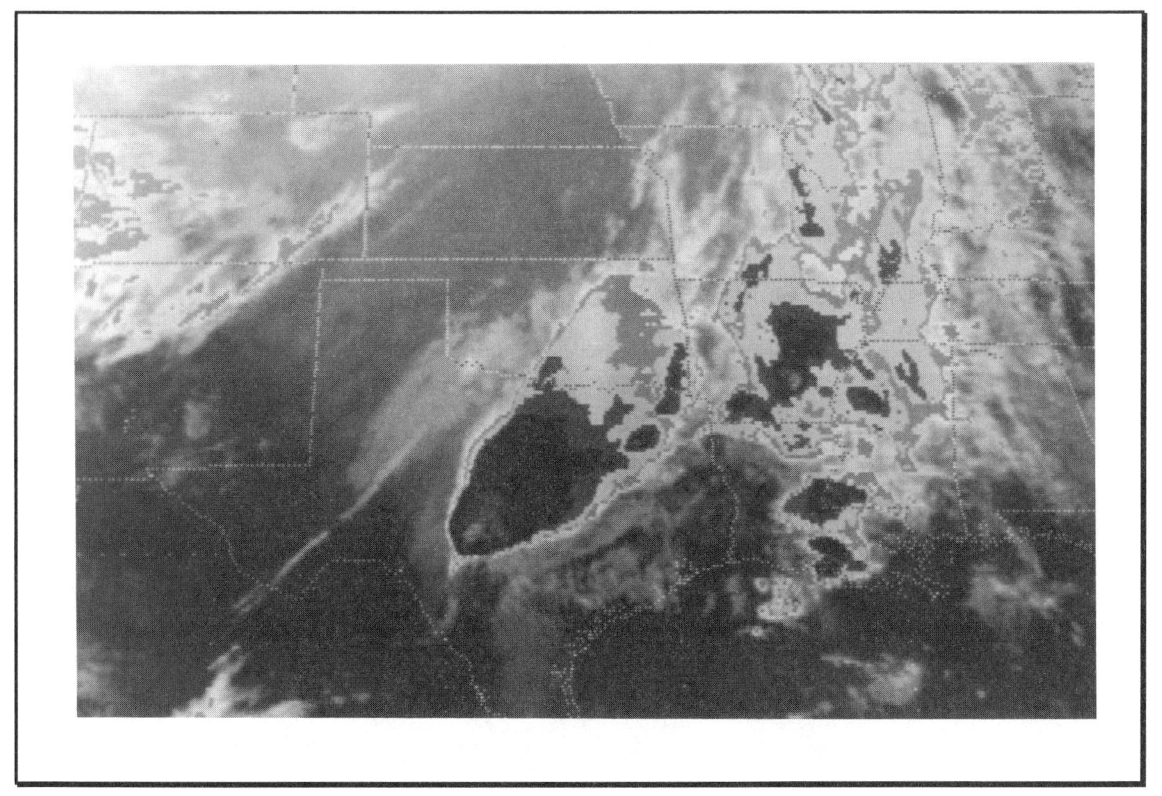

Figure 11.24. Large wedge-shaped thunderstorm over central Texas (GOES IR, MB enhancement, 1701 UTC, October 29, 1991)

Figure 11.25. Thunderstorm pair along the Gulf Coast
(GOES VIS, 1531 UTC, October 31, 1991)

Figure 11.26. Thunderstorm pair along the Gulf Coast
(GOES IR, MB enhancement, 1501 UTC, October 31, 1991)

Merging thunderstorms

Merging thunderstorms will also produce very heavy rain. This can be observed when two or more individual thunderstorms come together to form one large thunderstorm complex that is very intense. A storm of this type is often referred to as a **mesoscale convective system (MCS)**. To study merging thunderstorms, it is necessary to make repeated observations of the system that is producing the storm. Figures 11.27 and 11.28 consist of a sequence of four images of a thunderstorm merger. In figure 11.27 (top), a series of smaller thunderstorms can be seen at the tail of a mid-latitude oceanic cyclone. Five hours later, in figure 11.27 (bottom), these storms can be seen as they begin to merge. The combined energy of the thunderstorms and the heat from the warmer waters of the Gulf Stream form one very strong thunderstorm, shown in figure 11.28 (top). Eventually, the storm weakens and begins to dissipate (figure 11.28, bottom).

Figure 11.29 is another example of a thunderstorm merger. In figure 11.29A, several thunderstorms are located in a line across the central United States. The two larger storms in Arkansas merge together and form an MCS on the Tennessee border (figure 11.29B). The MCS continues to grow and takes on a comma shape in figure 11.29C; it produced very heavy rainfall and flash flooding along the Ohio River valley.

Moisture plumes from the tropics

The ITCZ is often the source of moisture for thunderstorms and other heavy rainfall events. Over the tropics, intense heating causes large amounts of water to evaporate. Along the ITCZ, this moisture is carried upward in the atmosphere, creating large areas of thunderstorms. This excess moisture and its associated thunderstorms can be seen in satellite imagery as a more or less continuous band of moisture near the equator. At times, large plumes of high-level moisture will extend from the tropics into mid-latitudes. These **water vapor plumes (WVP)** are seen best in water vapor imagery as northward surges of moisture from the ITCZ. Large thunderstorms and flash flood events often occur in storms located along the northernmost tip of a water vapor plume. Here, high-altitude moisture and upper-level winds interact with low-level moisture and often create conditions favorable for thunderstorms and heavy precipitation.

Figures 11.30 and 11.31 illustrate the relationship between water vapor plumes and heavy precipitation. Figure 11.30 is a GOES IR image that has been enhanced using the MB enhancement curve. A large thunderstorm cluster can be seen centered over Missouri. This was one of the many storms that contributed to the devastating midwestern floods of 1993. Figure 11.31 is a water vapor image taken at the same time as figure 11.30. A well-defined water vapor plume can be seen extending from the tropical Pacific Ocean, across Mexico, and into the central United States. The large thunderstorm is located at the northern tip of the water vapor plume.

Damaging winds and tornadoes

Although it is impossible to actually see winds on a satellite image, cloud top patterns often provide clues to identifying areas where high winds are likely. In this manner, satellite imagery can be used, along with other data such as radar and surface reports, to forecast damaging wind events such as windstorms, downbursts, and tornadoes.

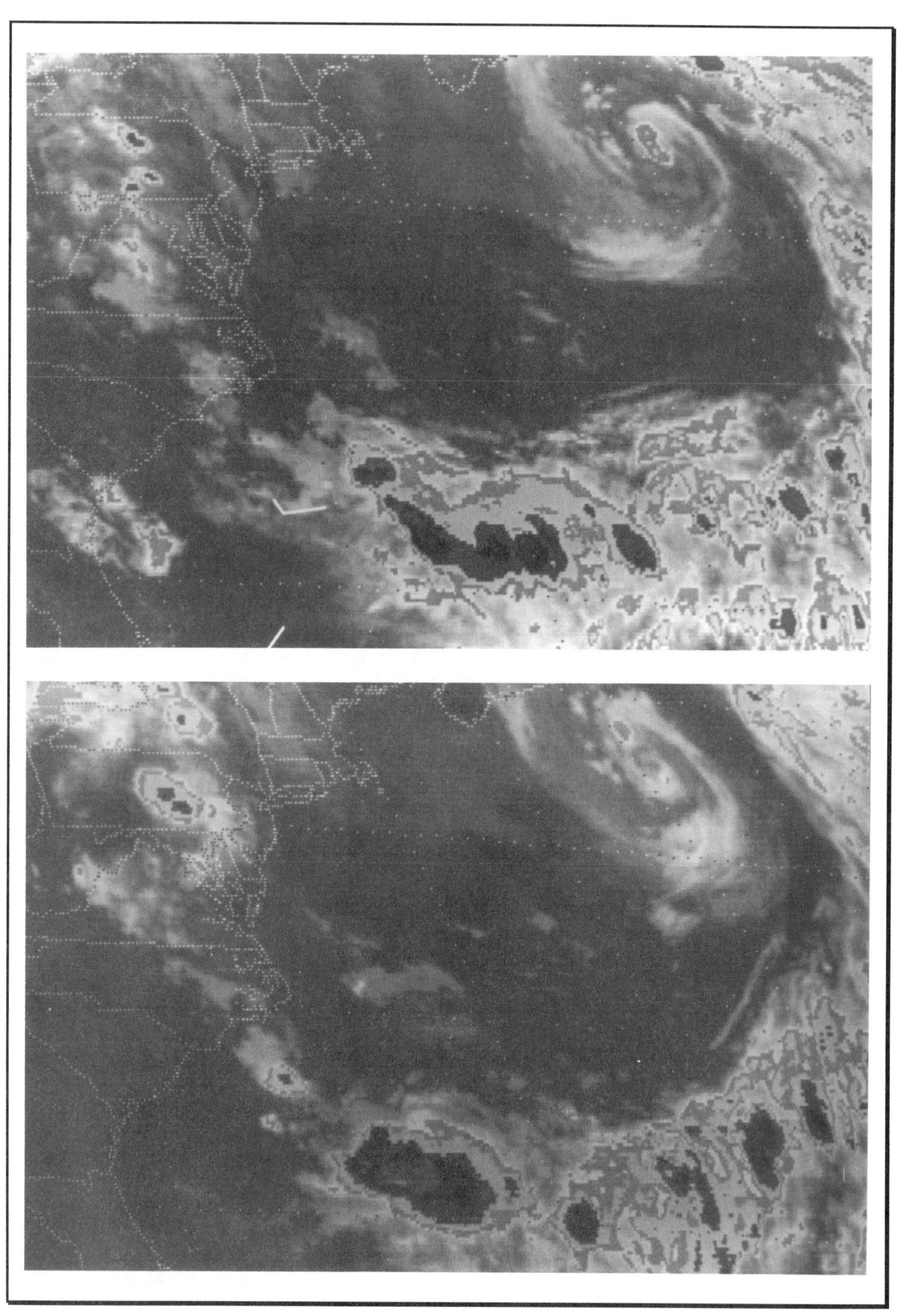

Figure 11.27 Thunderstorms merging into an MCS over the Gulf Stream (GOES IR, MB enhancement, [top] 0600 UTC, [bottom] 1101 UTC, July 20, 1986)

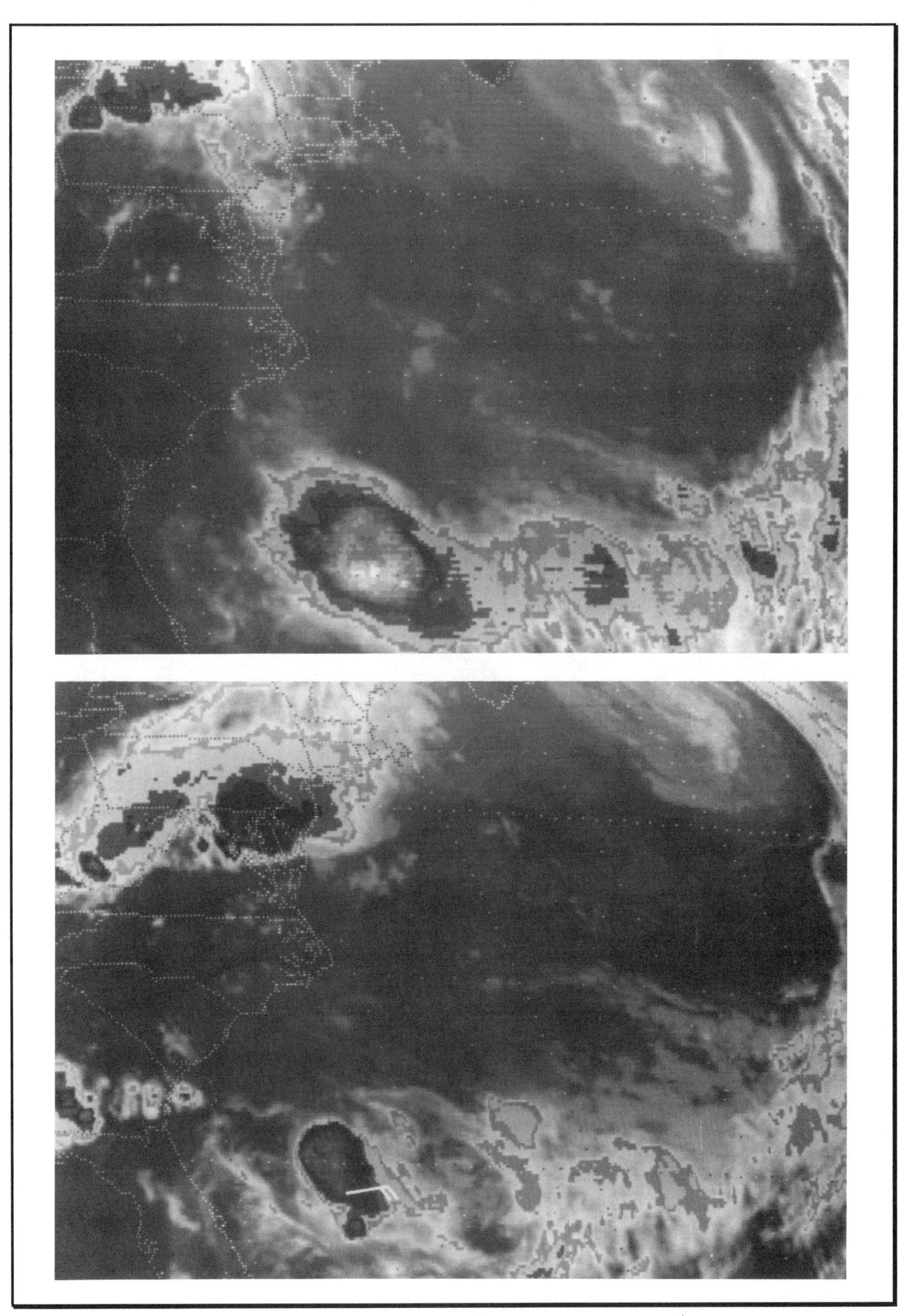

Figure 11.28. Thunderstorms merging into an MCS over the Gulf Stream (GOES IR, MB enhancement, [top] 1701 UTC, July 20, 1986; [bottom] 0100 UTC, July 21, 1986)

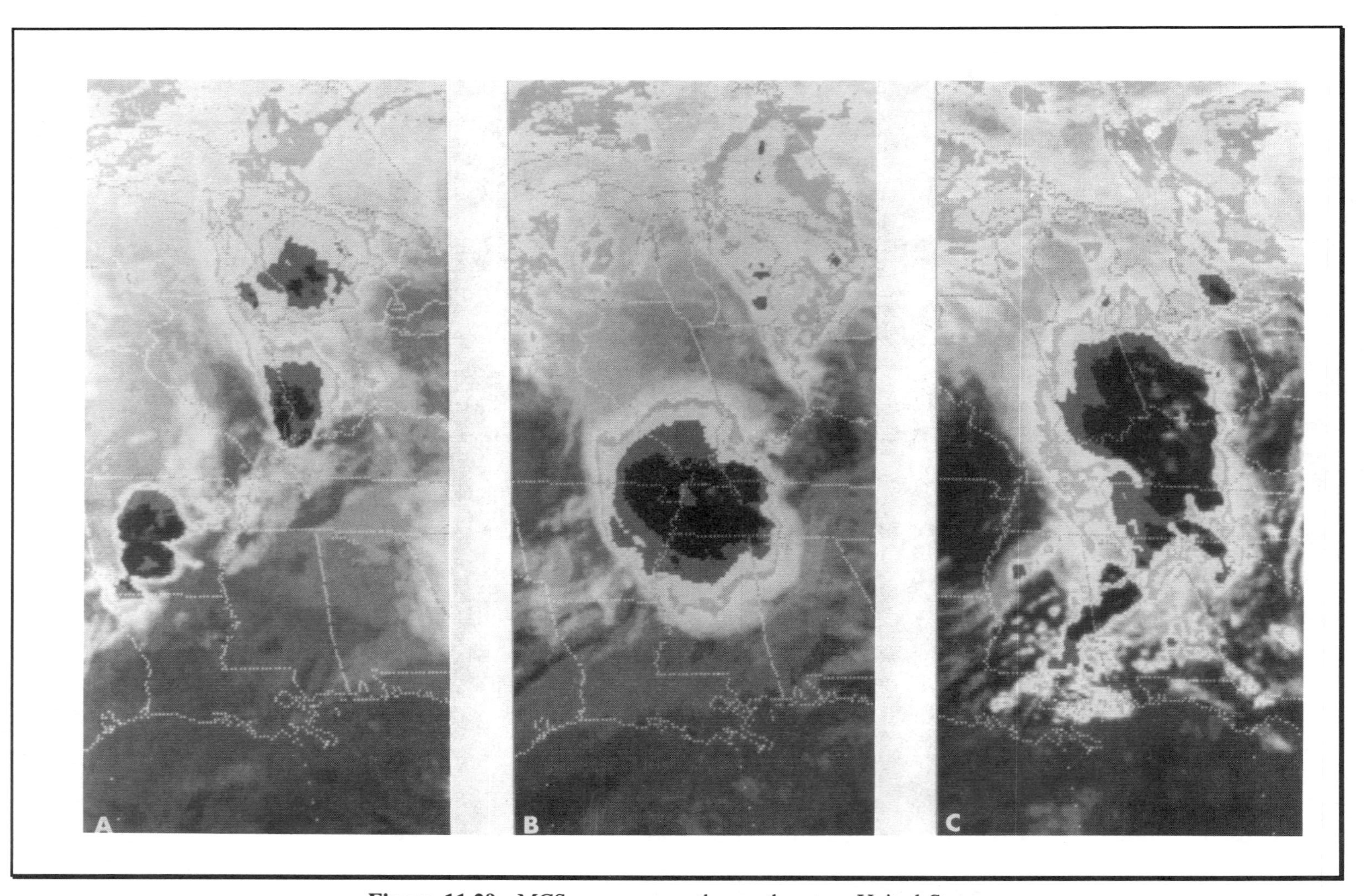

Figure 11.29. MCS merger over the southeastern United States
(GOES IR, MB enhancement, [A] 1201 UTC, [B] 1601 UTC, [C] 2001 UTC, April 9, 1991)

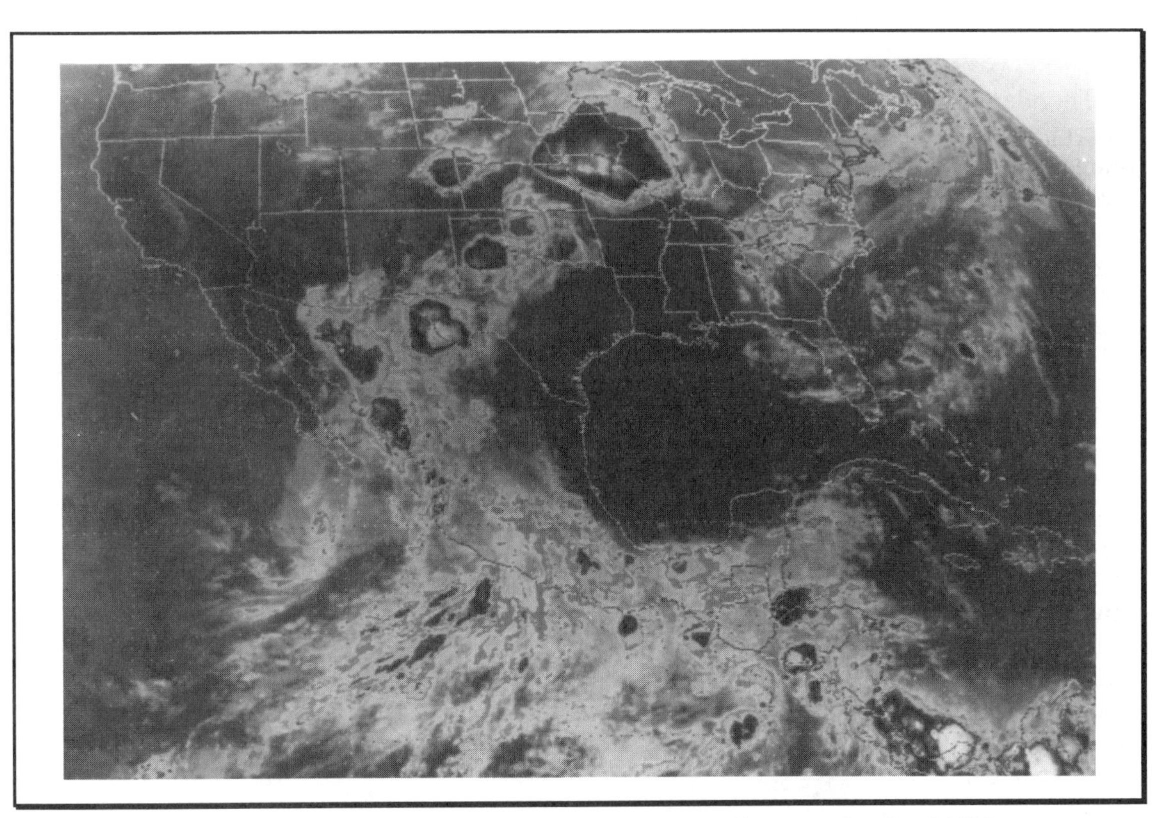

Figure 11.30. Thunderstorms associated with the midwestern floods of 1993 (GOES IR, MB enhancement, 0401 UTC, July 1, 1993)

Figure 11.31. Water vapor plume associated with the midwestern floods of 1993, not the same scale as figure 11.30 (GOES WV, 0401 UTC, July 1, 1993)

Windstorms

Thunderstorms that produce very strong, straight-line winds are known as windstorms. Often these storms can be identified by the cloud top temperature pattern in enhanced IR imagery. In a conventional thunderstorm, the anvil head of the storm spreads out in a wedge shape, with the apex pointing upwind and the anvil advancing ahead of the storm. The coldest cloud tops are usually located near the apex of the storm. During a windstorm, however, the outflow of the storm races ahead of the storm so fast that it outruns the upper-level cirrus shield. When the outflow moves this fast, it forces the air ahead of the storm to rise very rapidly, producing intense convection in the leading edge of the storm. Thus, in enhanced IR imagery, the colder cloud top temperatures appear along the leading edge of the storm, instead of in the apex or upwind region. This is known as a **leading-edge gradient**, and is often a signature of a storm with very heavy winds.

Figure 11.32 is a sequence of images of a windstorm traveling over northern Illinois. This storm exhibits a leading-edge gradient, since the front portion of the storm contains the whitest, coldest cloud tops. The leading edge of the storm is traveling at the same speed as the surface and low-level winds; therefore, the wind speed along the leading edge can be calculated by measuring the distance the leading edge has traveled over the time elapsed between images. This method only works on thunderstorms with a leading-edge gradient, since conventional storms are usually masked by the cirrus anvil. Figure 11.19 also shows a line of thunderstorms with a leading-edge gradient. These storms produced damaging winds in the Ohio and Tennessee Valleys and the mid-Atlantic regions of the United States.

Downbursts

During a strong thunderstorm, a large bubble of cold air may be supported in the upper atmosphere by strong updrafts. When the support for this air weakens, this cold, dense air may fall rapidly to the ground. When it reaches the ground it then spreads out, much as a drop of water does when it hits the ground. The effect on the ground is a very strong, destructive straight-line wind known as a **downburst** (or **microburst**). The damage from these downbursts is often mistakenly blamed on tornadoes, since both phenomena occur during severe thunderstorms. Downbursts are especially of interest to airport meteorologists, since they present a major hazard to aviation.

In satellite imagery, downbursts can often be detected because the area where a downburst occurs usually exhibits pronounced warming in the cloud tops. This will show up as a dark spot surrounded by cold cloud top temperatures in enhanced IR imagery. This pattern can be seen in figures 11.33 and 11.34. This sequence of images covers a two-hour period of a large MCS in the southeastern United States. In figure 11.33 (top), a small dark spot at point *A* indicates possible warming associated with a downburst. One half hour later, in figure 11.33 (bottom), the warming in this region expanded and a downburst was reported. In figure 11.34 (top), the cloud tops began to cool again and they appear brighter; however, a new downburst forms near the same area at point *A* in figure 11.34 (bottom).

Tornadoes

Tornadoes are the most powerful storms that occur in the atmosphere. Fortunately, they are very much smaller than other major storms such as hurricanes and mid-latitude cyclones. Within a tornado, winds circulate at speeds that can approach 500 km/hr

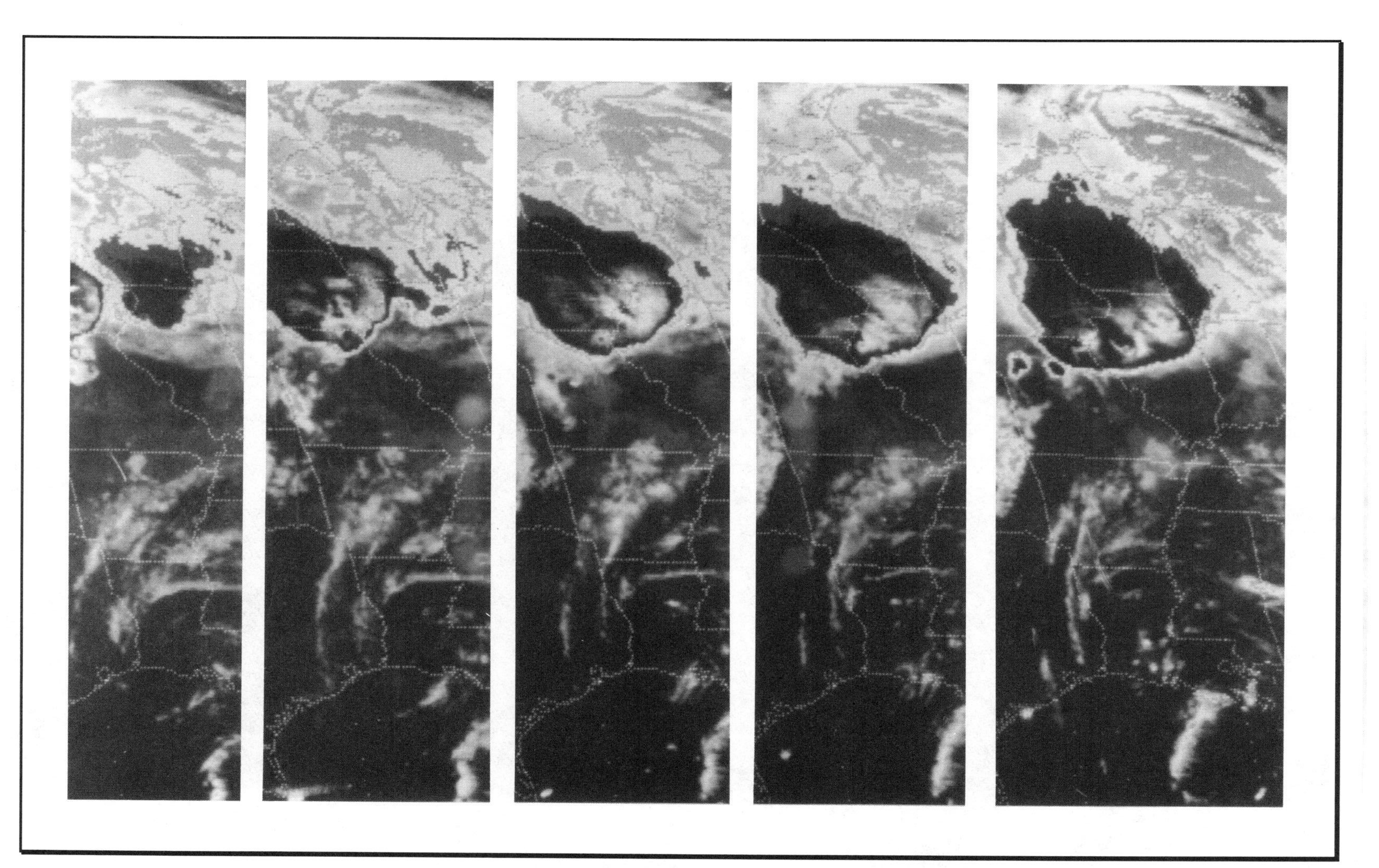

Figure 11.32. Chicago windstorm with a leading-edge gradient
(GOES IR, MB enhancement, 1601 UTC to 2001 UTC [hourly], July 2, 1992)

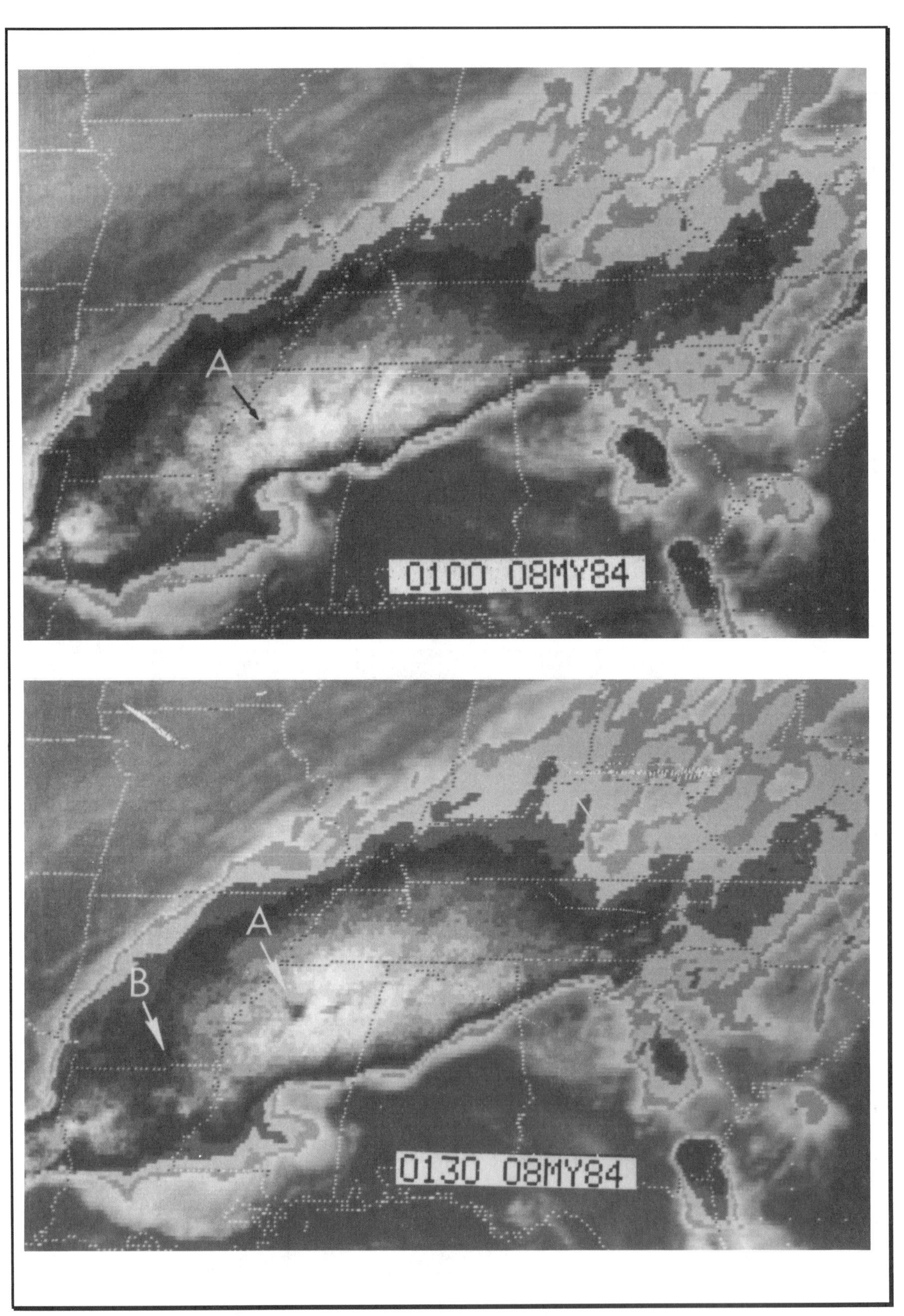

Figure 11.33. Large MCS with downbursts over the southeastern United States (GOES IR, MB enhancement, [top] 0100 UTC, [bottom] 0130 UTC, May 8, 1984)

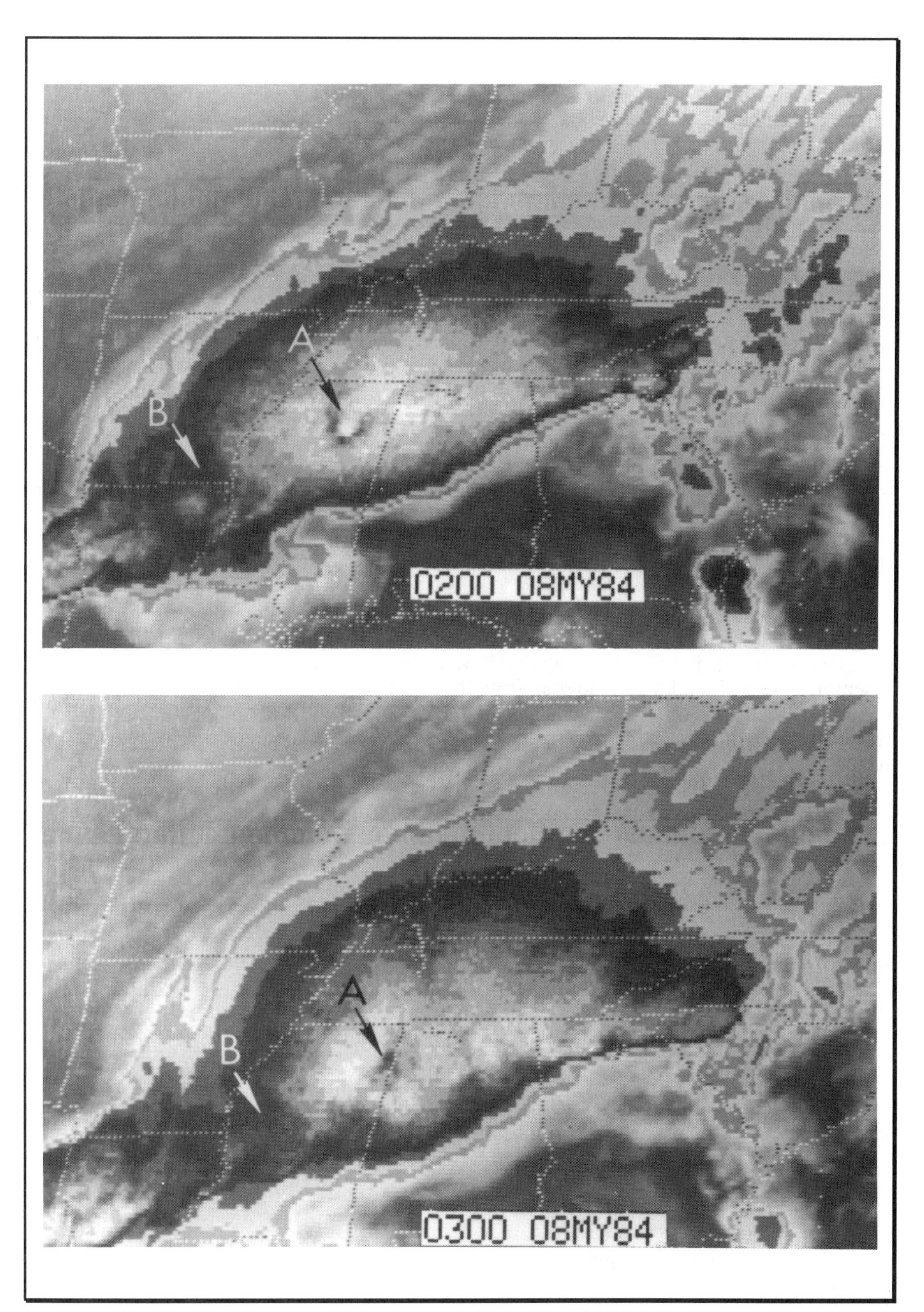

Figure 11.34. Large MCS with downbursts over the southeastern United States (GOES IR, MB enhancement, [top] 0200 UTC, [bottom] 0300 UTC, May 8, 1984)

(300 mph). A typical tornado is only a few hundred meters wide and will stay on the ground for about 10 km (6 miles), although some tornadoes can grow to more than 1 km in width and stay on the ground for several hundred kilometers. The high winds in a tornado can be extremely destructive. Some of the larger, more powerful ones have left almost total devastation in their wake. Each year many people are killed or left homeless by the destructive forces of tornadoes.

The United States has the distinction of having more tornadoes per year than any other country in the world, although tornadoes do occur in Europe, India, Australia, Japan, and parts of southern Africa and South America. They generally occur on landmasses affected by a jet stream, though they can form over bodies of water. In the United States, tornadoes can occur in virtually every state; however, they are rare in the Rocky Mountain states, along the western coast, and in Alaska and Hawaii. More tornadoes form in the central and eastern portion of the United States than anywhere else in the world. This is because the unique topography and pressure patterns that occur in the central plains region are ideal for tornado formation. The Rocky Mountains to the west act as a barrier to low-level winds; the lack of mountain barriers to the north allows cold air to move south from Canada; and southerly winds allow warm, moist air to move northward. With high pressure often anchored over the southeastern United States or the nearby Atlantic, it is not uncommon for troughs or low-pressure centers to develop over the plains. The cyclonic circulation of these lows tries to bring together these various air masses. As the air masses clash, severe thunderstorms often develop.

Tornadoes generally form in well-developed thunderstorms that exist as either individual storms or as a series of thunderstorms within a squall line. In satellite imagery, thunderstorms that spawn tornadoes often have a characteristic V, or wedge, shape that resembles the profile of a tornado seen from the ground. Tornadoes commonly occur near the apex (upwind) of the wedge and travel to the northeast. When thunderstorms with a V shape are located on a satellite image, they should be identified as areas of possible tornado formation.

Often radar is used to confirm whether a thunderstorm is a possible tornado-producing storm. Television newscasts will often show the conventional radar image of an area when severe weather is nearby. Tornado-producing thunderstorms may produce a classic **hook echo** pattern in radar imagery. This hook echo shows up as a curved band, hooking around a center of circulation in the southwest portion of the thunderstorm. A V-shaped thunderstorm in satellite imagery and a hook echo in the radar imagery provide strong indications that a storm is producing one or more tornadoes. **Doppler radar** now provides actual wind measurements inside thunderstorm clouds. Thus, the circulation of the tornado can often be detected before it reaches down to the ground. Using this data, tornado warnings for localized areas can be issued.

Figures 11.35–38 offer a view of a tornadic thunderstorm in Oklahoma on May 11, 1992, from both a synoptic and a mesoscale perspective. In figure 11.35, a large wedge-shaped thunderstorm can be seen in Oklahoma. The region behind this storm is relatively cloud free, indicating the presence of a dry air mass, one of the conditions necessary for tornadic thunderstorms.

Figure 11.36 is a water vapor image of the same scene. The jet stream is marked by a dark band of dry air sweeping across from Mexico, through Texas and into Arkansas. The large thunderstorm has formed on the front left quadrant of the jet. Jet streams often supply energy and rotation to thunderstorms, helping tornadoes to develop. It is also apparent in this image that the thunderstorm has formed to the east of the center of maximum vorticity in a mid-latitude cyclonic system.

Figure 11.35. Tornado-producing thunderstorm over Oklahoma (synoptic view) (GOES VIS, May 11, 1992)

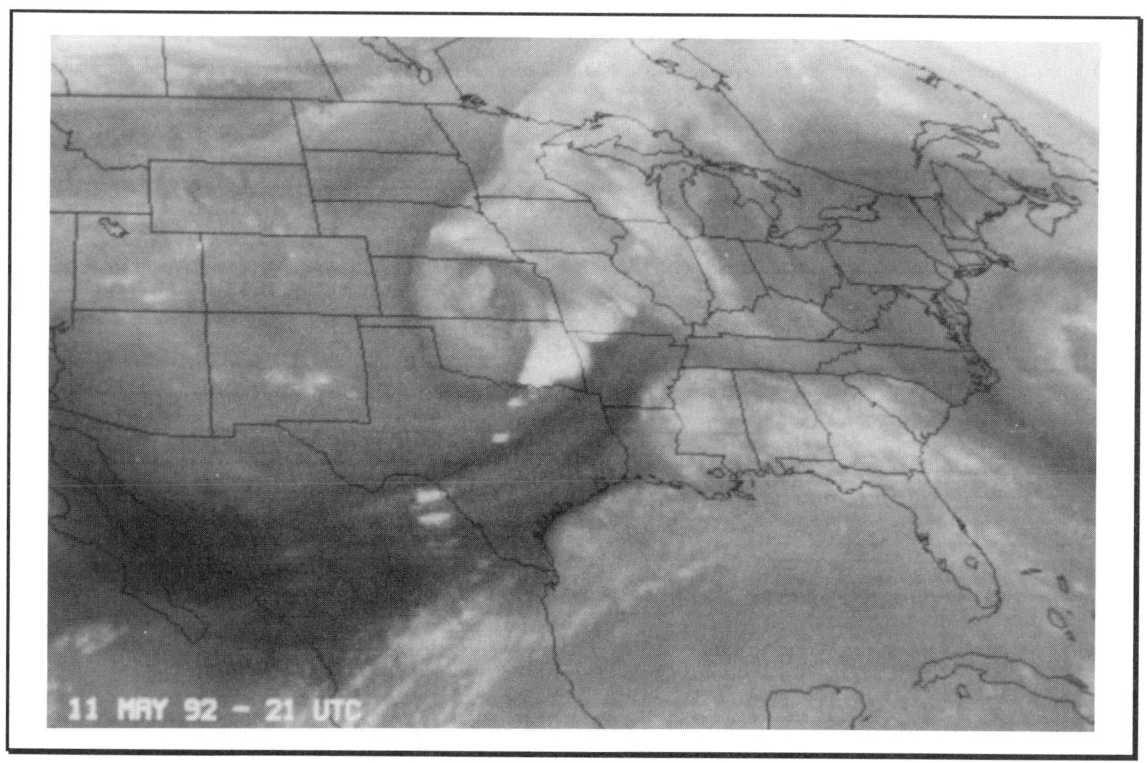

Figure 11.36. Tornado-producing thunderstorm over Oklahoma (synoptic view) (GOES WV, May 11, 1992)

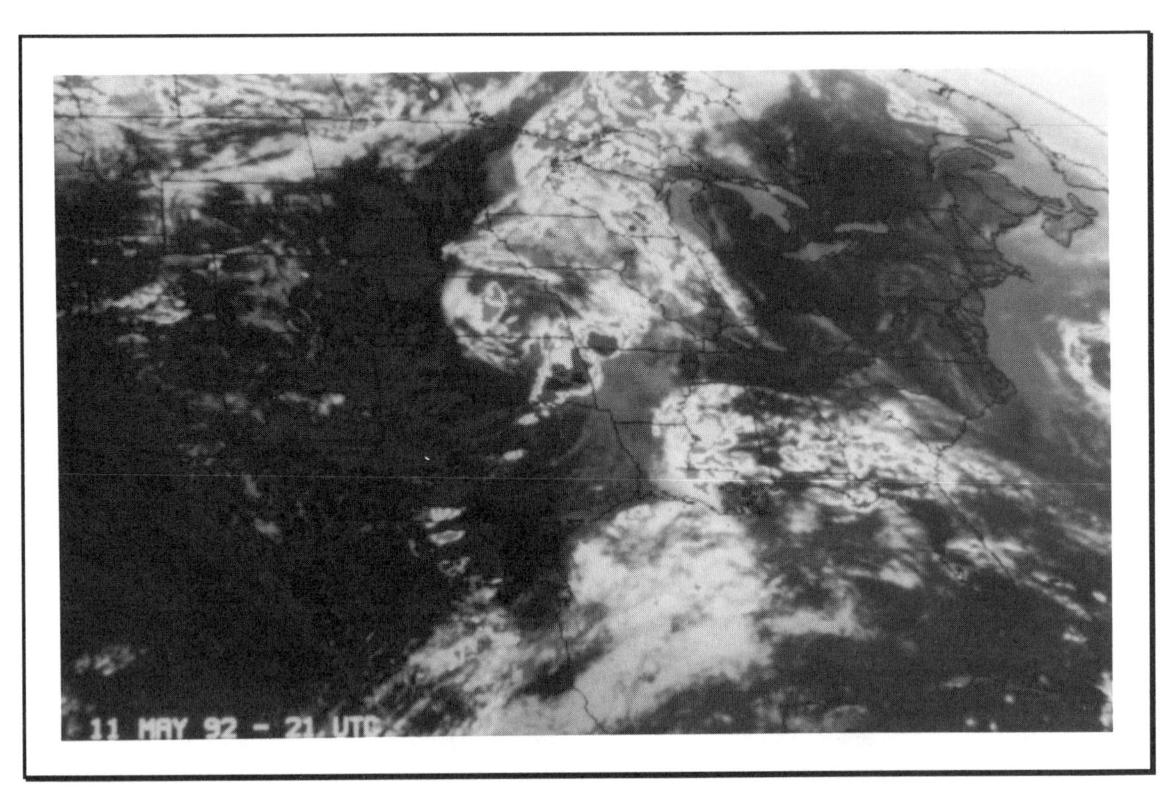

Figure 11.37. Tornado-producing thunderstorm over Oklahoma (synoptic view) (GOES IR, MB enhancement, May 11, 1992)

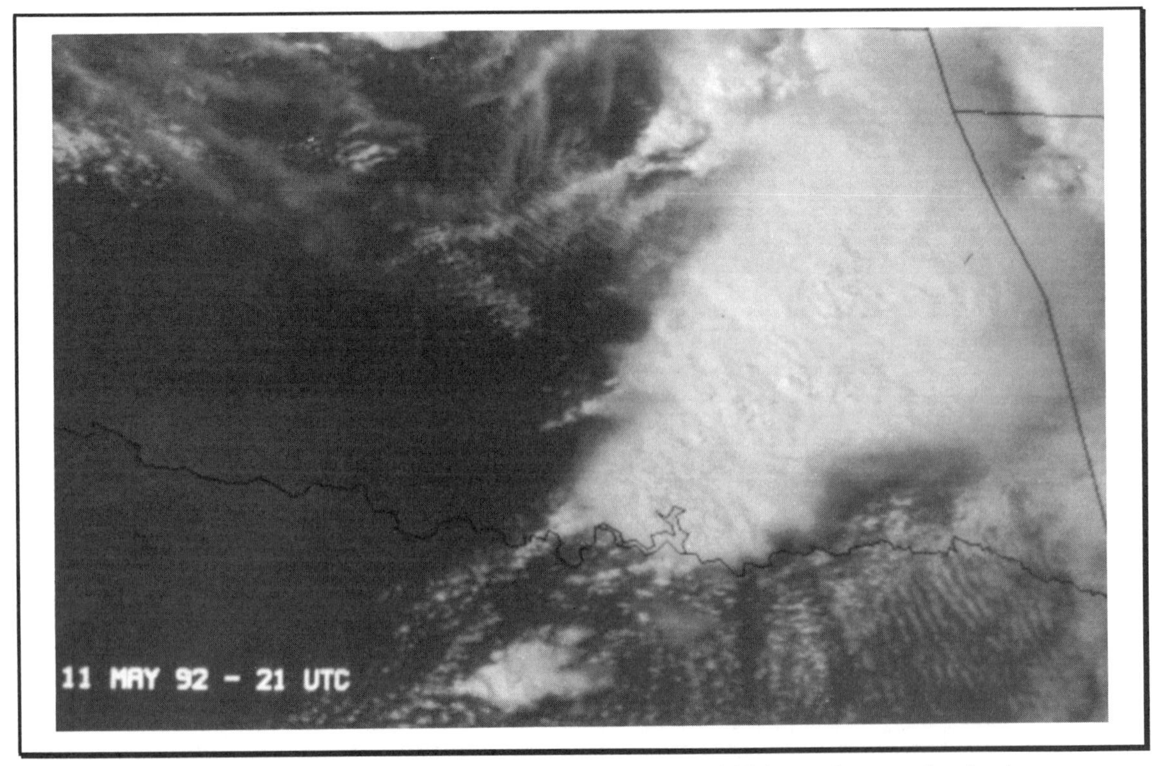

Figure 11.38 Tornado-producing thunderstorm over Oklahoma (mesoscale view) (GOES VIS, May 11, 1992)

Figure 11.37 is an enhanced IR image that identifies the location of the highest, coldest clouds. Finally, figure 11.38 is a close-up VIS image of the thunderstorm. This storm exhibits the classic wedge shape of tornadic thunderstorms. This storm spawned several tornadoes in southeastern Oklahoma.

Case Study: Thunderstorm development

Figures 11.39 and 11.40 include a sequence of VIS and enhanced IR images that depict a series of thunderstorms in the central plains region of the United States. These storms produced very heavy rainfall, flooding, and strong winds in and around Wichita, Kansas. Many of the concepts discussed earlier can be seen in this sequence of images; therefore, they serve as an excellent summary for this chapter.

In the first image (figure 11.39A), a line of thunderstorms can be seen extending from west to east in central Kansas, north of Wichita. The anvil cirrus clouds are being spread to the north, where they become very thin and ragged along the edges. These clouds extend well beyond the area of active convection. This storm is beginning to weaken; however, there are several indications that further thunderstorm development is likely in this region.

Just south of the thunderstorm line in Kansas lies a dense mass of cumulus clouds. This indicates a convectively unstable region. Along the southern boundary of the line an outflow boundary appears to be moving southward, out from under the storm, and toward the cumulus mass. This line exhibits a leading-edge gradient, with the coldest and highest tops (as seen in IR imagery) nearer its leading (southern) edge. New storm development is likely where the outflow of the older, weakening storm interacts with the unstable region to the south.

Further south, a dry line can be detected, starting at the Oklahoma panhandle and extending to the southwest into Texas. This boundary can be identified by a narrow line of cumulus clouds with an area of clear, dry air to the west. Further into Texas, this line appears to fork, with cumulus clouds and several small thunderstorms developing along the lines. This region is also a likely candidate for further thunderstorm development.

In figure 11.39B, at 2131 UTC, a clear zone can be seen forming to the north of the thunderstorm line in central Kansas. This indicates subsiding air associated with a mesohigh. A new thunderstorm can be seen forming in southern Kansas, where the outflow from the older storm has interacted with the unstable cumulus region. A second, rapidly growing thunderstorm can be seen along the dry line, inside the eastern portion of the Oklahoma panhandle. The enhanced IR image indicates that cloud top temperatures are quite cold in both storms, especially in the areas around their overshooting tops. The storm near Wichita has an enhanced-V signature, within 90 minutes of forming. These features are indications that these storms will continue to develop and will likely produce severe weather. In northern Texas, a line of storms ahead (east) of the dry line is also growing, and cloud top temperatures are cooling rapidly. Along the actual dry line, the cloud tops also exhibit development, though it is not as rapid.

By 2201 UTC, the line of storms in Kansas has weakened further, and the clear zone has continued to widen (figure 11.39C). The old outflow boundary from this storm is more clearly visible as this storm dissipates. The new Kansas and Oklahoma thunderstorms (actually clusters of storms) have continued to grow. Large anvils can be seen spreading downwind from both storms. Both exhibit the enhanced-V signature and overshooting tops. These storms also appear to be moving closer together, indicating a

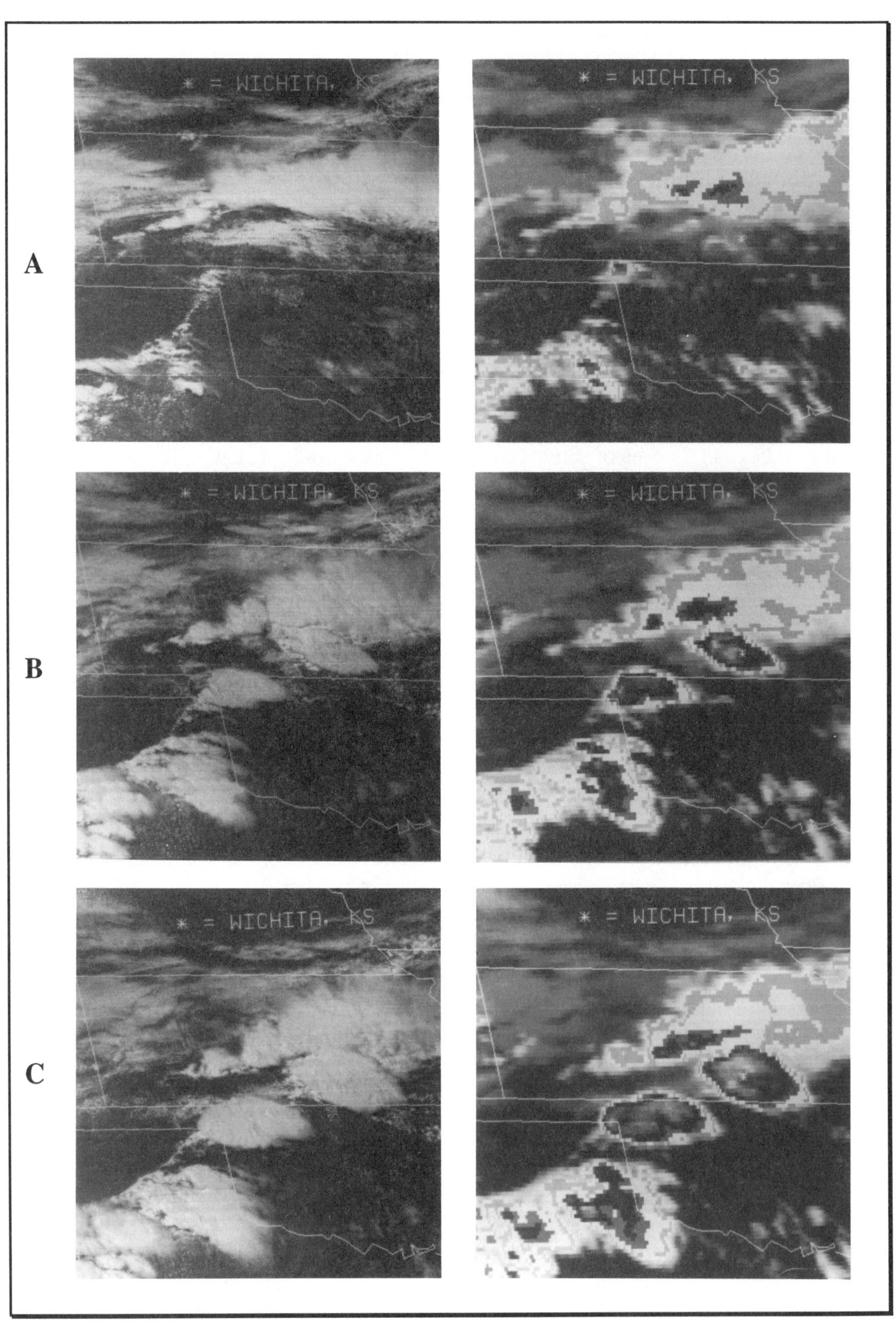

Figure 11.39. Thunderstorm development near Wichita, Kansas (GOES VIS/IR, MB enhancement, [A] 2001 UTC, [B] 2131 UTC, [C] 2201 UTC, July 8, 1993)

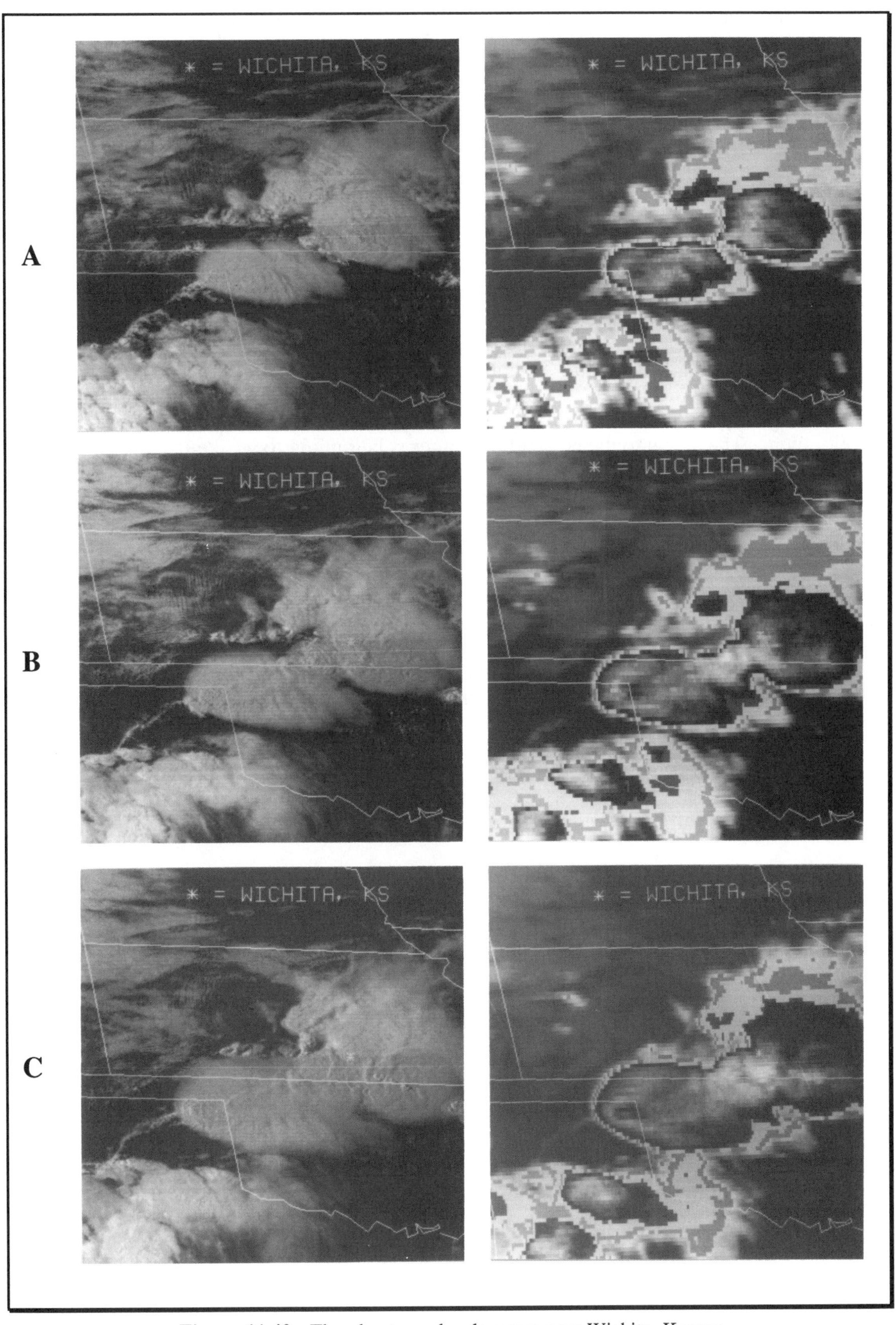

Figure 11.40. Thunderstorm development near Wichita, Kansas
(GOES VIS/IR, MB enhancement, [A] 2231 UTC, [B] 2301 UTC, [C] 2331 UTC, July 8, 1993)

possible future merger. Meanwhile, the storms in Texas have also grown, and appear to be moving closer to each other. As they continue to grow, outflow from these storms may interact and lead to new storm development in the region between them.

In figure 11.40A, a remnant of the outflow boundary from the original line of storms can still be seen oriented west to east, nearly parallel to the Oklahoma and Kansas border. The two younger, large storm clusters have matured and have nearly grown together. The enhanced-V signature is much more pronounced, and the rate of cloud top cooling is rapid in both. In the VIS image, interaction between outflow from these storms can be seen in the region between them, evidenced by the new cumulus development in this area. In Texas, new convective development is occurring in the zone between the two smaller storms located along the dry line. Outflow from these storms has resulted in a shifting of the dry line toward the northwest.

By 2301 UTC, the Kansas and Oklahoma storms have merged just southwest of Wichita, Kansas (figure 11.40B). Convection in the region between them is being enhanced by outflow from the two parent storm clusters and exhibits pronounced development. The storm to the east of the merger is beginning to weaken, as indicated by warming in cloud top temperatures. The storm to the west of the merger continues to get energy from the unstable air associated with the dry line, and therefore it still exhibits the enhanced-V cloud top pattern. In Texas, a new thunderstorm continues to grow as the earlier storms on either side weaken and begin to dissipate.

Finally, in figure 11.40C, the merger in Kansas continues to grow rapidly. Several new overshooting tops are easily seen in the VIS image. Cloud top temperatures are very cold at the point of the merger, while in both parent storms they are warming. The western storm still exhibits the enhanced-V that indicates strong convection and heavy rainfall. In Texas, the dry line is still apparent, though it has moved to the northwest a considerable distance. The region immediately to the east of the dry line remains very clear as a result of subsidence from the thunderstorm activity in that area. Though the original line of storms in Kansas dissipated several hours before this image was made, it still has an effect on the cloud patterns in figure 11.40C. Note the large, cloud-free area in central Kansas where the air continues to subside as a result of the dissipation of this storm.

Interactions between thunderstorms can affect local weather significantly. Sometimes the interactions act to weaken storms; at other times they create very intense storms. The result is often severe weather in one part of the state or county, and nicer weather nearby. Satellite, radar, and lightning data are among the new tools helping meteorologists better understand these effects.

CHAPTER 12

Tropical Cyclones

Characteristics of tropical cyclones

Meteorological satellites are uniquely equipped for the task of tropical cyclone surveillance. Satellites provide a great deal of valuable information about the position, intensity, and motion of these storms. This information is used to forecast, analyze, and provide warning of tropical cyclones around the world.

Tropical cyclones are storms that originate in tropical latitudes; they include tropical depressions, tropical storms, hurricanes, typhoons, and cyclones. These various types of storms are similar; their main difference is where they form. **Hurricanes** are tropical cyclones that occur in the Atlantic Ocean and the eastern and central Pacific Ocean, while **typhoons** are tropical cyclones that originate in the western Pacific Ocean. **Cyclone** is a more specific term that is often used to describe tropical cyclones that form in the Indian Ocean and near Australia. Though much of this chapter describes hurricanes (tropical cyclones with sustained winds of at least 120 km/hr, or 74 mph), the concepts can be equally applied to typhoons and cyclones.

Hurricane season

Tropical cyclones are most likely to occur at specific times during the year. For example, in North America, the **hurricane season** officially runs from June 1 through November 30. However, the majority of hurricanes occur during August, September, and October, when the ocean waters have had ample time to heat up. Hurricanes that strike the United States are usually generated in the Atlantic Ocean, the Caribbean Sea, and the Gulf of Mexico (between 10° N and 30° N).

Figure 12.1. Hurricane Hugo (GOES VIS, September 21, 1989)

Many hurricanes also form in the Pacific Ocean off the west coast of North America and move westward toward the central Pacific, though some of these storms circle back and strike the northwest coast of Mexico. During the hurricane season, the actual hurricane-forming region shifts, being slightly farther north at the beginning of the hurricane season than it is toward the end. The hurricane season in the eastern and central Pacific Ocean, the typhoon season in the western Pacific Ocean, and the cyclone season in the Indian Ocean and near Australia are slightly different.

The structure of a hurricane

A tropical cyclone is defined as a cyclonically rotating atmospheric vortex that ranges in diameter from a few hundred miles up to one or two thousand miles. It is associated with a central core of low pressure and convective clouds that are organized into spiral bands, with a sustained convective cloud mass at or near the center. Unlike mid-latitude systems, hurricanes and other tropical cyclones are storms without associated fronts. But like other storms, they are characterized by a central core of low pressure and winds that blow cyclonically (counterclockwise in the Northern Hemisphere; clockwise in the Southern Hemisphere) around that core.

The lowest core pressures in a hurricane typically range between 920 and 980 mb. Hurricanes normally have an eye in the central region where warm humid air is sinking toward the Earth. The eye may be as much as 50 km (30 miles) in diameter; it develops as winds increase and spin around the central core of low pressure. The eye of a hurricane may be cloud free or have an overcast of clouds (known as the **central dense overcast**) that produces much less rain than the surrounding bands of clouds. Weather within the eye of a hurricane can be deceptively calm and pleasant, often causing people to mistakenly believe that the storm has passed. In many cases, people have left their shelter during eye passage thinking the storm was over, only to be greeted by the second half of the storm. A clearly defined eye can be seen in figure 12.1, a VIS image of Hurricane Hugo before it struck Charleston, South Carolina, in 1989. The eye is relatively cloud free and is surrounded by dense clouds.

Air flows into the base of a hurricane and spirals around the eye of the storm. Water vapor in the air condenses, and thick rain bands form. These bands are seen as cloud arms that spiral around the central portion of the storm. They consist of very thick clouds and thunderstorms that can be very violent and bring very heavy rainfall, high winds, and occasional tornadoes. Dense rain bands and lines of thunderstorms can be clearly seen in figure 12.1, especially in the southeastern and northern portion of Hurricane Hugo. Also associated with a hurricane is a **storm surge**, which is an increase in water level in coastal areas due to the combined effect of wind, waves, and reduced central pressure. Storm surge and its associated coastal flooding are the deadliest parts of a hurricane.

The wind speeds vary within a hurricane, reaching a maximum just outside the eye, in a feature called the **eye wall**. The eye wall contains the heaviest rains, as well. A typical hurricane may only be 500 km (300 miles) in diameter; however, the winds can approach 320 km/hr (200 mph), although they more typically reach only 160 km/hr (100 mph), which is strong enough to take roofs off buildings and cause considerable damage. If a hurricane were symmetrical and if it were not moving, the wind speeds would be fairly consistent on all sides of the eye. But because hurricanes move, the winds on the righthand side of the path of motion tend to be stronger. The winds on the lefthand side of the eye tends to be slower. This is because the wind speeds are a result of the combination of the speed of rotation and the speed at which the hurricane travels.

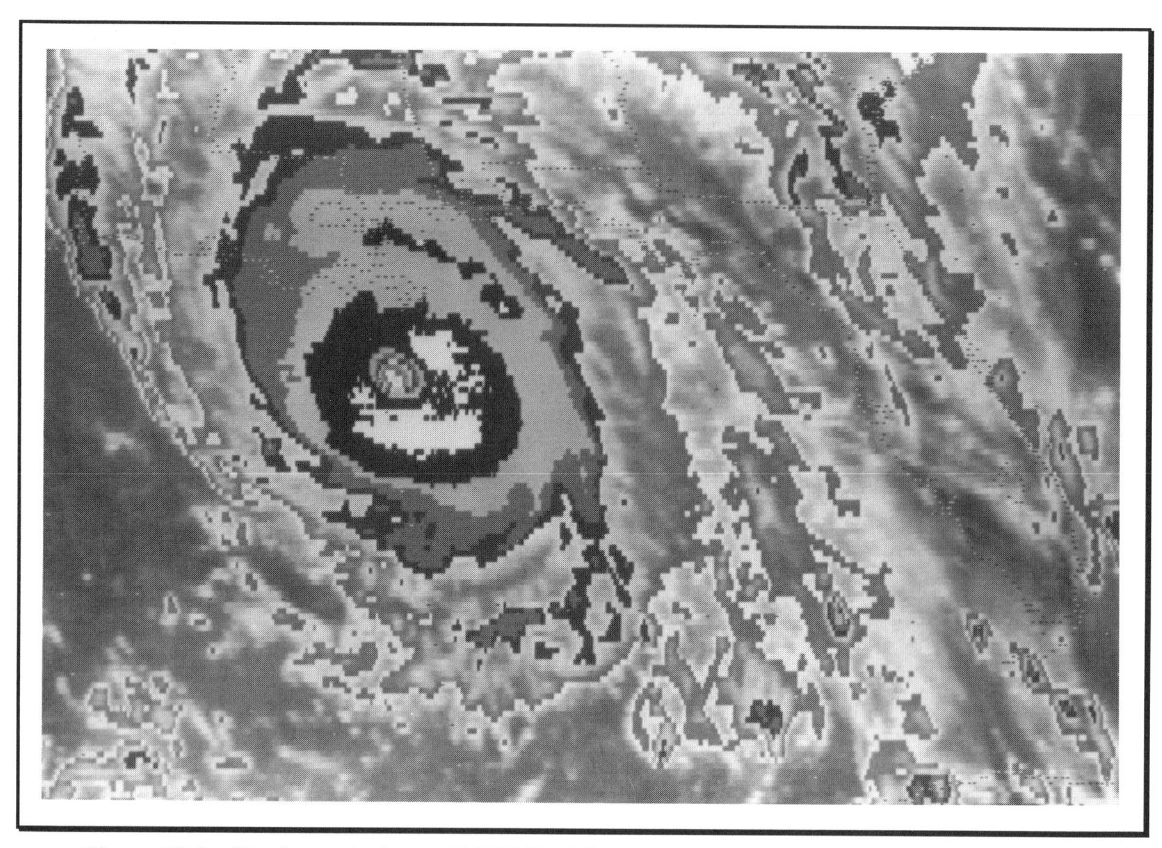

Figure 12.2. Hurricane Andrew (GOES IR, BD enhancement, 1831 UTC, August 25, 1992)

The temperature structure of a hurricane is such that the eye is warmer than the surrounding regions by several degrees. This difference in temperature is a very important characteristic in studying the intensity of a hurricane, since typically, the greater the temperature gradient, the more intense the storm. Meteorologists use the BD enhancement curve to observe the temperature gradient in the cloud tops surrounding the eye of the storm. Figure 12.2 is an enhanced image of Hurricane Andrew showing the temperature gradient within the storm. The coldest cloud top temperatures appear white in this image and are located on the southeastern portion of the storm just outside the eye wall. This would likely be the region where the storm is the strongest. At this point, Andrew exhibits a rather sharp temperature gradient between the eye and the surrounding regions, indicating that the storm is fairly strong. Other highlighted areas in the image help locate strong thunderstorms within the spiral rain bands.

The life cycle of a hurricane

Tropical cyclone development will only occur when very specific conditions exist. A hurricane originates as a **tropical disturbance** with relatively light winds, a weak area of low pressure, extensive cloudiness, and some precipitation. Many such disturbances exist at any given time in the tropics, but very few actually develop into hurricanes since the conditions required to do so are very specific (table 12.1). The hurricane's main source of energy is warm, humid air from over the ocean; therefore it requires warm ocean temperatures. The air over the ocean must also be very warm and moist. As the air rises through the storm, water vapor condenses into liquid water. Each droplet of water that condenses releases a certain amount of energy, known as latent heat, which fuels the hurricane. If a developing storm encounters colder ocean water or land, this energy source is lost and the storm will weaken. For a hurricane to form, the winds at all altitudes must be from the same direction. **Wind shear** refers to a condition in which wind direction and speed change throughout the lower 15 km (45,000 ft) of the atmosphere. When wind shear is present, the storm is frequently torn apart and prevented from developing into an organized system.

Conditions Leading to Tropical Cyclone Development
* Ocean surface temperatures of 27° C (80° F) or warmer.
* Very warm, humid air.
* Little **vertical wind shear** (wind direction and speed relatively constant throughout the lower 15 km of the atmosphere)

Table 12.1

Occasionally, when all the required conditions are present, a tropical disturbance evolves into a **tropical depression**, a closed low-pressure system. As the pressure drops, winds around the low-pressure core increase but remain less than 60 km/hr (39 mph). For a depression to reach the **tropical storm** stage, a distinct rotation must exist around the central area of low pressure and winds must reach speeds between 60 and 120 km/hr (39–74 mph). At this point, the tropical storm is given a name. To reach hurricane stage the storm must have a pronounced rotation around a central core of low pressure and wind speeds of at least 120 km/hr (74 mph). Once the storm is a hurricane, it can last for several days; however, as it ages and encounters land or cold ocean water, it loses its energy source and it begins to weaken. The storm may be downgraded to a tropical depression, and eventually it will die out, becoming a widespread area of heavy rains.

Observing hurricane development and estimating intensity

In satellite imagery, tropical cyclone development is analyzed by studying the cloud patterns and determining how they change with time. Repeated observations of a tropical cyclone provide information on the intensity and the rate of growth or decay of the storm. Figure 12.3 shows a modeled depiction of the developmental pattern of a tropical cyclone. This method of intensity analysis is based on the degree of spiraling in the cloud bands. The diagrams at the top of the graph illustrate the day-by-day changes in the shape of the cloud bands for a typical storm. The vertical axis of the graph is the **tropical number (T-number)** of the tropical cyclone. This number rates the intensity of the storm. Normally, a cyclone will exhibit a growth rate of 1 T-number per day. The straight line represents the intensity change and typical growth rate of a hurricane. The wavy line superimposed on the graph represents the degree of expected variability of intensity on a day-to-day basis.

This graph can be used as a conceptual model for estimating hurricane intensity and rate of development. When an observed cyclone shows the same daily increase in spiral banding as the diagram shows, the storm is developing at a typical rate. If the curved band spiral develops more quickly or more slowly, the rate of growth is considered to be fast or slow, respectively.

The associated central pressure and wind speeds are found below each T-number. The initial stage of tropical cyclone development is first recognized when curved cloud lines and

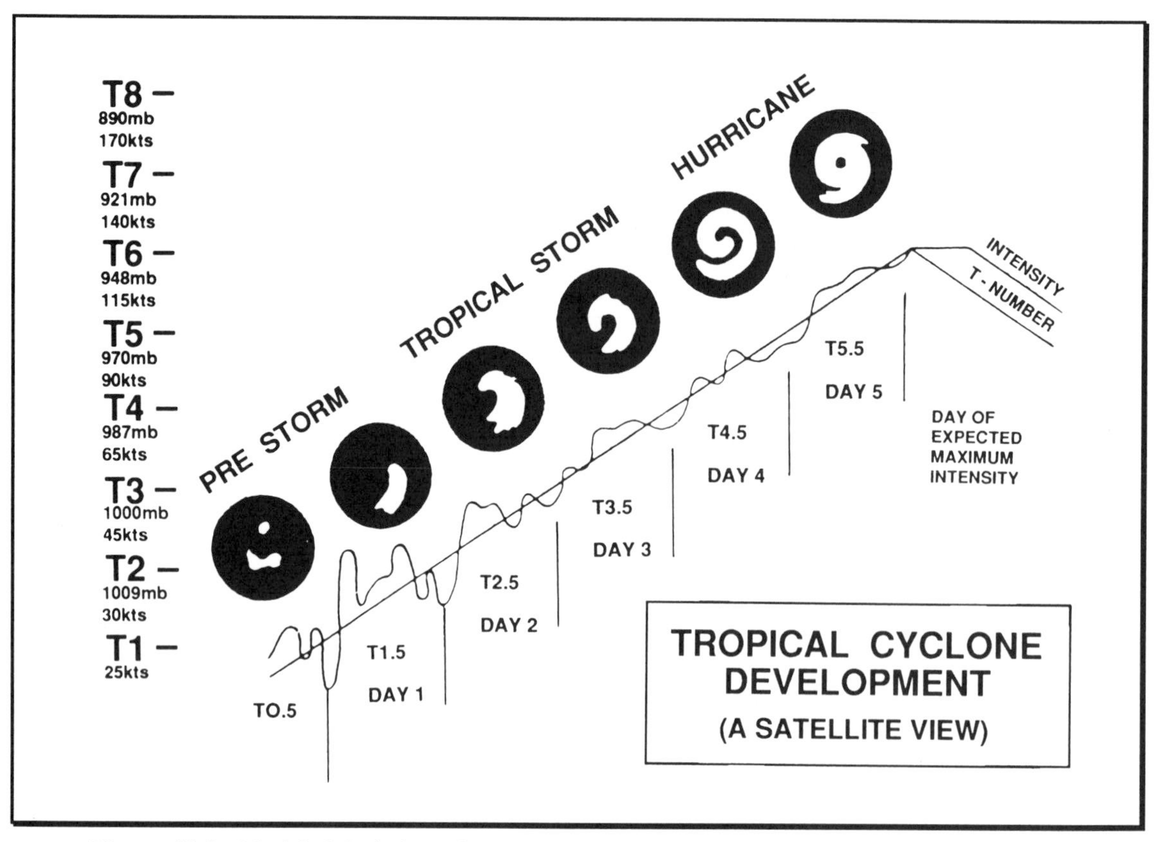

Figure 12.3. Modeled depiction of tropical cyclone development as observed from satellite

bands define a cloud system center near or within a deep cloud layer. This stage is referred to as the T1 stage. The T2 stage should appear about 24 hours later. When the curved band spirals halfway around the center of the disturbance, the weak tropical storm (T2.5) stage has been reached. The minimal hurricane stage (T4) is attained when the cloud band completely encircles the center. Once the eye is observed (T4.5), continued intensification is indicated by an increase in eye definition, increasing smoothness of the dense overcast, or embedding of the eye in the dense overcast.

As long as conditions remain favorable, a tropical cyclone should reach its maximum intensity four to six days after the T1 stage is seen. This time period varies according to the direction the hurricane is traveling. For storms that travel northward, a maximum intensity is expected in four days. A storm moving northwest is expected to reach a maximum intensity in five days, while a westward-moving storm is expected to reach a maximum in six days. The hurricane will reach its decaying stage when it moves out of the region where ocean temperatures are warm enough or when it moves across land.

Movement and tracking of hurricanes

Studying and predicting hurricane motion is extremely important since hurricanes are very destructive and, at times, deadly. Floods, high winds, ocean waves and swells, and tornadoes may all be part of the destructive forces associated with a hurricane. Studying the movement of a hurricane allows coastal areas and ships at sea to be warned in time to prepare for the storm. Accuracy is also very important. If storms are not accurately forecast, people may be caught unprepared. If repeated hurricane warnings are false, the warnings will not be taken seriously and people will begin to ignore them.

After its initial development, a hurricane typically shows its first movement to the west, guided by the trade winds. The combined effect of prevailing winds and the Coriolis effect will cause the storm to begin to take a more northwesterly direction. As it moves into and near the mid-latitudes, it is influenced more by the prevailing westerly winds and the jet stream. At this point the storm starts to curve to the north or the northeast. This curved path is typical of most hurricanes in the Northern Hemisphere, but many storms do not follow this classic path. Some move erratically, some loop, and some even move in unusual directions.

Using satellite imagery, tracking the position of a storm is fairly straightforward when an eye is present. When an eye is not discernible, it is much more difficult to locate the center of the circulation because it is probably covered up by high and middle-level clouds. Once the position of the storm is determined, it is recorded on a hurricane tracking chart

DATE	TIME	LAT	LONG	SPEED (Kt) & DIRECTION OF MOVEMENT	CENTRAL PRESSURE (mb)	MAX SUST. WINDS (Kt)	GUST (Kt)	COMMENTS
8/17/93	0900 Z	11.6 N	40.4 W	20 W	1007	30	35	Tropical depression
8/17/93	1500 Z	13.0 N	43.3 W	22 WNW	1005	35	45	Tropical Storm: named "Andrew"
8/17/93	2100 Z	13.4 N	45.3 W	20 WNW	1003	40	50	
8/18/93	0900 Z	14.3 N	48.9 W	18 WNW	1000	45	55	
8/18/93	1500 Z	15.3 N	51.2 W	20 WNW	1000	45	55	
8/18/93	2100 Z	15.9 N	52.8 W	17 WNW	1000	45	55	
8/19/93	0300 Z	16.6 N	54.4 W	16 WNW	1000	45	55	
8/19/93	0900 Z	17.6 N	56.3 W	18 WNW	1000	45	55	
8/19/93	2100 Z	19.2 N	59.5 W	18 WNW	1005	45	55	
8/20/93	0300 Z	20.1 N	59.8 W	14 NW	1009	40	50	
8/20/93	0900 Z	21.0 N	61.0 W	14 NW	1013	40	50	
8/20/93	1500 Z	21.3 N	61.3 W	9 NW	1015	40	50	
8/20/93	2100 Z	22.3 N	62.5 W	10 NW	1015	40	50	
8/21/93	0300 Z	23.7 N	63.0 W	12 NW	1013	45	55	
8/21/93	0900 Z	24.3 N	63.7 W	10 NW	1006	50	60	
8/21/93	1500 Z	24.7 N	64.6 W	9 WNW	1007	50	60	
8/21/93	2100 Z	25.2 N	65.4 W	8 WNW	1004	50	60	
8/22/93	0300 Z	25.6 N	66.4 W	9 WNW	1001	55	65	
8/22/93	0900 Z	25.8 N	67.5 W	9 WNW	994	65	75	Upgrade to Hurricane Andrew
8/22/93	1500 Z	25.9 N	69.0 W	12 W	974	80	95	Watch: NW Bahamas
8/22/93	2100 Z	25.9 N	70.4 W	13 W	974	85	105	Warn: NW Bahamas/ Watch: South Fla
8/23/93	0300 Z	25.6 N	71.9 W	12 W	959	95	115	
8/23/93	0900 Z	25.5 N	73.4 W	14 W	951	105	130	
8/23/93	1200 Z	25.5 N	74.1 W	14 W	951	105	130	Warn: South Fla/ Watch: W. coast Fla
8/23/93	1500 Z	25.4 N	75.0 W	14 W	930	115	140	Warn: West coast Florida
8/23/93	1800 Z	25.4 N	75.8 W	14 W	922	130	160	
8/23/93	2100 Z	25.4 N	76.5 W	14 W	923	130	160	
8/24/93	0300 Z	25.4 N	78.1 W	14 W	931	120	145	
8/24/93	0900 Z	25.4 N	80.3 W	16 W	932	120	145	
8/25/93	0300 Z	26.3 N	85.7 W	15 W	945	120	145	
8/25/93	0900 Z	26.8 N	87.4 W	15 WNW	949	120	145	Warning: Mississippi to Texas
8/25/93	1500 Z	27.5 N	89.2 W	15 WNW	944	120	145	
8/25/93	2100 Z	28.2 N	90.3 W	14 WNW	932	120	145	
8/26/93	0300 Z	29.0 N	91.0 W	11 NW	950	120	145	Warning: Mississippi to Louisiana
8/26/93	0900 Z	29.7 N	91.7 W	10 NW	955	100	120	
8/26/93	1500 Z	30.5 N	91.6 W	7 N	987	65	80	
8/26/93	1700 Z	30.7 N	91.6 W	7 N	995	65	80	Downgrade to Tropical Storm Andrew
8/26/93	2100 Z	31.2 N	91.5 W	7 N	992	45	55	Coastal warnings discontinued
8/27/93	0900 Z	32.2 N	90.1 W	8 NE	995	30	40	Downgrade to Tropical Depression
8/27/93	1500 Z	33.3 N	88.5 W	8 NE	998	30	40	Last advisory from Natl. Hurricane Cntr

Table 12.2. Storm track data collected during Hurricane Andrew (August 17–27, 1992)

and the date and time of the observation is noted. The direction and speed of the hurricane's motion can often be estimated and future positions fairly accurately predicted simply by extrapolating the previous path of travel. Table 12.2 is an example of information recorded while tracking a hurricane. This data was collected for Hurricane Andrew in August 1992. If the latitude and longitude of each observation are plotted on a map of the Atlantic Ocean and Gulf of Mexico, they will show Andrew's path.

Case study: Hurricane Andrew

The only major hurricane during 1992 was Andrew, a very strong hurricane that devastated south Florida as well as the gulf coast of Louisiana. The following sequence of images follows Andrew from the tropical storm stage through its final dissipation over the southeastern United States. Refer to table 12.2 for specific information about Andrew throughout its life cycle. Note that wind speeds are given first in knots (kt), since this is how they are issued by the National Weather Service.

Figure 12.4 is the first image in this sequence, taken at 1801 UTC on August 21, 1992. At this point, Andrew was a tropical storm (T3.5) that was heading west-northwest. The central pressure at this time was about 1007 mb and the maximum sustained winds were 50 kt (58 mph), with gusts up to 60 kt (69 mph). At this point in its development, the storm exhibits a pronounced rotation; however, the spiraling of the thunderstorm clouds is not completely closed, and the cloud bands are somewhat narrow.

Figure 12.4. Tropical Storm Andrew (GOES VIS, 1831 UTC, August 21, 1992)

Figure 12.5. Hurricane Andrew (GOES VIS, 1831 UTC, August 22, 1992)

Figure 12.5 shows Andrew 24 hours later. At this point Andrew had officially been a hurricane for 9 hours and was at the T4 stage of development. The spiral cloud pattern was a closed circulation, and an eye began to develop near the center of the storm. The central pressure was approximately 974 mb; maximum sustained winds were at 85 kt (98 mph), with gusts up to 105 kt (121 mph). The northern Bahama Islands, immediately to the west of the storm in this image, were under a hurricane warning, and southern Florida was put under a hurricane watch.

Over the next several hours, Andrew continued to strengthen at the expected rate of approximately 1 T-number per day. Figure 12.6 shows Andrew as it approaches the Bahama Islands at 1401 UTC on August 23. The spiraling of the clouds had increased, and at this point the eye was very well defined, indicating a T5.5 stage of development. The central pressure was 930 mb; winds were at 115 kt (132 mph) with gusts up to 140 kt (161 mph). The storm was still moving in a westerly direction, and southern Florida was put under a hurricane warning at about this time.

Before Andrew struck the east coast of Florida it reached its lowest central pressure of 922 mb (T6.5) on August 23 at 1800 UTC. Andrew caused widespread destruction and loss of life in southern Florida. As it passed over the land, it weakened considerably. By 2001 UTC on August 24, it had passed over southern Florida and was moving in a west-northwesterly direction again (figure 12.7). At this time, its central pressure was approximately 940 mb. Back over the warm water of the Gulf of Mexico, Andrew began to reintensify. Wind speeds at this time were 120 kt (138 mph), with gusts up to 145 kt (167 mph), and the Gulf Coast from Mississippi to Texas was under a hurricane warning.

Figure 12.6. Hurricane Andrew (GOES VIS, 1401 UTC, August 23, 1992)

Figure 12.7. Hurricane Andrew (GOES VIS, 2001 UTC, August 24, 1992)

Figure 12.8. Hurricane Andrew (GOES VIS, 1831 UTC, August 25, 1992)

Figure 12.9. Tropical Storm Andrew (GOES VIS, 1801 UTC, August 26, 1992)

Andrew continued to strengthen again over the warm waters of the Gulf of Mexico, and at 1831 UTC on August 25 (figure 12.8) the central pressure was approximately 940 mb (T5.5). At this point, the eye was once again very well developed and the degree of spiraling was high. Andrew was now traveling in a northwesterly direction with winds at 120 kt (138 mph) and gusts up to 145 kt (167 mph). The Mississippi coast and the Louisiana coast were under a hurricane warning, and the hurricane warning for the Texas coast was reduced to a hurricane watch.

After landfall along the Louisiana coast, Andrew began to weaken without warm ocean waters to supply energy to the storm (figure 12.9). By 1801 UTC on August 26, the central pressure of the storm had risen to 995 mb, its winds had fallen to 45 kt (52 mph), and Andrew was downgraded to tropical storm status. Now heading due north, Andrew continued to weaken and become less organized. It was further downgraded to tropical depression status at 0900 UTC on August 27.

By 1801 UTC on August 27 (figure 12.10), Andrew had become a widespread area of cloudiness and rainfall that was incorporated into a frontal system from the west. At this point, the central pressure had risen to approximately 1000 mb, and wind speeds were less than 30 kt (35 mph). Now heading in a northeasterly direction, Andrew traveled up the east coast and ended as a disorganized and rainy cloud system.

Figure 12.10. Tropical Depression Andrew (GOES VIS, 1801 UTC, August 27, 1992)

CHAPTER 13

Monitoring Air Quality

Introduction to air quality monitoring

Another important application of satellite imagery is in the monitoring of air quality around the globe. Using satellite data, it is often possible to locate source points for large amounts of pollution or smoke entering the atmosphere. The data collected by satellites also offer graphic display of the aerial extent of pollutants and can be used to track and predict their movement throughout the atmosphere. The role of pollution in regional cloud and weather modification can be assessed from satellites, as well.

Smog and haze, smoke, blowing sand and dust, and airborne volcanic material are the pollutants most frequently observed in satellite imagery. Collectively, these pollutants are often referred to as **aerosols**. Aerosols are solid particles that remain suspended in the atmosphere. When aerosols are introduced into the atmosphere, the process is referred to as **atmospheric loading**. Aerosols play an important role in the atmosphere. They block and scatter incoming solar radiation, thus affecting the climate of the Earth and the colors we observe in the sky. They interact with water droplets in the atmosphere and influence precipitation. They also pose a variety of health problems in heavily populated areas, especially for elderly people or individuals with respiratory problems.

Haze and pollution

Automobile exhaust and industrial activity add aerosols to the atmosphere; sometimes these aerosols become trapped and increase in concentration. Stable meteorological conditions can contribute to the trapping of particulates in the lower atmosphere. Water droplets can form around these suspended solid particles and grow, giving the sky a milky white appearance. This condition, known as **haze**, usually occurs when a high-pressure system with light winds stagnates over a portion of the eastern United States. Eventually, hazardous air quality can require the issuance of air pollution advisories and alerts.

Satellite observations provide information about the extent and the movement of haze. Additionally, approaching weather systems can be monitored in order to forecast whether air quality conditions will improve. The suspended particulate matter appears as a lighter gray area, especially in VIS satellite imagery. As the air mass over the affected area becomes more moist, the hazy areas appear increasingly light due to the growth of the particulates in the presence of water droplets. If the haze thickens, land details such as rivers and lakes may become obscured in the satellite data. In figure 13.1, stable atmospheric conditions have led to the formation of haze over the east coast of the United States (the haze is marked with "∞"). Fog (marked with "=") is also present in West Virginia, indicating light, low-level winds that permit the formation of thick haze. An upper-level low can be seen centered over the lower Mississippi River Valley. The cyclonic winds flowing around this system have spread the haze westward, through northern Kentucky and across much of the upper Mississippi River Valley.

Smoke

One of the most useful applications of satellite technology is in the monitoring of forest and brush fires. From the vantage point of space, fires in remote areas are easily spotted, and the proper fire-fighting personnel can then be notified to take action. Illegal slash burning in the Pacific Northwest, as well as in tropical and subtropical slash areas in Central America and South America, can be monitored using satellite data. These data also help in tracking the smoke and particulates released into the atmosphere, which can reduce visibility and degrade air quality.

Figure 13.1. Haze and pollution detection in satellite imagery (GOES VIS, August 15, 1984)

In addition to reducing air quality, smoke has a strong effect on local weather patterns. Regions on the Earth that are covered with smoke are not heated as rapidly as smoke-free regions; therefore, cumulus cloud formation can be limited. Adjacent smoke-free areas may heat up faster, resulting in a temperature gradient along the boundary between the smoke-covered and smoke-free areas. This can lead to thunderstorm development along the temperature gradient.

The smoke from forest and brush fires is often visible in both VIS and IR satellite imagery as a light gray plume. At the actual point where the fire is located, the smoke is seen as a dense, narrow plume that spreads out downwind. The smoke plume often has a wedge shape that points upwind, caused by the spreading of the smoke. Once it is under the influence of the prevailing winds, smoke thins and spreads out considerably. Smoke can often be seen hundreds of kilometers downwind of the original source point as a milky gray area in VIS imagery.

In the summer of 1988, extensive forest fires in and around Yellowstone National Park destroyed over 2 million acres of vegetation. In figure 13.2, taken on September 8, 1988, smoke plumes can be seen extending toward the east in northern Wyoming and Idaho. The smoke plumes are detectable in IR imagery since they are cold, but they are not as cold as cloud tops in Kansas and Nebraska. Thus, the plumes are primarily low level. This indicates that low-level winds are westerly in this area. Note that at the origin of the smoke cloud the plume is thicker, and therefore it appears brighter. As the smoke cloud is carried downwind, it thins and spreads out, and its brightness decreases.

Figure 13.2. Yellowstone fires (GOES IR, MB enhancement, September 8, 1988)

Blowing sand and dust

Dust storms frequently occur in areas where exposed soils dry out and are subsequently lifted into the atmosphere by strong local winds. This phenomenon is most often observed in desertlike regions where vegetation is sparse and soils remain dry throughout the year. Dust storms are commonly observed in the Sahara desert region and in the southwestern portion of the United States, where these conditions are often present. Blowing dust will generally appear as a milky gray area or band in VIS imagery. It is most apparent in the early morning and the late afternoon, when the sun is low on the horizon and scattering of light is at its maximum.

In the United States, strong low-level winds associated with late winter and early spring storm systems frequently cause large-scale dust storms in the Great Plains region. During this time of year, much of the land is barren of crops, as farming activity is just beginning. As solar radiation heats the lower atmosphere and the underlying ground, warmed air begins to rise and cooler air sinks. This local turbulence, along with the strong storm system winds, causes the soil particles to be lifted skyward. Once in the air, these soil particles can remain in suspension and be observed as they move hundreds of kilometers across the country. The blowing dust results in poor visibility (both at the surface and aloft), major soil erosion, soil drifting, and plant damage. People with respiratory problems are particularly vulnerable to such a degradation of air quality.

Figure 13.3 is a VIS image of Texas and the central United States. A mid-latitude storm system is located in the central portion of the country. Soil and dust particles have been lifted into the atmosphere at points *A* and *B* and carried by the winds circulating around the low-pressure system. In this figure, the dust can be seen as a light gray area over Texas and the central plains. The cyclonic circulation around the storm center is evident in the orientation of the two separate dust storms.

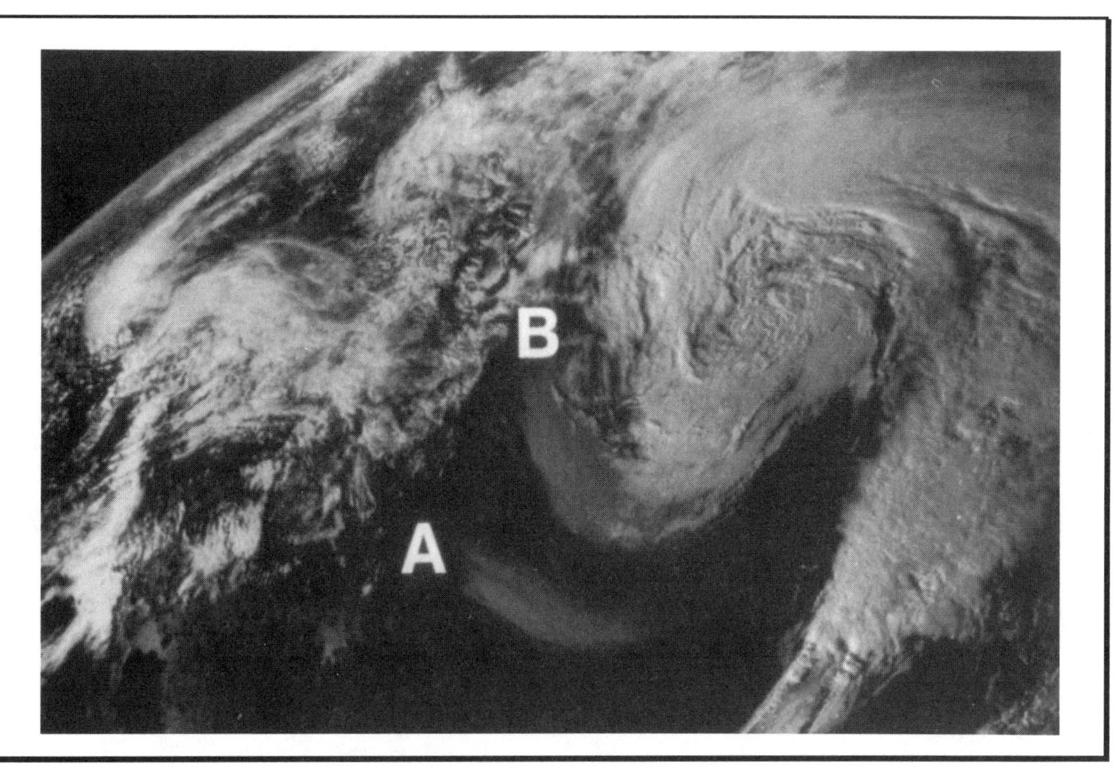

Figure 13.3. Dust storm over Texas and the central plains region of the United States (GOES VIS, April 2, 1982)

Volcanic activity

Volcanic eruptions add extremely large amounts of aerosols to the atmosphere in a very short period of time. The dense clouds of airborne volcanic particulate material are known as **ash clouds**. Ash clouds associated with volcanic eruptions can cause a variety of air quality problems. Areas close to a volcanic eruption can be covered by several inches or even several feet of volcanic ash. Further from the volcano, ash clouds can cause air quality to decline so severely that people with respiratory problems and other health conditions are in danger. Additionally, ash clouds are a hazard to aviation. On a global scale, aerosols from volcanoes that persist in the upper atmosphere are thought to spread around the entire planet and partially block solar radiation from reaching the lower atmosphere. The result is a short-term cooling of the Earth. Thus, not only do volcanoes affect the areas immediately surrounding them but they can also play a role in regulating the global climate.

Ash ejected from volcanoes can be seen under clear sky conditions in satellite imagery. Ash clouds generally have a high albedo and can be detected in VIS satellite imagery, especially where the plume is very thick. The debris can form a large, light-colored cloud. As the ash plume spreads and thins, it develops a hazy appearance. Material from the ash cloud can often be tracked for hundreds of kilometers as it moves under the influence of the prevailing winds.

IR imagery can also be used to locate and monitor volcanic activity. Since volcanic ash clouds often extend very high in the atmosphere, they tend to be cold, and they appear bright in IR satellite imagery. Figure 13.4 is an IR image of Alaska as Mount Redoubt erupted on March 23, 1990. The brilliantly bright white spot in the image is the ash plume

Figure 13.4. Eruption of Mount Redoubt, Alaska (HRPT IR, March 23, 1990)

Figure 13.5. Mount St. Helen's eruption (GOES VIS, 1615 UTC, May 18, 1980)

Figure 13.6. Mount St. Helen's ash cloud (GOES VIS, 2345 UTC, May 18, 1980)

Figure 13.7. Mount St. Helen's ash cloud, spreading (GOES VIS, 2331 UTC, May 19, 1980)

from the volcano. The water feature to the right of Mount Redoubt is Cook Inlet. Anchorage is located at the head of this inlet.

Figures 13.5–7 illustrate how VIS imagery can be used to monitor the spread of volcanic material in the atmosphere. Figure 13.5 shows the ash cloud associated with the eruption of Mount St. Helen's in Washington on May 18, 1980. The ring-shaped ash cloud can be seen as it spreads away from the volcano. Several hours later, westerly winds carried the ash from the eruption several hundred kilometers. In figure 13.6, material from the eruption can be seen spreading as far away as southwest Montana. This cloud buried some towns with several inches of ash. By the next day, the winds spread the ash cloud as far as Oklahoma and the northeast Texas panhandle. This can be seen in figure 13.7 as a wide area with a hazy appearance over the central plains states. Note that the plume extends to the northwest across Wyoming, Idaho, and northwest Oregon.

CHAPTER 14

Oceanography

Oceanography from space

Nearly 75% of our planet is covered by water. The oceans influence almost every aspect of life on Earth, making it vital to understand their role in the global environment. Oceanographic studies today focus on the study of the oceans as an integral component of a global system governed by the interactions of air, water, and land. These studies lead to improved understanding about the role of the oceans in governing long-term climatic change, short-term weather patterns, and distribution of living resources in the ocean.

Until the advent of satellite technology, global-scale oceanographic studies were not possible. The oceans are simply too large for human researchers to use traditional sampling techniques. Historically, information on ocean surface features was recorded on ocean-going vessels and was centered among the most heavily traveled shipping routes. Most notably, Ben Franklin mapped the **Gulf Stream** in 1777 using shipboard observations. This method of studying the oceans is not very practical for today's purposes since it requires long, costly missions and does not include measurements in remote areas. Furthermore, the spatial and temporal resolution of data collected during these missions is usually inadequate to resolve many ocean features.

The use of environmental satellites has greatly improved our ability to study the characteristics of the oceans. From a satellite, large areas of the oceans, including remote regions, can be observed at the same time. This allows entire oceans to be monitored and studied on a useful time scale and at a lower cost. This chapter will discuss several applications of weather satellite imagery to oceanographic studies.

Sea surface temperature

One of the most useful data sets offered by remote sensing of oceans is **sea surface temperature (SST)**. IR sensors on environmental satellites can be used to measure the temperature across large expanses of the ocean surface. This data has many important applications. Figure 14.1 is an example of an SST map produced daily by oceanographers for NOAA. These maps permit scientists to use ocean temperatures to observe ocean circulation and locate major ocean currents. Ocean current analysis can facilitate ocean transportation, much as jet stream analysis is used for routing aircraft. Additionally, by using SST, scientists can monitor changes in ocean temperatures and relate these to weather and climate changes. SST can also aid in the detection of newly formed sea ice, which might otherwise go unnoticed. The same data can also be used to monitor the amount of ice on inland bodies of water such as the Great Lakes and Hudson Bay. Finally, SST can be used to help locate living resources that are associated with specific thermal features in the oceans, such as fish that prefer a specific temperature range.

Oceanic circulation

The oceans are an important part of the Earth's heat exchange system. Ocean water at the equatorial regions of the Earth absorbs heat from the sun. These warm ocean currents then flow toward the poles, carrying heat away from the equator and distributing it to higher latitudes. Cold-water currents travel from the polar regions toward the equator, where they become heated again. This circulation is mainly wind driven and often matches the wind patterns across the globe. In the Atlantic and Pacific Basins in the Northern Hemisphere, currents generally flow clockwise (anticyclonically) as persistent high pressure near 30° latitude forces ocean circulation. In the Southern Hemisphere the circulation is mainly counterclockwise. These circulation patterns are known as **gyres**.

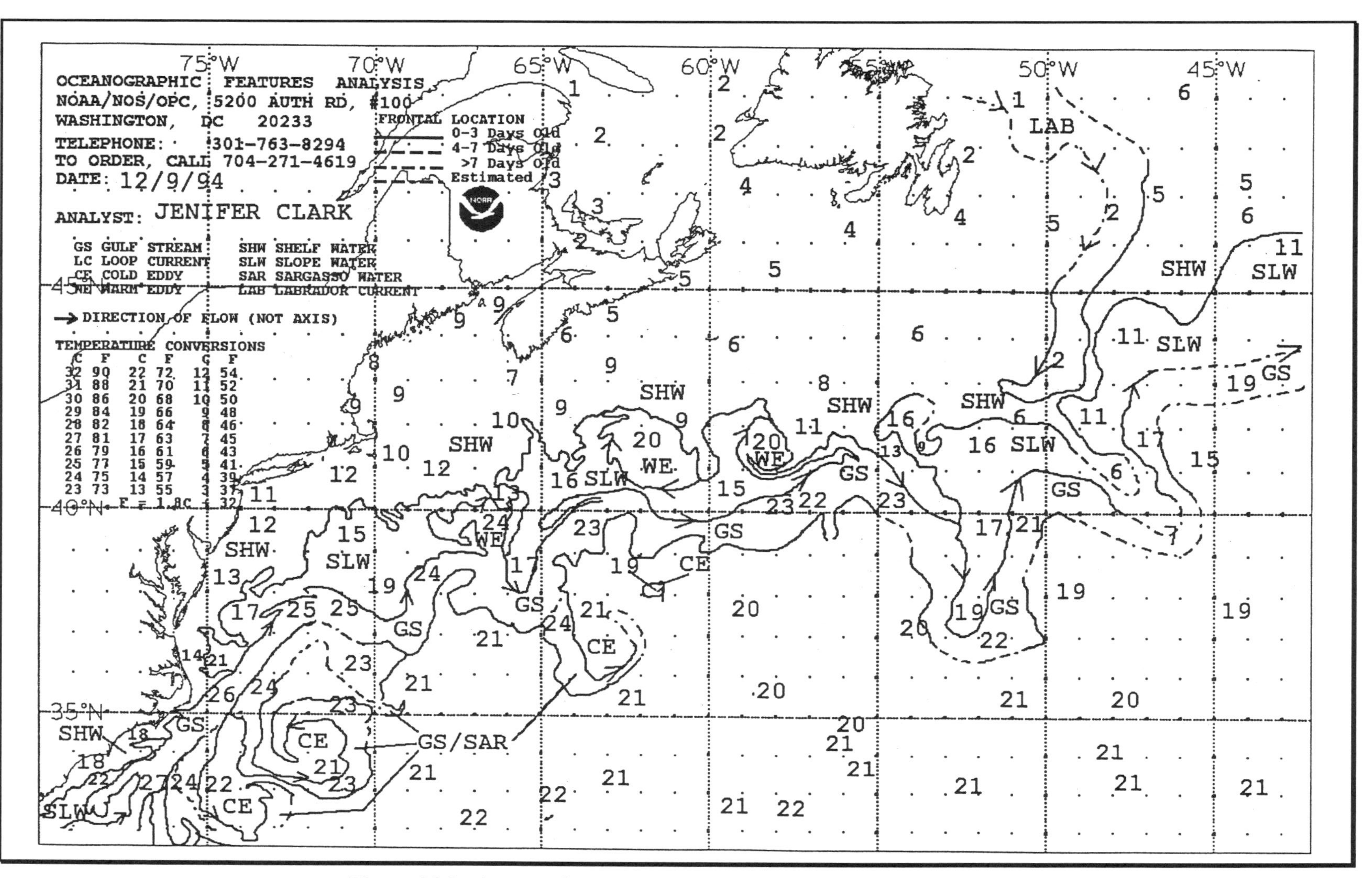

Figure 14.1. A sea surface temperature (SST) map produced by NOAA

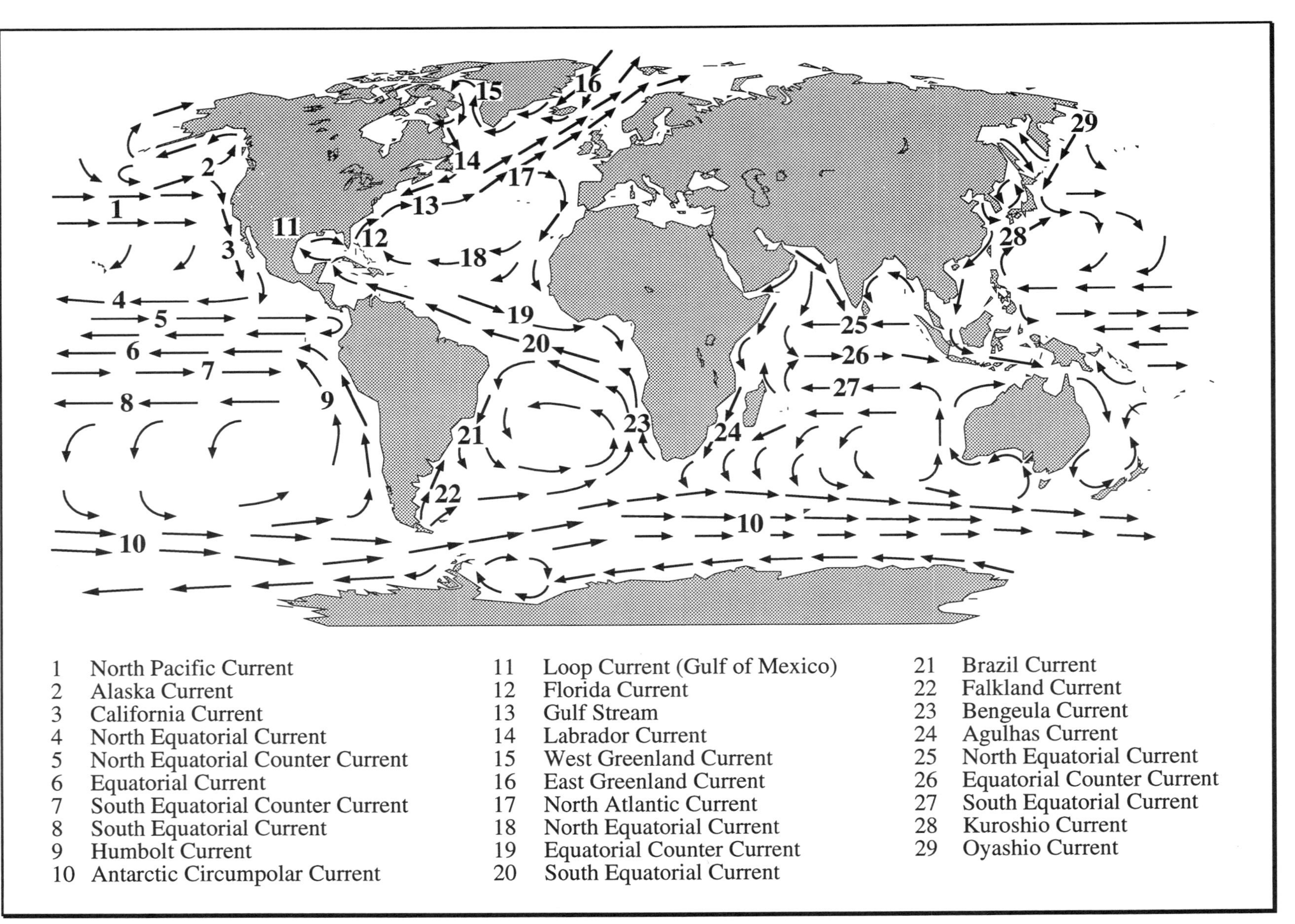

Figure 14.2. Major circulation patterns in the world's oceans

The map in figure 14.2 illustrates a generalized circulation pattern for the world. Typically, the currents that flow poleward are warm-water currents and the currents flowing toward the equator are cold-water currents. It is important to understand that ocean circulation is very dynamic, and this map only shows the average position of these currents. Additionally, this map does not show subsurface and deep ocean circulation, both of which are very important in ocean dynamics.

Oceanic fronts

Oceanic fronts are boundaries between water masses of different density. Density is a function of temperature and **salinity** (the amount of dissolved salts in water); therefore, both **thermal (temperature) fronts** and **haline (salinity) fronts** exist in the ocean. A thermal front is a zone with a pronounced horizontal temperature gradient, while a haline front exhibits a horizontal salinity gradient. Ocean fronts can extend from the surface to the very deep layers of the ocean, often separating very large volumes of ocean water.

Using IR satellite imagery on a relatively cloud-free day, it is possible to detect ocean thermal fronts in the surface layers of the ocean. This is accomplished by locating in an image a distinct gray shade difference that results from the horizontal temperature gradient across a thermal ocean front. Since these temperature differences are often on a magnitude of 2–5° C, image enhancement techniques are often used to highlight the temperature range of the ocean surface and improve the contrast between small differences in sea surface temperature.

Oceanic fronts can be permanent or transient features. Permanent oceanic fronts include the **Gulf Stream front** (figure 14.3), located off the east coast of North America, and the **Kuroshio Current front**, located off the east coast of Asia. These frontal boundaries always exhibit a pronounced horizontal temperature gradient and can be up to 1000 meters deep. Transient oceanic fronts usually occur seasonally and are generally weaker, with more diffuse boundaries. Transient fronts may only appear in the ocean for a few weeks during the year; however, they are important components of the ocean system.

One limitation in studying ocean thermal fronts using satellite imagery is the fact that satellites can only detect the temperature characteristics in the surface layer of the ocean, leaving the thermal characteristics of deeper waters undetected. This problem is especially pronounced in summer months, when intense solar heating tends to make the surface layer **isothermal** (characterized by equal temperatures throughout). This can obscure the surface expression of permanent ocean features from the view of a satellite. For example, the Kuroshio Current can lose its thermal surface expression during the summer months, even though the front is still well defined at deeper depths.

The Gulf Stream

The Gulf Stream is a strong warm-water current that generally flows northward, nearly parallel with the Atlantic coast of the United States. The **Florida Current** is the southern portion of the Gulf Stream that extends from the southeastern tip of Florida to Cape Hatteras, North Carolina. At Cape Hatteras, the Gulf Stream turns eastward and flows into the northern Atlantic Ocean, where it slowly cools. This current is studied on a regular basis, and its position is charted regularly to aid shipping in the oceans around North America. The Gulf Stream is also an important component of weather over the oceans. As air comes into contact with this warm-water current, it is heated. Thunderstorms are often seen forming over the Gulf Stream, making it especially important to transatlantic shipping

Figure 14.3. The Gulf Stream and Florida Current along the east coast of North America (HRPT, enhanced IR, April 24, 1990)

and air travel (see chapter 11). This current is also responsible for bringing warmer air temperatures to the United Kingdom, moderating the climate in several cities, including London. In figure 14.3 the Florida Current and the Gulf Stream are visible off the east coast of the United States. Since it is warmer than the surrounding waters, the frontal system appears as a region in the water that is a darker shade of gray.

The Gulf Stream can often be observed as it interacts with the various masses of ocean water that surround it. The ocean floor slopes very gently for the first few hundred kilometers off the eastern coast of North America, forming a feature known as the **continental shelf**. This water is relatively shallow and therefore experiences a wide temperature range throughout the year. Beyond the continental shelf, the ocean floor drops very sharply in a region called the **continental slope**. This slope leads to the deep ocean floor. The water over each of these regions can often be discerned in satellite imagery, since the temperature and/or salinity characteristics differ in each region. To the west of the Gulf Stream lie the colder **shelf water** from the continental shelf and the slightly warmer **slope water** from the continental slope. These waters are lower in salinity and temperature than the Gulf Stream. They can be seen to the west of the Gulf Stream in figure 14.3 as the waters that appear as a lighter gray shade. The boundary between these cooler waters and the warm Gulf Stream is the **Gulf Stream front**. In satellite imagery this front is often very sharp, since the temperature gradient can be very steep across this boundary. At times, cooler shelf or slope waters will flow into the Gulf Stream and interact with it. This is known as a **cold-water intrusion.**

To the east of the Gulf Stream in figure 14.3 is the **Sargasso Sea**. The Sargasso Sea is a pool of warm water, several hundred feet deep, that forms the upper layer of the central North Atlantic Ocean. It is slightly cooler than the Gulf Stream water, and the salinity of its water is higher. The boundary between these two bodies of water is often less sharp than the boundary between the Gulf Stream and the slope and shelf waters; however, it can often be detected in satellite imagery.

Gulf Stream flow is characterized by large-amplitude meanders that often break off and form eddies. These patterns are similar to those observed in the atmosphere. **Cyclonic eddies** are formed by southward meanders of the Gulf Stream that become very large, break away from the Gulf Stream, and develop separate, closed circulations (counterclockwise in the Northern Hemisphere; clockwise in the Southern Hemisphere). Cyclonic eddies are often termed **cold-core eddies**, since the central area of the eddy contains cooler slope water surrounded by a ring of warmer Gulf Stream water. A portion of a cold-core eddy can be seen at the extreme eastern edge of figure 14.3. It can be identified by a ring of warmer water surrounding a pool of cooler water.

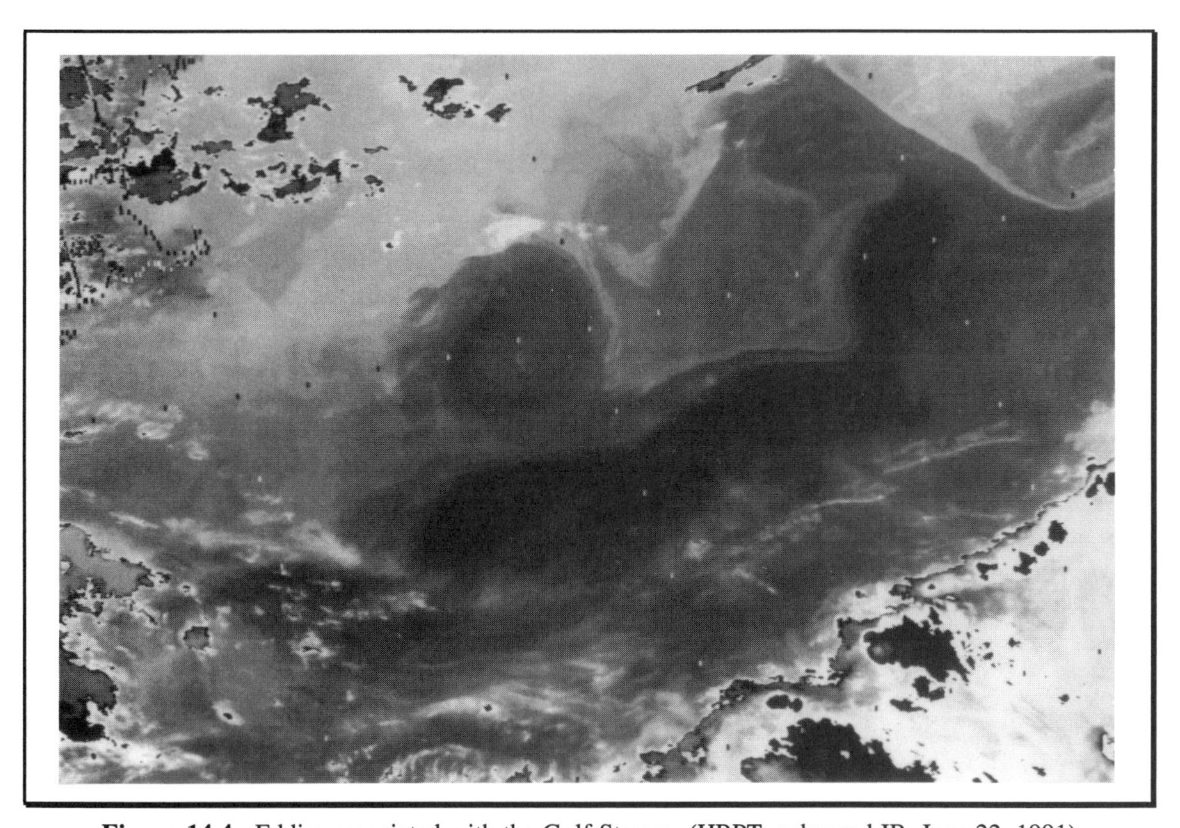

Figure 14.4. Eddies associated with the Gulf Stream (HRPT, enhanced IR, June 22, 1991)

Anticyclonic eddies are formed by northward meanders of the Gulf Stream that break away from the Gulf Stream and develop separate, closed circulations. Anticyclonic eddies are often called **warm-core eddies** since the central area of the eddy contains warmer water from the Gulf Stream surrounded by a ring of cooler shelf water. These eddies often form in conjunction with the bottom topography of the ocean, especially along the edge of the continental shelf boundary. Once they form, they spin off by themselves and decay slowly. In figure 14.4, several warm-core eddies can be seen along the northern edge of the Gulf Stream. The most distinct eddy is located east of Cape Cod. It is easily identified by the ring of warm water surrounded by a ring of cooler water. To the east of this eddy is a large-amplitude meander that may become cut off as a warm-core eddy.

The Gulf of Mexico Loop Current

The **Gulf of Mexico Loop Current** is a warm-water current that flows through the Gulf of Mexico. The Loop Current enters the Gulf of Mexico between Cuba and the Yucatan Peninsula, circles through the gulf, and exits through the Florida Strait, where it joins the Florida Current. Once in the Gulf of Mexico, the Loop Current forms an anticyclonic (clockwise) loop which can extend in a northward direction until it breaks off as a large eddy. This eddy will then slowly drift off toward the western portions of the gulf. Figure 14.5 shows the Loop Current as it extends northward into the Gulf of Mexico. Colder water can be seen as it interacts with the warmer water of the Loop Current. Often, water flowing from the Mississippi River delta can be seen interacting with the warmer water of the Loop Current. Although the Loop Current is a permanent feature in the gulf, it is usually only detectable by satellite in the colder months. In the warmer months, the surface layer in the Gulf of Mexico becomes isothermal due to solar heating, obscuring the current's surface expression from the satellite.

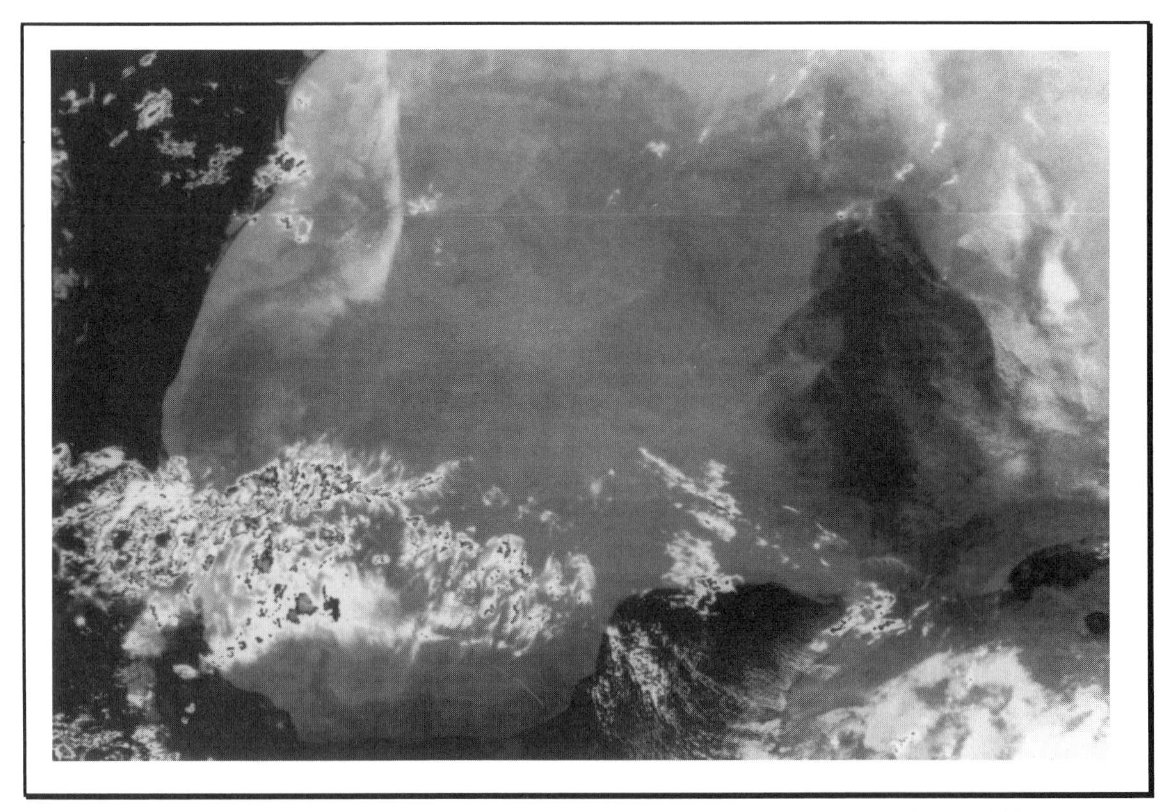

Figure 14.5. Gulf of Mexico Loop Current (HRPT, enhanced IR, March 11, 1988)

The California Current

In the higher latitudes of the North Pacific Ocean, a cool current called the **North Pacific Current** flows eastward and approaches the west coast of North America. **The North Pacific Current front**, which separates colder Arctic water from warmer water from the south, is a permanent ocean feature that extends several hundred meters into the ocean. Although it can be difficult to detect in warmer months, it remains fairly well defined below 100 meters throughout the year.

As the North Pacific Current reaches North America, it splits into the northward-flowing **Alaska Current** and the southward-flowing **California Current**. The California Current flows parallel to the North American coast from British Columbia to California. It eventually feeds into the **North Equatorial Current**, a warmer-water current that flows from east to west near the equator. The California Current is a weak and cold current that is spread out over a much broader area than the Gulf Stream. The surface expression of the California Current can also be difficult to detect owing to seasonal fluctuations in solar heating; therefore its appearance will vary considerably in space, time, and intensity.

Upwelling

Upwelling occurs in a body of water when subsurface water rises toward the surface. Since the temperature of water generally decreases with depth, upwelled water is colder than the surface water it replaces. Thus, in an upwelling area, surface temperatures are usually colder than surrounding temperatures. IR satellite data can provide information on the position and strength of the surface temperature gradient associated with upwelling. Areas of upwelling usually appear as bands of lighter gray shades (colder temperatures) along coastlines. Upwelled water is often seen interacting with near-shore currents, and

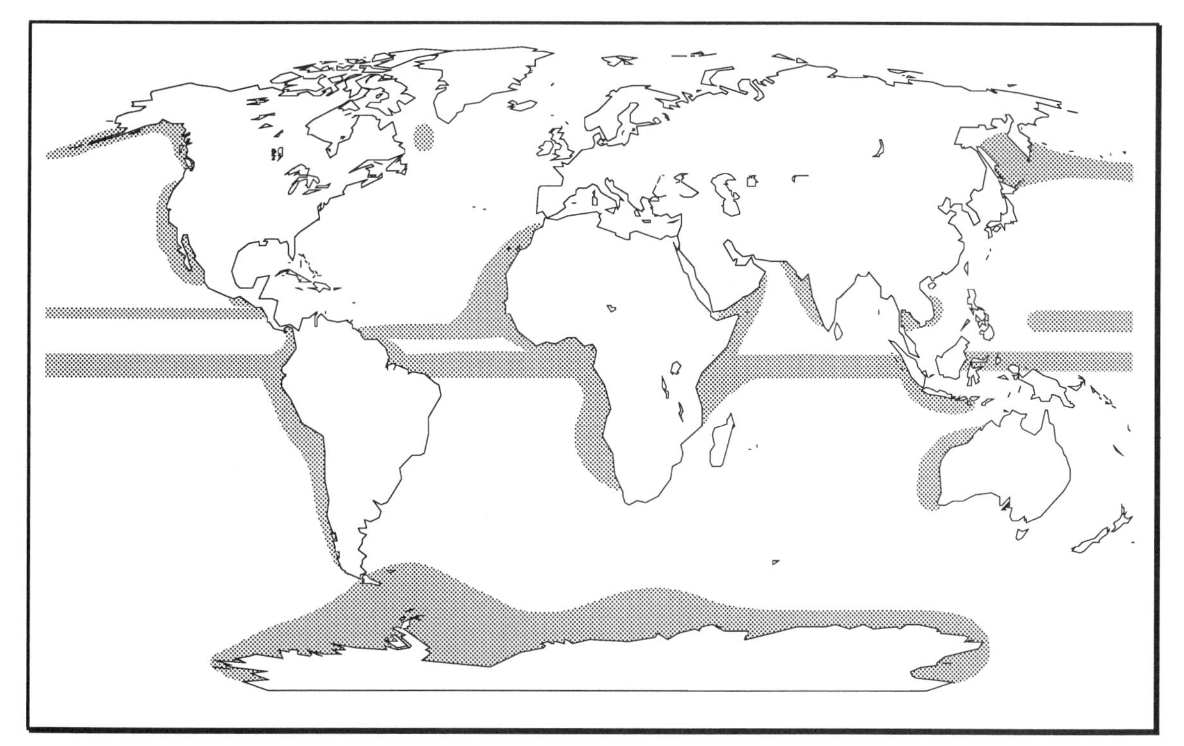

Figure 14.6. General areas of upwelling in the world

oceanic thermal fronts are often observed between upwelled and non-upwelled water. Figure 14.6 is a map showing the general areas of upwelling around the world.

Upwelling can significantly influence the meteorological conditions in the region surrounding the upwelling area. Cold waters associated with upwelling can significantly inhibit thunderstorm development. If upwelled water is cooler than the air, it can cool the air sufficiently for fog and stratus clouds to develop. This can create areas of unequal surface heating which can lead to the development of surface fronts in the atmosphere along the boundary between upwelled and non-upwelled water, or between upwelled water and adjacent land. Cold, upwelled waters off the western coast of Africa generally prevent tropical disturbances from forming into stronger tropical storms and hurricanes that could eventually affect North America.

Upwelling is also capable of influencing biological productivity in the oceans and along the coastline. Upwelled water is rich in nutrients; it thus helps **phytoplankton** (marine organisms that use light and nutrients from the water) to grow, providing a food source for various marine organisms. The areas in figure 14.6 where upwelling commonly occurs are some of the most productive regions in the world. Some of the world's largest fisheries are dependent on seasonal upwelling and the tremendous amount of production in these regions.

Causes of upwelling

Coastal upwelling is the most commonly observed type of upwelling. It is caused by surface **wind stress** (the force of winds pushing on the water surface) in combination with the effect of the Earth's rotation (the Coriolis effect). These two forces produce a net transport of surface water in an offshore direction (figure 14.7). The divergence of surface water away from the coast causes an upwelling of cooler, subsurface water.

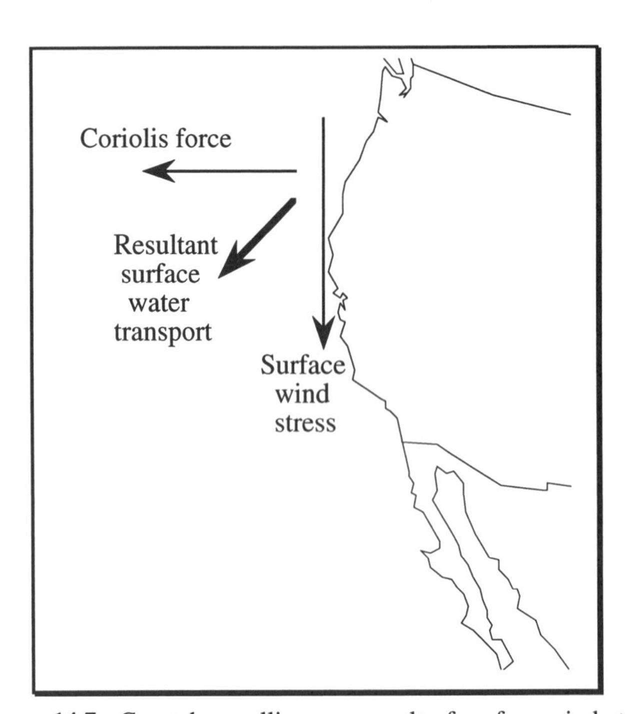

Figure 14.7. Coastal upwelling as a result of surface wind stress

Wind direction is essential in determining whether coastal upwelling will occur. Coastal upwelling will occur in the Northern Hemisphere when the open ocean is to the right of an observer who is looking in the direction toward which the wind is blowing. In the Southern Hemisphere, the ocean has to be to the observer's left for upwelling to occur. For example, along the west coast of North America, a northerly wind blowing parallel to the coastline can result in coastal upwelling (figure 14.7), while along the west coast of South America, a southerly (south to north) wind is required for upwelling. The migration and strength of global wind and pressure patterns such as the subtropical high are important in causing coastal upwelling. Upwelling is especially common along the west coast of continents where persistent high-pressure systems are located, resulting in winds that flow toward the equator parallel to the coastline.

The strength of upwelling depends on wind characteristics such as speed, duration, fetch, and direction. Since all of these fluctuate throughout the year, upwelling varies with the seasons. Coastal upwelling is most pronounced off the western U.S. coast and the northern African coast from April through August and off the Peruvian and Chilean coasts from March to May. The upwelling season lasts for months; however, upwelling does not always occur during the entire season. Short-term increases in winds produce upwelling events occurring on a time scale of days to weeks, while short-term changes in wind direction may prevent upwelling from occurring.

During an upwelling event, cold, upwelled water is revealed by gray shade patterns in IR imagery. Upwelling can be observed in figure 14.8. Northerly winds associated with the Pacific high have resulted in upwelling conditions along the California coast. The light gray shades just offshore indicate coastal upwelling of cooler bottom water. The upwelled water in this image can also be seen interacting with the cool waters of the California

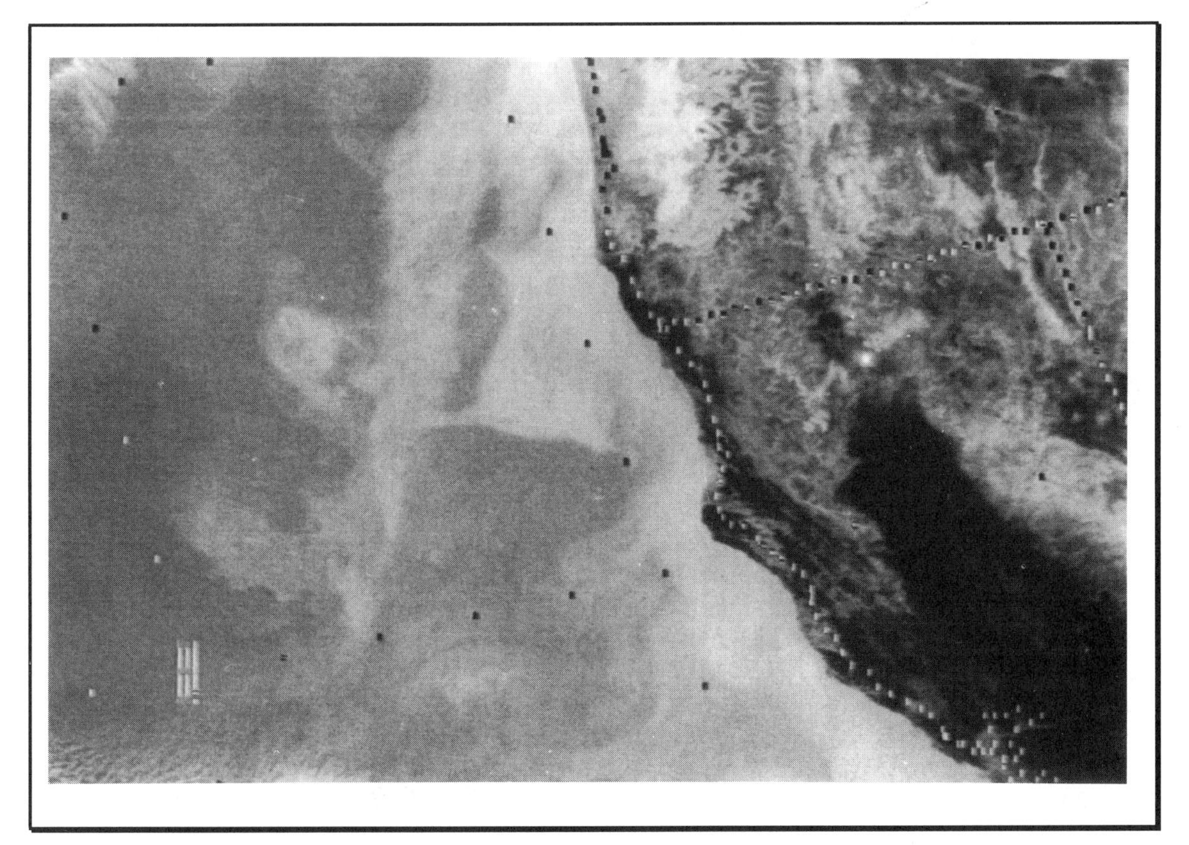

Figure 14.8. Upwelling off the California coast (HRPT IR, September 9, 1991)

Current as it flows in a southerly direction. Although the temperature gradient is very small and contrast is not very apparent, several spiral eddies can be seen where the upwelled water is influenced by the California Current.

The **bathymetry** (bottom topography) of the ocean floor can also influence upwelling. For example, a more steeply sloped ocean floor produces more upwelling than a gentle slope under the same wind conditions. **Topographically induced upwelling** occurs when an ocean current flows over a bottom projection that forces the current to rise and transport subsurface water to the surface. Upwelling of this type also can be observed on the lee (downstream) side of islands, on the lee side of large land promontories, and in eddies where surface waters are forced to diverge.

Dynamic upwelling occurs when the flow of oceanic currents causes surface water to diverge in the open ocean. Where adjacent ocean surface waters flow away from each other, deeper water rises to replace them. Upwelling of this type is associated with the Equatorial Current and the northern boundary of the Equatorial Counter Current.

El Niño

El Niño is an unusual warming of ocean surface waters in the eastern Pacific Ocean. Although El Niño events are not yet fully understood, it is known that they generally occur every 3–5 years. Scientists are currently studying this phenomenon to determine its effects on the global climate and to develop ways of minimizing the economic impact of such an event.

Under normal conditions, the Pacific Ocean waters off the west coast of South America are cooler than the waters of the central and western Pacific. Steady easterly trade winds cause surface water in the eastern Pacific to flow toward the west, causing cooler, deeper water to upwell. The cooler waters affect local air temperatures and inhibit the growth of thunderstorms; therefore, this region is generally less convectively active than the central or western Pacific. Figure 14.9 is an IR image of the tropics during normal conditions. Notice that very little convective activity takes place off the coast of South America.

During an El Niño event, however, the trade winds weaken and the coastal upwelling diminishes. This allows warmer surface waters in the eastern Pacific. The increase in sea surface temperature during an El Niño year allows increased convection to occur, and strong thunderstorms develop off the west coast of South America. The overall temperature pattern in the region changes, resulting in changes in upper-level winds. This shift in atmospheric conditions affects airflow and moisture flow, which can result in changes in weather patterns in regions all around the world. Figure 14.10 is an IR image from the 1983 El Niño event. Note the greater amount of tropical convection off the South American continent that occurs as a result of abnormally warmer ocean waters.

Not only are meteorological and climatic issues centered around El Niño but this phenomenon has important social and economic implications. The seasonal upwelling in the eastern Pacific Ocean provides nutrient-rich water that forms the basis for many major fisheries in smaller countries such as Peru. During an El Niño event, inhibited upwelling limits the success of the fishing industry in these countries. Since fishing is often the only major industry in these countries, an El Niño event can mean disaster to a country's economy. Scientists are currently trying to better understand El Niño in order to predict when an El Niño event is likely to occur. This would allow these countries to prepare in advance and prevent a major economic impact.

Figure 14.9. Normal atmospheric conditions in the eastern Pacific (non–El Niño year) (GOES IR, March 24, 1984)

Figure 14.10. Eastern Pacific atmospheric conditions during an El Niño event (GOES IR, March 24, 1983)

Ice detection and forecasting

For any company or organization that is involved in ocean transportation in polar seas, ice is a concern. It can block shipping and create dangerous conditions for ocean transportation and it is dangerous to structures such as oil rigs that are located in polar oceanic regions. Ice can also modify submarine acoustics, making it a concern for vessels dependent on sonar. This is not a problem limited to the higher latitudes, however, as ice, and especially icebergs, can extend far into mid-latitude bodies of water. In addition, ice can form on inland bodies of water and create transportation problems. This makes ice detection and forecasting an important application of satellite imagery.

Figure 14.11. Ice cover on Lake Erie (HRPT VIS, March 22, 1994)

Ice on inland bodies of water

Lakes and rivers in higher latitudes are often partially or completely covered with ice during the winter months. Examples include the St. Lawrence Seaway and the Great Lakes. These bodies of water form an important shipping lane for the central portion of the United States and Canada. When covered with ice, this seaway must be closed to shipping. Prior to the advent of satellite technology, it had to be closed for the most of the winter. Often it remained closed all winter, even when the ice retreated for several weeks. Using satellite imagery, however, ice analyses and forecasts can be made which can help keep the seaway open for a greater amount of time. Ice can also dam up rivers and lake outlets during winter months. A rapid melting or breaking up of the ice can result in a surge of water similar to the flood released by a breaking dam. Satellite imagery can help identify areas that may be prone to flooding in the event of a rapid melting of the ice.

In VIS satellite imagery, ice cover on lakes and rivers usually appears light gray to white. Thick ice generally will appear brighter than thin ice. When covered by snow, a frozen water surface will usually appear very bright, since there is no ground cover (such as trees) to block the ice and snow from the satellite's view. Figure 14.11 depicts ice cover on Lake Erie. In this image, the ice appears as various shades of light gray and white, depending on the thickness of the ice and the amount of snow cover on the ice. Snow can also be seen on the southeastern portion of this image. Notice that the shallow western portion of the lake is not covered with ice. Ice patterns on the lakes can often be related to the depth of the water.

Sea ice

Large regions in the polar latitudes are covered by **sea ice** throughout the year. Sea ice forms from ocean water; therefore, it is characterized by a **bulk salinity** (the amount of salt in the water). The salinity of sea ice is greatest when it is newly formed. As sea ice ages, it **rejects** (or loses) salt and increases in tensile strength. The bulk salinity of sea ice ranges from 20 ppt (parts per thousand) to a low of 5 ppt after the first winter season.

Sea ice plays an integral role in regulating the global climate. The presence of ice prevents the exchange of heat between the ocean and atmosphere, thus reducing the temperature extremes in the polar regions. Globally, sea ice reduces the total amount of solar heating of the Earth by reflecting solar radiation away from the surface.

Sea ice also has a strong influence on the global ocean circulation. When ice forms in the oceans during the fall and winter months, salt is rejected from the ice, increasing the salinity of the surface water. As the ice melts during the spring and summer months, a relatively fresh surface water layer creates a strong vertical density gradient, affecting surface water circulation. As sea ice migrates southward and subsequently melts, it brings relatively colder and less saline water to lower latitudes. The net equatorward transport of ice therefore contributes to increasing water salinity in the polar regions and decreasing salinity in subpolar regions.

Stage of Development	Types	Thickness	Characteristic Albedo
New ice	Frazil, slush shuga, grease	0–10 cm	Very low; ice appears dark against water surface
	Dark nila	0–5 cm	Very low; ice appears dark against water surface
	Light nila	5–10 cm	Low; ice is dark gray
Young ice	Gray	10–15 cm	Moderately low; ice appears gray
	Gray-white	15–30 cm	Moderate; ice appears light gray
First-year ice	Thin	30–70 cm	High; very light gray compared to snow-covered ice
	Medium	70–120 cm	Very high; ice appears white
	Thick	120–200 cm	Very high; ice appears white
Old ice	Second year	2.0–2.5 m	Very high; ice appears white
	Multi-year	> 2.5 m	Very high; ice appears white

Table 14.1. Classes of sea ice based on age and associated albedoes

Sea ice also affects the distribution of living organisms in the ocean. In the winter months, ice acts as an insulator between warmer ocean waters and extremely cold polar air. This allows the ocean to remain warm enough to support abundant wildlife. The edge of the ice zone is also known to be an area high in biological activity. Winds that blow off ice packs toward the ocean can induce upwelling of nutrient-rich water, which stimulates the growth of phytoplankton and attracts numerous larger organisms to the regions around the ice edge. In this manner, the distribution of sea ice can affect the location of marine resources in polar regions.

Sea ice formation

As sea ice ages, it becomes thicker and its albedo increases. Young, thin sea ice usually has a very low albedo and may even go undetected in a VIS satellite image if it is thin enough. Old ice or snow-covered ice has a very high albedo and tends to be very bright in VIS imagery. Table 14.1 categorizes the different types of sea ice according to age and albedo.

Sea ice formation is affected by a variety of factors, including air temperature, winds, currents, and water salinity. As a result, many types of sea ice have been defined. In its earliest stages, sea ice is known as **new ice**. New ice ranges from 0 to 10 cm (0 to 4 in) in thickness. As sea water first begins to freeze, it is called **frazil ice**. This ice consists of individual ice crystals suspended in water. Under calm conditions, the crystals often form a solid sheet, but in waves or currents they form a layer on the surface known as **grease ice**. **Slush**, by contrast, is a viscous floating mass of ice formed from a mixture of snow and water. **Shuga** is an accumulation of spongy ice pieces a few centimeters across that often forms from grease ice or slush under agitated ocean conditions. **Nila** is the initial stage of layering in sea ice. At times ocean waves will cause these layers of nila to break into circular pieces of ice known as **pancake ice**.

As sea ice consolidates into a layer with a thickness of 10–30 cm (4–12 in.), it is referred to as **young ice**. Young ice is further classified according to thickness into **gray ice** (10–15 cm, or 4–6 in.) and **gray-white ice** (15–30 cm, or 6–12 in.). At a thickness of 30 cm (12 in.), sea ice is termed **first-year ice**, and it remains first-year ice until it melts or survives a summer melt period. First-year ice rarely reaches an average thickness greater than 2 meters (6.7 ft.). Ice that has survived one melt season is termed **second-year ice**, and ice that has survived multiple melt seasons is **multi-year ice**. These types of ice differ substantially from first-year ice. During the melt season, melt water ponds form on the ice and water percolates through the ice, causing changes in the bulk salinity. These types of ice are also characterized by a rough, rolling surface and a very high albedo.

Sea ice distribution

The areas of sea ice coverage are divided into three regions. The **permanent ice zone** consists of the areas of the ocean that are always covered with sea ice. The **seasonal ice zone** is a zone in which sea ice is present for part of the year, but not throughout the entire year. The seasonal ice zone is marked by the **maximum winter extent** of the ice on one side and the boundary of the permanent ice zone on the other side (i.e., the minimum summer extent). A third zone, the **marginal ice zone**, extends from the ice edge to a point sufficiently away from open ocean that the ice is not affected by the ocean (i.e., no influence from wave action or swell). This zone is usually 100–200 km (60–120 miles) wide. Figure 14.12 shows the minimum and maximum extent of sea ice in the Arctic region, and figure 14.13 shows the same for sea ice around Antarctica.

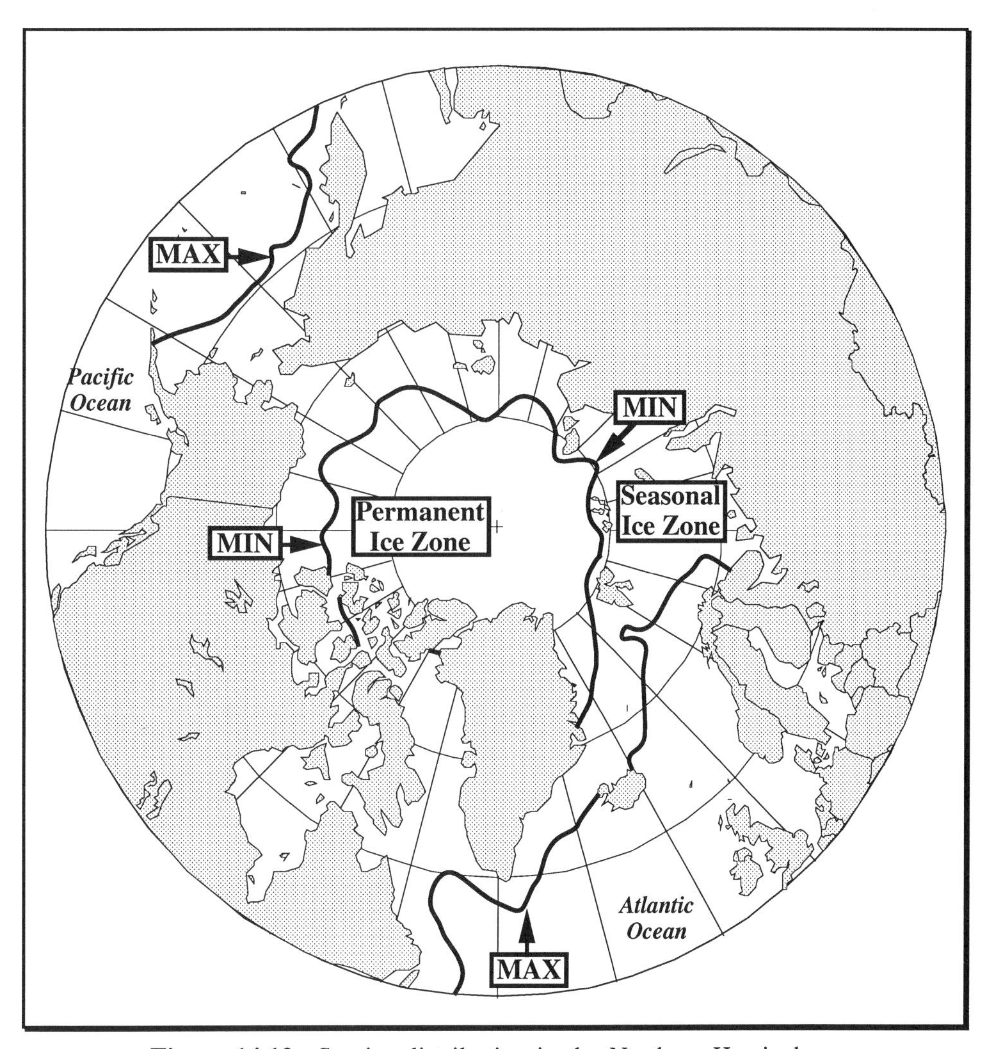

Figure 14.12. Sea ice distribution in the Northern Hemisphere

Sea ice movement

Sea ice can be divided into two categories, **fast ice** and **drift ice**. Fast ice is sea ice that grows outward from the coast and is anchored to the sea bed. Drift ice is sea ice that is not considered fast ice. Drift ice, as its name implies, floats freely and is in continual motion. The motion of sea ice is directly related to surface wind speed and direction. The general rule is that sea ice drifts at 2% of the surface wind speed and 15–30° to the right of the surface wind direction (in the Northern Hemisphere). Sea ice can also be categorized on the basis of the density of the **ice pack** (the clustered drift ice). The concentration of sea ice can range from open water with very little ice to compact or consolidated ice. Using a chart similar to the one in figure 14.14, the ice concentration can be estimated from satellite imagery. Although this method is qualitative in nature, it can be helpful while making an ice forecast. Simply observe the ice pack and make an estimate of the percentage of the water covered by the ice pack, using the chart as a guide.

Sea ice can appear very compact; however, it never completely covers the ocean because of motion, fracturing, and melting. These processes can lead to the formation of open water areas that include narrow, linear **leads** (ranging from several meters to kilometers) and **polynyas** (ranging in area to thousands of square kilometers). These features are

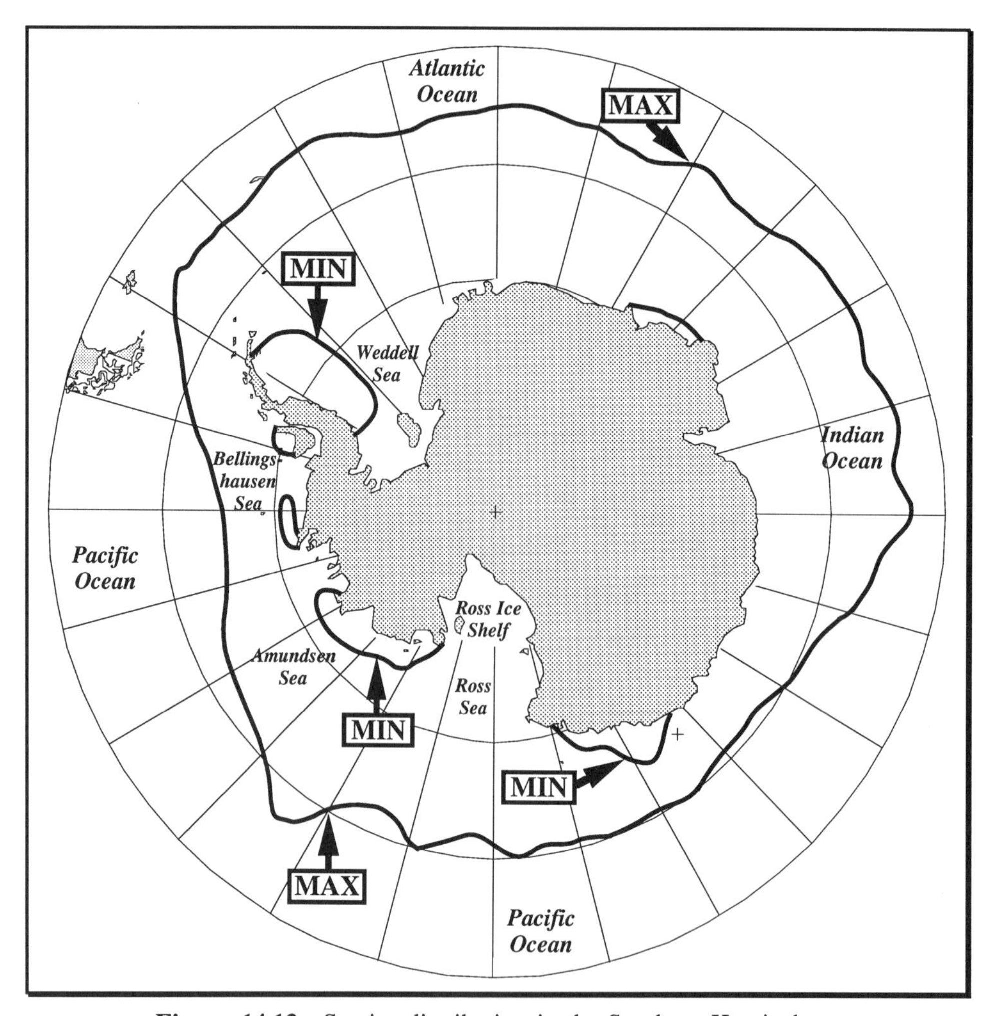

Figure 14.13. Sea ice distribution in the Southern Hemisphere

especially common along the coastline, where persistent offshore winds tend to break up the ice. Frequent deformation of the ice pack tends to break uniform layers of ice into irregular shapes called **floes**. **Ridges** can form on ice floes when movement and collisions of the floe cause ice to be compressed with enough force to crumble into broken piles. A **rafted floe** can form when one ice floe overrides another. **Fissures** are cracks in the ice that form due to the stress of movement within an ice floe.

In the Arctic Basin, there are two major circulatory features that characterize the movement of sea ice. The first is a large clockwise circulation in the Beaufort Sea north of Alaska known as the **Beaufort Gyre** (figure 14.15F). The second is known as the **Transpolar Drift Stream (TDS)** (figure 14.15G). This drift originates in the East Siberian Sea and exits the Arctic Basin via the Davis Strait, east of Greenland. Over 90% of the sea ice that exits the Arctic Ocean does so through the TDS. Much of this ice continues to drift toward the equator in the **East Greenland Current** (figure 14.15H). Using a series of polar orbiter satellite images, one can monitor these migrations of sea ice over a period of several days once a piece of ice with a recognizable shape has been identified. Direction and speed of movement can be calculated, and this information can be used to alert ships in the area.

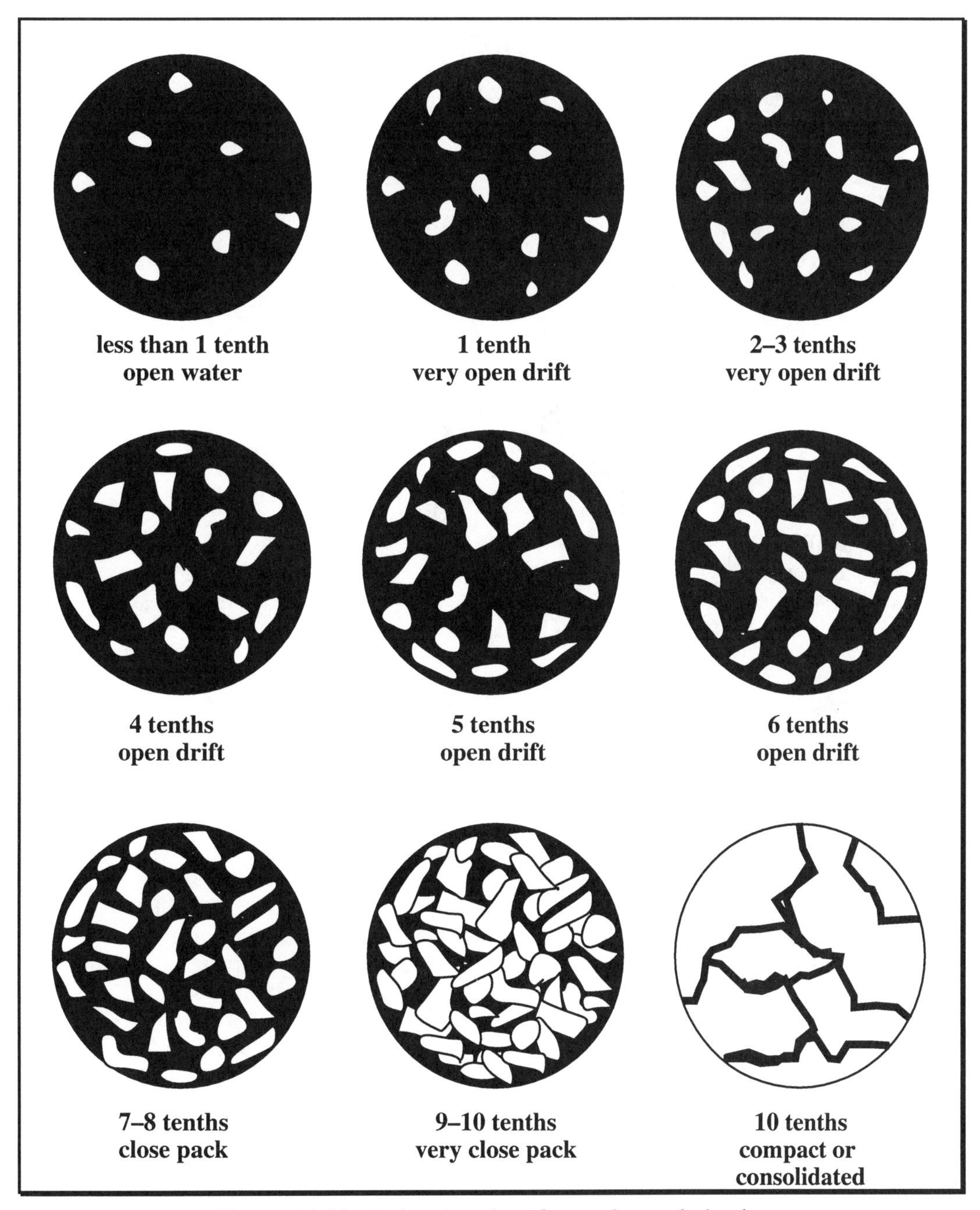

Figure 14.14. Estimation chart for sea ice pack density

Figure 14.16 is an HRPT image of an Arctic sea ice pack off the northern coast of Alaska. Barrow, Alaska is located nearly in the center of the image, at *B*. The Alaskan coastline runs diagonally from the lower left corner to the upper right corner. Fast ice is attached to the coastline and extends out into the ocean (as indicated by the arrows). There is a slight difference in the shade of gray between the land and the fast ice that makes it possible to locate the shoreline. Off the coast of Alaska is a dense ice pack. Many fissures and cracks can be seen in the ice. The ice density changes from consolidated at the top of the image to roughly 7–8 tenths packed at the bottom of the image. When this image was taken, the region made national news because several migrating whales had become trapped in the fast ice (in circled area) and were unable to participate in the annual migration. International rescue efforts eventually released the whales safely.

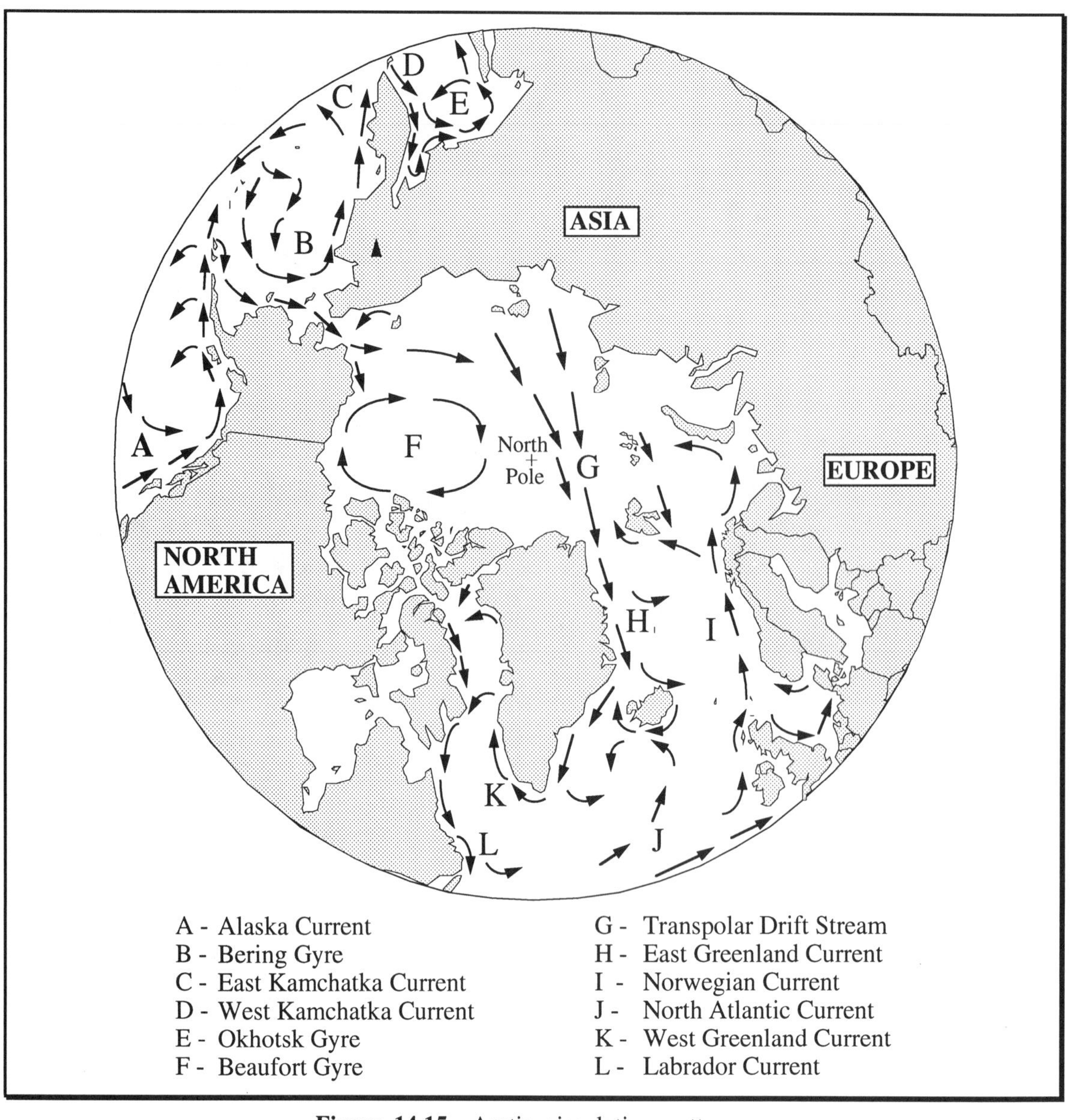

Figure 14.15. Arctic circulation patterns

Icebergs

Icebergs are different from sea ice in that they are formed as they break off **(calve)** from glaciers that originate on land. Therefore, they form from fresh water and retain different physical characteristics than sea ice. They have a different density than sea ice, and they reach different depths. Icebergs are typically a greater danger to shipping than are free-floating ice floes because 90% of an iceberg's size remains below the water. The iceberg season actually occurs during spring and summer months, when glaciers are weakened and calve more frequently.

The principal region of iceberg formation in the world is along the west coast of Greenland, where approximately 10,000 new icebergs form each year. Icebergs also form from glaciers in the area surrounding Prince William Sound (Alaska) and in areas around Antarctica. Antarctic icebergs tend to be larger than their Arctic counterparts. In 1991,

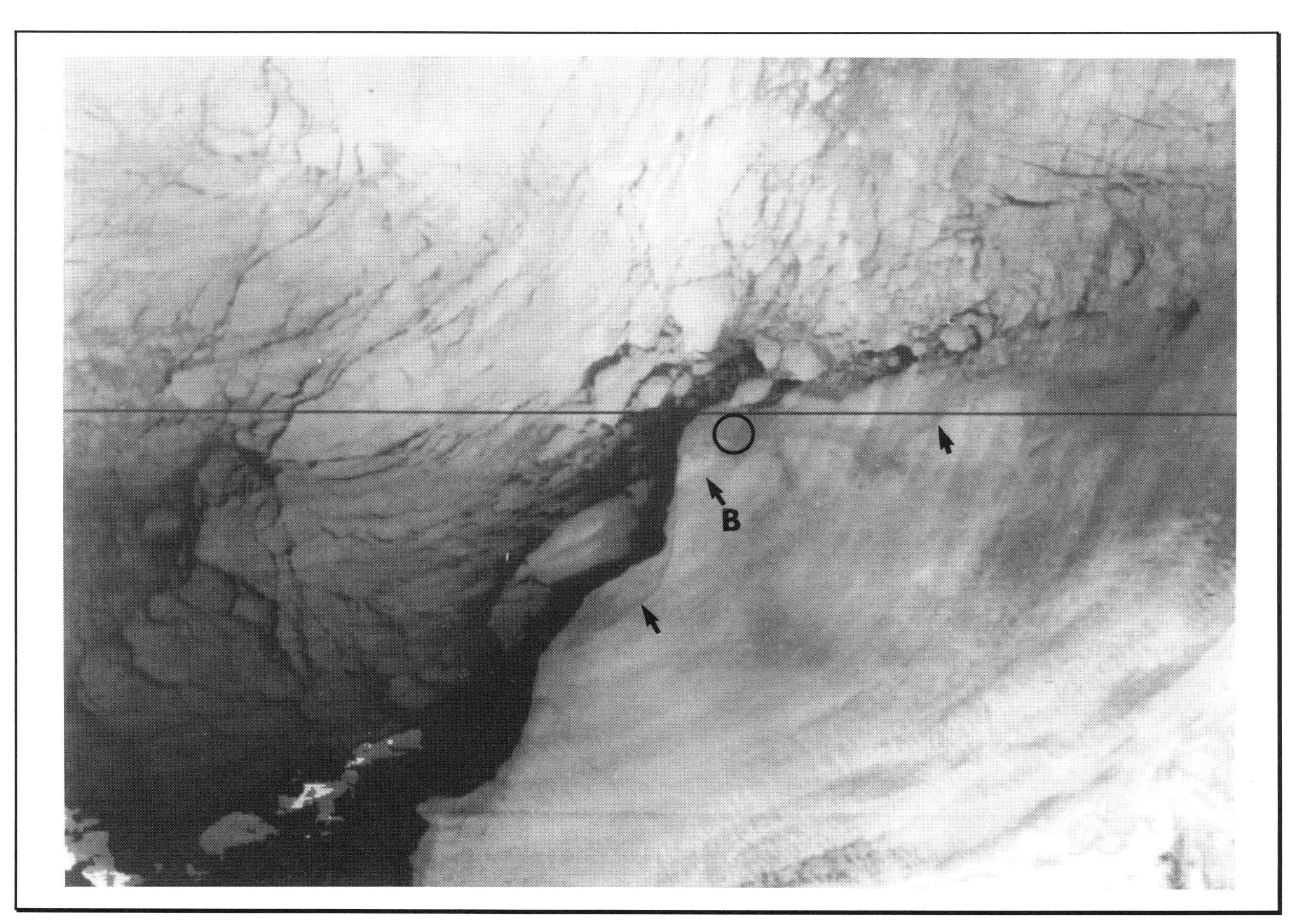

Figure 14.16. Sea ice near Barrow, Alaska (HRPT VIS, October 21, 1988)

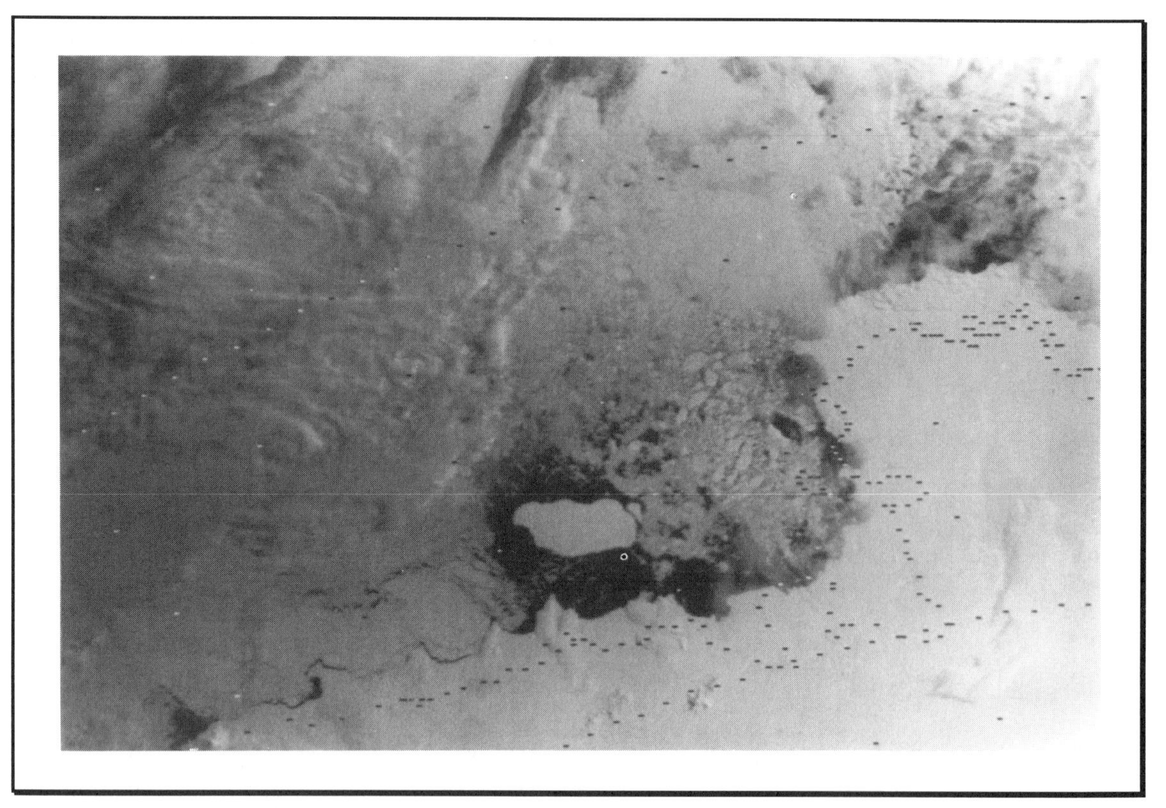

Figure 14.17. Iceberg A-24 in the Weddell Sea, Antarctica (HRPT VIS, November 26, 1991)

iceberg A-24 broke free from the sea ice in the Weddell Sea, Antarctica, and floated into the South Atlantic. The initial size of A-24 was estimated as three times the size of Rhode Island. This iceberg can be seen in figure 14.17, an HRPT image of Antarctica. The iceberg appears as a large white object just off the Antarctic coast and is surrounded by a sea ice floe.

Appendix A

Conversions and Constants

Length

1 micrometer (μm) =	0.0000394 inch 0.001 millimeter 0.0001 centimeter
1 millimeter (mm) =	0.0394 inch 0.1 centimeter 0.001 meter
1 centimeter (cm) =	0.394 inch 0.01 meter
1 meter (m) =	39.4 inches 3.28 feet 0.001 kilometer
1 kilometer (km) =	0.621 miles 0.540 nautical mile 1000 meters
1 nautical mile (nm) =	1.85 kilometers 1.15 miles 1 minute of latitude

Area

1 square kilometer =	0.39 square mile 247.1 acres

Pressure

1 atmosphere =	14.7 pounds/square inch 1.013 bar 1013 millibar 760 millimeters of mercury at 0° C

Volume

1 milliliter (ml) =	0.06 cubic inch 1.0 cubic centimeter 0.001 liter
1 liter (l) =	1.06 quarts 33.8 fluid ounces 1000 milliliters

Speed

1 centimeter/second =	0.394 inch/second 0.02 mile/hour (mph) 0.036 kilometer/hour
1 kilometer/hour =	0.621 mile/hour 27.8 centimeters/second
1 knot (kt) =	1.85 kilometers/hour 1.15 miles/hour 1.0 nautical mile/hour 51.5 centimeters/second
1 meter/second =	2.23 miles/hour 3.6 kilometers/hour

Temperature

Fahrenheit to Celsius:	°C = (°F – 32) ÷ 1.8
Celsius to Fahrenheit:	°F = (°C x 1.8) + 32
Celsius to Kelvin:	K = °C + 273

Wind Symbols

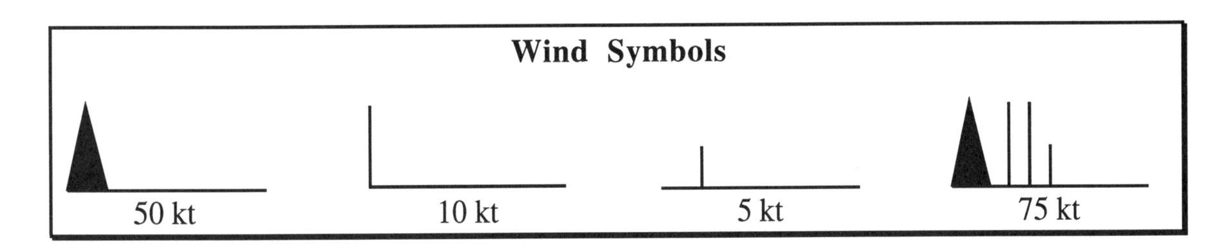

Appendix B

Internet Resources for Satellite Imagery

The Internet has become an excellent source for satellite data, with sites offering imagery covering all areas of the earth. In addition to imagery, these sites often include detailed weather data such as surface maps, radar summaries, computer model forecasts, and upper-air charts that can aid in satellite image interpretation. The following is a partial list of some Internet sites that offer satellite imagery and related data. This list assumes that the reader is familiar with basic Internet navigation and file transfer procedures. You may want to refer to an Internet resource guide if you are unfamiliar with these procedures and wish to obtain imagery in this manner.

For each site, the connecting address and a brief description of relevant features is given. The addresses, or Uniform Resource Locators (URLs), for some sites are given in several different formats. These address formats and the associated type of server are listed in the table below.

Address format	Type of server
ftp://host.name.domain/directory/	FTP site
http://host.name.domain/directory/	WWW server
telnet://host.name.domain	telnet site
gopher://host.name.domain	Gopher server
wais://host.name.domain	WAIS server
news:newsgroup.name	newsgroup

Most sites on the internet include "README" files (or similar files) with instructions and information about the various data available at the site. It is a good idea to read these files before attempting to access data provided at a site with which you are not familiar. Many sites also include connections to other sites, which often provide additional data. Take some time to explore what each site has to offer in order to determine which is most useful for your purposes.

For those with internet access, an on-line document that is very helpful is "Resources for Weather Related Resources on the Internet." This document lists all known Internet sites that post weather-related information, including satellite imagery from all over the world. It also details various sources for software and CD-ROMs that are very useful. The document is updated every two weeks and can be obtained via anonymous FTP, Gopher, or World Wide Web (WWW) browsers from the following locations:

ftp://rtfm.mit.edu/pub/usenet/news.answers [filename = weather/data/part1]
ftp://vmd.cso.uiuc.edu/wx.sources.doc [filename = weather/data/part1]
http://www.cis.ohio-stat.edu/hypertext/faq/usenet/weather/top.html
ftp://kestrel.umd.edu/pub/wx [filename = weather/data/part1]
gopher://wx.atmos.uiuc.edu

Or, to obtain a copy of the document through e-mail, send a message to mail-server@rtfm.mit.edu. Leave the subject line blank. The message text should read:

send /pub/usenet/news.answers/weather/data/part1

At each update, the document is also posted to the following newsgroups:

sci.geo.meteorology
news.answers
sci.answers

Site: **Bureau of Meteorology, Melbourne, Victoria, Australia**
Address: gopher://babel.ho.bom.gov.au

Features: Satellite imagery of Australia from GMS satellite, Australian weather forecasts.

Site: **Florida State University Weather Gopher**
Address: gopher://metlab1.met.fsu.edu

Features: Source for daily satellite imagery, surface plots, surface analysis, computer model forecasts, radar summary, upper-air maps.

Site: **Ohio State Weather Gopher**
Address: gopher://geograf1.sbs.ohio-state.edu

Features: Satellite imagery from GOES 7 and GOES 8, updated hourly; VIS, IR, and WV imagery of North America; Antarctica imagery; satellite information and prediction charts; tropical disturbance information; upper-air maps; and U.S., Alaska, and international surface charts.

Site: **Oregon State University Atmospheric Sciences Weather Information**
Address: ftp://ats.orst.edu/pub/weather

Features: Latest GOES VIS, WV, and IR imagery of the U.S. and the northwest region of the U.S.; national weather summaries; individual city surface maps and forecasts; global composite water vapor imagery. "Andrew" and "Emily" directories contain archived hurricane imagery.

Site: **Purdue University Weather Processor Server**
Address: gopher://thunder.atms.purdue.edu
http://thunder.atms.purdue.edu
ftp://thunder.atms.purdue.edu

Features: VIS, IR, and WV GOES imagery updated hourly, composite VIS/IR imagery, composite IR/radar summary and IR/surface maps, surface maps, upper-air maps, radar summaries, computer model output, archived imagery.

Site: **University of Edinburgh (Scotland) Department of Meteorology**
Address: ftp://ftp.met.ed.ac.uk
ftp://cumulus.met.ed.ac.uk

Features: Daily forecasts and Meteosat imagery of Europe and the U.K., Atlantic, Africa; weather animations.

Site: **University of Illinois Weather Machine**
Address: http://www.atmos.uiuc.edu
gopher://wx.atmos.uiuc.edu

Features: Very comprehensive weather information resource, including real-time IR, VIS, and WV imagery from GOES 7 that is updated hourly; archived imagery; state and international forecasts; satellite image discussion; severe weather information; hourly surface maps; upper-air maps; computer model forecasts. Includes links to other useful weather-related servers.

Site: **University of Maryland Department of Meteorology**
Address: ftp://kestrel/umd.edu

Features: Surface analysis/radar summary for U.S. cities; VIS, IR, and WV imagery of the U.S.; archives including eclipse imagery, Hurricane Andrew, and 1993 Mississippi River flooding.

Site: **University of Wisconsin–Madison Space Science and Engineering Center**
Address: gopher://gopher.ssec.wisc.edu
http://www.ssec.wisc.edu

Features: Good source for satellite imagery, including daily GOES 7 and GOES 8 browse imagery, global composite imagery, daily full-disk GOES VIS and IR imagery, Southern Hemisphere and Antarctica imagery, GOES user's guide, software.

Glossary

The number at the end of each definition indicates the page number of the first use of the term.

Absolute zero. The theoretical temperature of a substance characterized by a complete absence of heat (equal to –273 ˚C). (2)
Absorption band. A range of wavelengths characterizing electromagnetic energy that is absorbed by the atmosphere. Absorption bands exist in the short-wavelength portion of the spectrum and in narrow bands within the infrared. (8)
Advanced Very High-Resolution Radiometer (AVHRR). The primary sensor on board U.S. polar-orbiting meteorological satellites. Five channels detect visible, near infrared, and thermal infrared radiation. (8)
Aerosols. Solid particles that remain suspended in the atmosphere: e.g., pollutants, smoke particles, sand, dust, and airborne volcanic material. (196)
Air mass. A large body of air with consistent temperature and moisture characteristics throughout. (126)
Alaska Current. A northward-flowing current along the northwestern coast of North America. (211)
Albedo. A property that describes the amount of light a surface reflects. (4)
Altocumulus clouds. Mid-altitude clouds with a "puffy" (cumuliform) shape. (76)
Altostratus clouds. Mid-level clouds with a flat, sheetlike (stratiform) shape. (76)
Amplitude. A measure of the amount of displacement in a wave. (2)
Anemometer. An instrument that measures wind speed. Small cups on the end of a rotor catch the wind, spinning the rotor and determining the wind speed. (41)
Anticyclone. A high-pressure region characterized by sinking air and anticyclonic air circulation. (103)
Anticyclonic. A term used to describe clockwise circulation (in the Northern Hemisphere) or counterclockwise (in the Southern Hemisphere). (103)
Anticyclonic eddy. An eddy that forms during a northward meander of the Gulf Stream and is characterized by a clockwise circulation. It has a warm core and is surrounded by a ring of cooler water. (Also called a warm-core eddy.) (210)
Applications Technology Satellite (ATS 1). The first combined geostationary communications and weather satellite, launched December 2, 1966. (17)
Arc-cloud line. A bow-shaped line of convective clouds that forms along the leading edge of the cold air outflow of a thunderstorm. (94)
Ascending node. The south-to-north pass of a polar orbiting satellite over a ground station. (12)
Ash cloud. A large cloud of particulates released during a volcanic eruption. (199)
Atmosphere. The layer of gases that surrounds a planet. The Earth's the atmosphere is very thin (about 800 km, or 500 miles) and composed primarily of nitrogen (79%) and oxygen (20%). (56)
Atmospheric loading. The process by which aerosols are introduced into the atmosphere. (196)
Atmospheric pressure (air pressure). The force caused by the weight of the atmosphere. At sea level, average atmospheric pressure is 14.7 lbs per square inch (1013 mb). As altitude increases, atmospheric pressure decreases. (64)
Atmospheric window. A range of wavelengths characterizing electromagnetic energy that is allowed to pass through the atmosphere. Atmospheric windows are found in the radio and visible portions of the electromagnetic spectrum. (8)
Aurora Australis. An auroral display visible in the Southern Hemisphere. (64)
Aurora Borealis. A brilliant display of light, known as an auroral display, that can be seen in the Northern Hemisphere. It is caused by charged solar particles interacting with air molecules in the upper atmosphere. (Also known as the Northern Lights.) (64)
Automatic Picture Transmission (APT). The continuous transmission of AVHRR data from polar orbiting satellites. Daytime transmission includes near-infrared and thermal infrared channels. Nighttime transmission includes two thermal infrared channels. (13)
Autumnal equinox. The position in earth's orbit in which the sun's direct rays are on the equator during their southward migration. On the Autumnal equinox, day and night are equal. Marks the first day of autumn in the Northern Hemisphere. (60)
Baroclinic zone cirrus. The cirrus cloud shield associated with a jet stream. (111)
Bathymetry. The study of the underwater topography of the oceans. (214)
BD enhancement curve. An infrared enhancement curve used in hurricane analysis; highlights cloud top temperatures surrounding the eye of the hurricane. (36)
Beaufort Gyre. A large-scale drift pattern of sea ice in the Beaufort Sea, north of Alaska. (220)
Bermuda high. A high-pressure region in the Atlantic Ocean, located near 30˚ N. (103)
Black body. A theoretical object that absorbs radiation at all wavelengths. (6)
Breeze front. A boundary between warmer and cooler air masses, typically along a land, sea, or lake breeze; usually characterized by a line of convective

clouds, and considered a favorable location for thunderstorm development, especially within a warm, humid air mass. *See also* lake, land, and sea breezes. (159)
Brightness value. An indicator of the energy level sensed by a satellite. It is represented by varying gray tones on a satellite image. In VIS imagery, bright tones indicate a high albedo; in IR imagery, bright tones indicate colder temperatures. (27)
Bulk salinity. A measure of the amount of salt in sea water or sea ice. (217)
California Current. A cold-water current that flows southward along the west coast of North America. (211)
Calving. The process by which a large piece of ice breaks off a glacier to become an iceberg. (222)
CD-ROM. A storage device (used with a personal computer) which is capable of holding large quantities of data and software. The contents of a CD-ROM disk can be read but not altered. CD-ROMs are used extensively to store datasets, including satellite imagery, from various scientific organizations. (22)
Central dense overcast. A layer of high-altitude clouds that forms over the eye of a hurricane. (185)
Channel. A specific band of wavelengths sensed by a satellite sensor. (8)
Cirriform clouds. High-altitude ice clouds, characterized by a very thin, wispy appearance. (76)
Cirrocumulus clouds. Cirrus-type clouds that exhibit a cellular structure. (77)
Cirrostratus clouds. Cirrus (ice crystal) clouds that form a thin haze or sheet across the sky. (77)
Cirrus clouds. High-altitude ice crystal clouds. They appear very thin and wispy due to upper-level winds spreading the cloud across the sky. (76)
Cirrus shield. A band of cirrus clouds that is usually present adjacent to the main axis of a jet stream. (113)
Cirrus streaks. Thin lines of cirrus clouds that often form along a jet stream when little water vapor is present in the atmosphere. They can be used to locate troughs and/or ridges in a jet stream. (115)
Closed-cell convection. Convection that results in cumulus clouds that appear solid, with clear rings surrounding the cells. (94)
Cloud band. A nearly continuous cloud formation with a distinct long axis with a length-to-width ratio of at least 4 to 1 and a width greater than 1° of latitude. (68)
Cloud bank. A region of clouds that has been formed or maintained as air is dammed up by topography. (96)
Cloud displacement. The apparent repositioning of a cloud on a satellite image, caused by the effect of the viewing angle of a satellite sensor. (Also known as parallax.) (39)
Cloud element. The smallest cloud form that can be seen in a satellite image. (68)
Cloud finger. An extension from the forward side of a frontal cloud, usually extending from a frontal band toward the equator. (68)
Cloud line. A narrow cloud band in which the individual cells are connected and the line width is less than 1° of latitude. (68)
Cloud shield. A broad cloud pattern that is no more than four times as long in one direction as it is wide in the other. (68)
Cloud streets. Parallel lines of convective clouds, usually oriented parallel to the wind direction and often caused when cold air flows over a warmer surface and convection is initiated. (68)
Coastal upwelling. Upwelling caused by a combination of wind stress on the water surface and the rotation of the Earth (Coriolis effect); the most commonly observed type of upwelling. (212)
Cold-core eddy. *See* cyclonic eddy. (209)
Cold front. The boundary between an advancing cold air mass and a warmer air mass. A cold front is characterized by cumulus-type clouds and thunderstorms. Colder, fair weather often follows the passage of the front. (126)
Cold-front type occlusion. An occluded front in which the coldest air lies behind the cold front. The warm front is completely lifted from the ground. (130)
Cold-water intrusion. Colder ocean water flowing into and interacting with a warmer-water ocean current. (209)
Comma cloud. A cloud or cloud system with a distinct comma shape due to rotation of the clouds. Usually it is associated with large, mid-latitude cyclones but it can be present in smaller-scale circulation patterns. (68)
Condensation. The process of a gas changing into a liquid. This process releases energy. (57)
Condensation nuclei. Tiny particles in the air onto which water vapor condenses, forming water droplets or ice crystals. These particles most often include sea salt, smoke, dust, and aerosols. (67)
Conduction. The transfer of heat through contact. (58)
Continental (c) air mass. An air mass that originates over land. It is characterized by low humidity. (126)
Continental polar (cP) air mass. An air mass originating over land in higher latitudes. Its humidity is low and its temperatures are cold. (126)
Continental shelf. A gently sloping region of the ocean floor adjacent to the continental coastline. (208)
Continental slope. A steep zone of transition between the continental shelf and the deep ocean. (208)

Continental tropical (cT) air mass. An air mass originating over land in lower latitudes. Its humidity is low and its temperatures are warm. (126)
Contrast. The degree of tone difference, on a gray scale, between two features on a satellite image. (27)
Convection. The rising and sinking action of a fluid caused by differential heating. In the atmosphere, heated air rises and cooled air sinks, forming convection cells. This process distributes heat away from the Earth and throughout the atmosphere. (58)
Convergence. The process by which winds come together. (85)
Convergence zone. An area in which convergence is taking place. Convergence zones are often present along frontal boundaries and breeze fronts and are favorable locations for thunderstorm development. (146)
Coriolis effect. An apparent force caused by the rotation of the Earth. It causes an object moving across the Earth in the Northern Hemisphere to appear to have been deflected to the right of its path of travel. In the Southern Hemisphere, objects appear to be deflected to the left. (102)
Cosmos. A series of Russian polar orbiting oceanographic satellites. (22)
Country breeze. A breeze that blows into a city as a result of the urban heat island effect. (50)
Crest. The maximum upward displacement of a wave. (2)
Cumuliform clouds. Clouds that appear "puffy" and exhibit vertical development. (73)
Cumulonimbus clouds. Cumulus clouds that can grow to 60,000 ft. They usually produce thunder and lightning, heavy rain, and hail, and may produce tornadoes. (74)
Cumulus clouds. Common fair-weather clouds. They are low-altitude clouds, usually characterized by limited vertical development, and are popularly known as "cotton puff" clouds. (73)
Cut-off, cold-core low. A low-pressure system that becomes cut off from its associated jet stream. (134)
Cyclogenesis. The process that leads to the development of cyclones (or low-pressure systems). (130)
Cyclone. A storm system that is centered around an area of low pressure and characterized by cyclonic air circulation. (103) Also: a tropical storm formed in the Indian Ocean (*see* hurricanes). (184)
Cyclonic. A term describing a counterclockwise circulation (in the Northern Hemisphere) or clockwise circulation (in the Southern Hemisphere). (103)
Cyclonic eddy. An eddy that forms during a southward meander of the Gulf Stream and is characterized by a counterclockwise circulation (in the Northern Hemisphere). It has a cold core and is surrounded by a ring of warmer water. (Also known as a cold-core eddy.) (209).
Dendritic pattern. The branching pattern that river valleys form on the land. The name is derived from "dendros," meaning treelike. (72)
Descending node. The north-to-south pass of a polar orbiting satellite over a ground station. (12)
Dew point. The temperature at which air becomes saturated with water vapor and condensation begins. (57)
Diffluent winds. Winds that are spreading away from each other (usually at an upper level). (153)
Direct readout data. Satellite data that is collected directly from the satellite transmission, by means of antennas and radio receivers that are sensitive to the satellite's transmission frequency. Imagery and data collected from direct readout is usually displayed on a personal computer. (13)
Divergence. The process by which winds flow away from each other. (85)
Doppler radar. A radar system used to locate areas of precipitation and measure wind patterns within a storm. (176)
Downburst. A very strong localized downdraft within a thunderstorm that often creates damaging winds. (The term is used interchangeably with "microburst".) (172)
Downdraft. A current of sinking air. (68) Also- downward currents of rain-cooled air within a thunderstorm. (146)
Downloading. The process used to transfer a digital file from one computer to another. (25)
Downwind. A term describing the direction toward which the wind is blowing. (92)
Drift ice. Sea ice that is not attached to the land. (219)
Dry line. A boundary between a dry air mass and a humid air mass. In the United States, a dry line often forms in the area from west Texas northward to western Nebraska when a dry tropical air mass (cT) from the southwestern deserts meets a humid tropical air mass (mT) from the Gulf of Mexico. It is a prime location for severe thunderstorm development. (153)
Dry slot. The region in a comma cloud where the jet stream crosses the cloud system. It appears relatively cloud free in IR and VIS imagery, and is dark in water vapor imagery. (132)
Dynamic upwelling. Upwelling caused by diverging currents at the surface of the ocean. (214)
East Greenland Current. A southward-flowing oceanic current off the eastern shore of Greenland. (220)
Edge definition. The appearance of the cloud edges in a satellite image. The edges can be ragged or well defined. (70)
Electromagnetic spectrum. The full range of wavelengths of electromagnetic energy, arranged according to wavelength and frequency. Includes

radio, microwave, infrared, visible light, ultraviolet, x-ray, and gamma ray energy. (3)
El Niño. An unusual warming of tropical oceans off the northwest coast of South America. Generally occurs on a 3–5 year cycle and is believed to affect the weather patterns over a large area of the world. (214)
Enhancement. Alteration of the gray scale of a satellite image in order to highlight specific features. (36)
Enhancement curve. An enhancement scheme that is developed for a specific purpose, such as thunderstorm or hurricane analysis. (36)
Equatorial low. A persistent low-pressure region near the equator where heated air rises. (103)
Evaporation. The process by which a liquid changes to a gas. This process consumes energy. (57)
Eye. The central portion of a hurricane, in which air is sinking; therefore, it remains relatively cloud free. (36)
Eye wall. The area surrounding the eye of a hurricane; the region in which the heaviest rainfall and strongest winds are located within the storm. (36)
Fast ice. Sea ice that is formed at the shoreline and is attached to the bottom of the sea. (219)
Feng Yun. A series of Chinese polar orbiting weather satellites. (22)
Fibrous texture. A term used to describe the appearance of clouds that appear to be streaked or fiberlike. (70)
Filling. A late phase of a storm system's development, in which central pressure is rising and the storm is weakening. (136)
First-year ice. Sea ice that has reached a thickness of 30–200 cm (12–80 in.). (218)
Fissures. Cracks in sea ice caused by differential movement within the ice floe. (220)
Flash flooding. Rapid local flooding during heavy precipitation events, ice jam breakups, and dam failures. (162)
Florida Current. A warm-water current in the Atlantic Ocean that flows from the southern tip of Florida northward to join the Gulf Stream. (207)
Floe. A large area of irregularly shaped pieces of sea ice. (220)
Fog. A stratus cloud that has formed at ground level. (67)
Frazil ice. Newly formed sea ice that consists of ice crystals suspended in water. (218)
Frequency. The number of waves that pass a point in a given time period. Frequently is inversely proportional to wavelength (long wavelengths are associated with low frequencies, and short wavelengths with high frequencies). (3)
Front. A boundary that separates two air masses. (126)
Frontal boundary. The point of contact between two different air masses. A frontal boundary can be found at the surface or high in the atmosphere. (126)
Gamma rays. Electromagnetic energy with the shortest wavelengths. Gamma rays are produced by the radioactive decay of various elements. (3)
Geostationary Operational Environmental Satellite (GOES). A series of geosynchronous satellites operated by the United States. (16)
Geostationary satellite. A type of satellite that is in a geosynchronous orbit and therefore remains over the same point on the Earth during its orbit. This type of satellite allows for constant monitoring of one region of the Earth. (10)
Geosynchronous Meteorological Satellite (GMS). A series of geostationary satellites operated by Japan. (22)
Geosynchronous orbit. An orbit in which a satellite remains stationary with respect to the Earth. This is because its angular velocity is the same as the Earth's. (16)
Global wind movement. Air circulation on a global scale that is a result of differences in incoming solar radiation across the Earth. Global winds cover larger areas than local winds and are usually more persistent in direction and speed. (85)
GOES East. A U.S.-operated geostationary weather satellite located over 75° W longitude. (17)
GOES West. A U.S.-operated geostationary weather satellite located over 135° W longitude. (17)
GOMS. A Russian-operated geostationary weather satellite that provides coverage of central Asia and the Indian Ocean. (23)
Gray ice. Young ice with a thickness of 10–15 cm (4–6 in). (218)
Gray scale. A scale of gray tones used in satellite imagery. It consists of 256 different tones that range from white to black. (27)
Gray tone display. A type of image display in which up to 256 shades, ranging from white to black, are used to represent energy levels sensed by a satellite. (27)
Gray-white ice. Young ice with a thickness of 15–30 cm (6–12 in). (218)
Grease ice. Newly formed ice that consists of an irregular layer of ice crystals broken up by waves or currents. (218)
Great Lakes Effect. A modification of the climate in the region surrounding the Great Lakes due to the increased moisture and heat from the lake surfaces. The term typically refers to enhanced downwind snow shower action during colder months. (88)
Ground station. A facility on Earth that is capable of receiving satellite transmissions. (13)
Gulf of Mexico Loop Current. A warm-water current that circulates through the Gulf of Mexico. (210)

Gulf Stream. A warm-water current that flows northward along the Atlantic coast of the United States and curves northeast across the North Atlantic toward Europe. (204)
Gulf Stream Front. The boundary line between warm Gulf Stream water and cool continental shelf and slope water. It is found along the eastern coastline of the United States. (207)
Gust front. The boundary between cold air flowing out of a thunderstorm and warmer air surrounding the storm. The leading edge of this front is a preferred location for new thunderstorm development. (147)
Gyre. An anticyclonic circulation pattern in the ocean basins, which results in clockwise flow in the Northern Hemisphere oceans and counterclockwise flow in the Southern Hemisphere oceans. (204)
Haline front. The boundary between ocean waters with different salinity levels. (207)
Haze. A phenomenon that occurs when aerosols from pollution and automobile exhaust are trapped in the atmosphere. Water droplets form around the suspended particles, giving the sky a milky appearance. Haze can lead to hazardous air quality conditions. (196)
Heat island. An area that is warmer than the surrounding regions. The term often refers to urban locations. (49)
High-pressure system. A region in which atmospheric pressure is higher than in the surrounding regions. Subsiding air in the system creates weather conditions characterized by few clouds and little precipitation. (111)
High-Resolution Picture Transmission (HRPT). Continuous transmission of high-resolution (1 km) data from the AVHRR on U.S. polar-orbiting meteorological satellites. (13)
Hook echo. A hook-shaped feature that is observed on conventional radar. It indicates the presence of a strong center of circulation in a thunderstorm and suggests the possibility of tornadoes. (176)
Hurricane. A tropical cyclone formed in the Atlantic Ocean or the eastern and central Pacific Oceans. Hurricanes are characterized by a strong central low pressure, heavy precipitation, and sustained winds at least as fast as 120 km/hr (74 mph). (184)
Hurricane season. The time period during which hurricanes are most likely to form. For North America, the season is officially from June 1 to November 30; however, most hurricanes form during August, September, and October, when the current and cumulative effects of atmospheric heating are at a maximum. (184)
Icebergs. Floating pieces of ice that break off from fresh-water glaciers over land. (222)
Ice pack. A cluster of drift ice. (219)
Inflection point. The point on a comma cloud at which the cloud curvature changes from cyclonic (counterclockwise) to anticyclonic (clockwise) in the Northern Hemisphere. It is often located where a jet stream intersects with the cloud system and near a region where warm and cold fronts meet. (132)
Inflow. Warm, moist air that flows into a thunderstorm. (146)
Infrared (IR). A portion of the electromagnetic spectrum that is felt as heat. (3) IR sensors on meteorological satellites are used to detect thermal properties of the Earth's atmosphere, oceans, and land surfaces. (8)
INSAT. An Indian geostationary meteorological and communication satellite. (23)
Instrument response delay time. The time it takes for a satellite sensor to respond to a sudden, sharp increase or decrease in brightness. Due to the delay, objects in a satellite image can appear to be displaced from their actual locations. (41)
Internet. An international network linking computers used by scientists, researchers, and educators for data and information exchange. (25)
Intertropical Convergence Zone (ITCZ). A region near the equator in which surface trade winds from the Northern and Southern Hemispheres converge. It forms a more or less continuous belt of clouds and precipitation across the equatorial region. (105)
Inversion. An atmospheric condition in which warmer air lies over colder air. This prevents air from rising, resulting in a very stable atmosphere with little vertical motion. (63)
Island eddies. Swirling cloud patterns downwind of an island caused by disruption of airflow by the island. (98)
Island lee lines. Long lines of convective clouds that form downwind of tropical islands. They are caused by air flowing around the island and converging on the downwind side of the island. (100)
Island wave pattern. A wave pattern in the clouds caused by disruption of air as it flows over an island. The waves resemble mountain waves and tend to be perpendicular to the wind direction. (98)
Isothermal. Characterized by uniform temperatures throughout. (207)
Jet stream. A relatively narrow and vertically thin region of rapidly moving air at high altitudes. In middle latitudes, the jet stream typically moves from a westerly direction with speeds that can exceed 200 km (120 mph) in winter and 100 km (60 mph) in summer. (65)
Jet stream axis. The location of the higher-velocity winds within the jet steam. (111)
Katabatic winds. Local winds that form as a result of cold, dense air flowing downhill. (94)

Kuroshio Current. A permanent warm-water current that flows along the eastern Asian coastline. (207)
Lake breeze. A local wind that forms when the land surrounding a lake is heated more rapidly than the air over the lake. Warm air over land rises, and cooler air over the water flows onshore. (89)
Land breeze. A local wind that forms at night and in early morning hours when the land cools below the temperature of adjacent water. As warm air over the water rises, air flows offshore (from land) to replace it. (91)
Lapse rate. The change in temperature with altitude in the atmosphere. (63)
Latent heat. Energy that is released as water vapor condenses into water droplets. In the atmosphere, latent heat is an important source of energy for storms such as hurricanes and thunderstorms. (57)
Lead. A narrow region (on a scale from several meters to kilometers) of open water in a sea ice floe. (219)
Leaf cloud. A leaf-shaped cloud associated with high-amplitude jet stream ridges. Leaf clouds often form into larger, more organized storm systems. (130)
Leaf-to-comma cyclogenesis. The process in which a leaf cloud develops into a comma cloud system. (130)
Lee high cirrus clouds. Cirrus clouds that form as air is forced to rise over mountains. (97)
Leeward. A term that refers to being downwind from a particular geographical point, such as a mountaintop or an island. (92)
Limb darkening. An error on the outer edges of a satellite image caused by the sensor angle as it scans these portions of the Earth. Objects on the outer edges of a satellite image appear to have a distorted shape and appear colder than they actually are in IR imagery. (40)
Local wind flow. A wind that forms as a result of local differential heating. It usually affects small areas and may change direction and speed often. (85)
Loop Current. *See* Gulf of Mexico Loop Current. (210)
Low-pressure system. A region where air pressure is lower than in nearby areas. As the air rises in these systems, clouds and precipitation form. Most storms and precipitation form in low-pressure regions. (111)
Lumpy texture. A term describing the appearance in VIS imagery of clouds that are characterized by a rough cloud top surface. This causes many shadows to be seen on the cloud, giving it a "lumpy" appearance. (70)
Marginal ice zone. The sea ice edge that is sufficiently away from open ocean water so that the ice is not affected by the ocean's wave action, swell, etc. (218)
Maritime (m) air mass. A humid air mass whose origin is over the oceans. (126)
Maritime polar (mP) air mass. An air mass that originates over the oceans in higher latitudes. Its humidity is high and its temperatures are cold. (126)
Maritime tropical (mT) air mass. An air mass that originates over the oceans at lower latitudes. Its humidity is high and temperatures are warm. (126)
Maximum vorticity. The central point in a storm system, where wind speeds are at their maximum and the storm exhibits the greatest degree of spiraling. (132)
Maximum winter extent. The farthest extent of sea ice during the winter months. (218)
MB enhancement curve. A type of infrared enhancement especially useful for thunderstorm analysis. It highlights cold cloud tops in order to locate areas of intense convection and thunderstorm development. (36)
Meander. To travel in a wavelike pattern. This pattern is often observed in the jet stream and in ocean currents. (110)
Megahertz (MHz). A unit used to measure the frequency of a wave. (13)
Meridional flow. Jet stream flow that is characterized by high-amplitude meanders. It allows a great deal of transport of warm and cold air masses and development of more intense storms. (111)
Mesofront. A small-scale boundary between air masses of different temperatures. Mesofronts include thunderstorm outflow boundaries and breeze fronts. (147)
Mesohigh. A region of high-pressure rain-cooled air at the ground underneath a thunderstorm. (147)
Mesopause. The boundary between the mesosphere and the thermosphere; located at an altitude of 50–90 km (30–54 miles). (64)
Mesoscale. A term describing events in the atmosphere that occur on a local scale (less than 250 km or 150 miles). Mesoscale forecasting includes the study of localized atmospheric conditions to determine the locations most favorable for storm development. (146)
Mesoscale Convective System (MCS). A single organized system formed by the merger of two or more thunderstorms. (167)
Mesosphere. This third major layer of the atmosphere lies above the stratosphere and extends from about 50 km to 90 km (30–54 miles). Temperatures decrease with altitude in this layer, and the air is extremely thin. No significant weather activity occurs here. (64)
METEOR. A series of Russian polar orbiting environmental satellites that transmit using APT format. (22)
METEOSAT. A series of geostationary satellites operated by the European Space Agency (ESA). (17)

Microburst. *See* downburst. (172)
Micron. A unit of measurement for length, equal to 10^{-6} meters. (4)
Microwave. A type of longer-wavelength electromagnetic energy best known for its usefulness in cooking. Also used by satellite sensors to locate water vapor in the atmosphere and assess soil moisture. (3)
Mid-latitude cyclone. A storm in the mid-latitudes characterized by a well-defined surface low-pressure area and associated warm, cold, and occluded fronts. (130)
Millibar (mb). A metric unit to describe air pressure. At sea level, average air pressure is 1013.25 mb. (64)
Monsoon. A seasonal shift in wind direction. Most notable are the monsoon seasons in India, although monsoons also occur in Australia and North America (near the Gulf of California). These winds bring moisture from the oceans during the wet season and dryer air from the land during the dry season. Their circulation is an important source of water for the regions that experience monsoons. (107)
Mountain convection. Convection that occurs when air is forced to rise over mountains in unstable atmospheric conditions. It can lead to thunderstorm formation over the mountains. (162)
Mountain upslope wind. A local wind that occurs as the sides of a mountain heat up during the day. Air rises up the mountain slope, leading to possible cloud and thunderstorm development. (92)
Mountain wave pattern. A wavelike cloud pattern caused by the disruption of air as it flows over mountains. The waves form downwind of the mountains and are generally oriented perpendicular to the wind direction. (96)
Multi-spectral imaging tool. A satellite sensor that is capable of detecting energy in a variety of wavelengths in each scan line. The VISSR (Visible and Infrared Spin Scan Radiometer) on board the GOES satellites in an example, since it detects energy in both VIS and IR wavelengths. (17)
Multi-year ice. Sea ice that has survived more than two melt seasons. (218)
National Environmental Satellite Data and Information Service (NESDIS). The branch of NOAA that is responsible for weather-related research, data collection, and data distribution.
National Oceanic and Atmospheric Administration (NOAA). The U.S. government agency (under the Department of Commerce) that conducts research on the earth's atmosphere and oceans. It is the agency responsible for operating the U.S. weather satellite program. (10)
National Weather Service (NWS). The branch of the U.S. government responsible for weather forecasting across the country. It is responsible for issuing short-term weather forecasts, weather summaries, and severe weather warnings across the United States.
Near infrared. Electromagnetic energy that mainly consists of solar radiation reflected by the Earth. (4)
New ice. Newly formed sea ice with a thickness ranging from 0 to 10 cm (0 to 4 in.). (218)
Nila. A later stage of new ice in which the ice sheet has consolidated into a thin sheet 1–10 cm (0.4–4.0 in.) in thickness. (218)
NOAA satellites. A series of polar orbiting environmental satellites operated by NOAA. (11)
Normal lapse rate. Condition when temperatures in the atmosphere decrease with increasing altitude. (63)
North Equatorial Current. A warm-water current that flows from east to west near the equator. (211)
Northern Lights. *See* Aurora Borealis. (64)
North Pacific Current. A cool current that flows eastward in the North Pacific Ocean and approaches the west coast of North America. (211)
North Pacific Current front. A permanent oceanic front in the North Pacific Ocean associated with the North Pacific Current. It separates colder Arctic water from warmer waters to the south. (211)
Occluded front. A front that has been lifted by a second, faster-moving front that has caught up to it. Cold fronts most often overtake warm fronts and lift them, creating an occluded front. (126)
Oceanic front. A boundary between ocean waters of two different densities. It can be caused by differences in temperature, salinity, or both. (207)
OKEAN. A series of polar orbiting environmental satellites operated by Russia that transmit using APT format. (22)
Opaque. Impenetrable to light. (4)
Open-cell convection. A process by which cloud features that form over warmer oceans as localized rain showers dissipate and leave a ring of clouds that mark the outer edge of precipitation-cooled air. The center portion of the "cell" is open, or cloud free. This signifies an unstable air mass. (94)
Outflow. The outward flow of air from beneath a thunderstorm. (147)
Outflow boundary. The boundary line between cooler thunderstorm outflow and warmer air surrounding the storm. (147)
Overshooting top. A phenomenon that occurs when towering convective clouds break through the cirrus top of a thunderstorm and give the cloud top a small area of a very lumpy texture. Overshooting tops occur in regions where the updraft region of the storm is intense enough to break through the cloud top. (74)
Pancake ice. Circular pieces of ice shaped by ocean waves. (218)

Parallax. The apparent displacement of a cloud due to the viewing angle of the satellite sensor. (39)
Permanent ice zone. Parts of the ocean that are permanently covered by sea ice. (218)
Phytoplankton. Marine organisms that are photosynthetic (use sunlight to make energy) and make up the base of the oceanic food chain. (212)
Pixel (picture element). The smallest element in a satellite image. Thousands of pixels make up each image, and when viewed together, they make a picture. (27)
Polar (P) air mass. An air mass that originates at higher latitudes and is characterized by colder temperatures. (126)
Polar easterlies. A band of easterly winds between the poles and 60° latitude on either side of the equator. (105)
Polar high. A persistent region of high pressure over the polar regions caused by very cold dense air sinking toward Earth. (105)
Polar jet stream. The branch of the jet stream that is closest to the poles on either side of the equator. (109)
Polar orbiting satellite. A type of satellite that orbits the Earth in a nearly pole-to-pole manner, at a relatively low altitude. (10)
Polynya. Large area (several thousand kilometers in area) of open water in a sea ice floe. (219)
Precipitation. Water droplets or ice crystals that become too large to remain suspended in the atmosphere and subsequently fall to the ground as rain, snow, hail, or sleet. (57)
Pressure gradient. The change in atmospheric pressure across a horizontal distance on the earth's surface. (85)
Prevailing Pacific high. A persistent region of high pressure in the Pacific Ocean located near 30° North. (103)
Prevailing westerlies. A band of westerly winds between 30° and 60° on either side of the equator. (104)
Radiation. Energy that is emitted from all objects in the universe and exhibits wavelike properties. (2)
Radio. The type of electromagnetic energy characterized by the longest wavelengths. A very important type of energy for communications. (3)
Radiometer. An instrument that measures the intensity of electromagnetic radiation at specific wavelengths. Satellite sensors are radiometers since they measure visible and infrared energy from the Earth. (8)
Radiosonde. An instrument package used with weather balloons to collect temperature, moisture, and wind data for use in analyzing the vertical structure of the atmosphere. (63)
Rafted floe. An ice floe that has overridden another ice floe. (220)
Remote sensing. The study of something without making actual contact with the object of study. (2)
Resolution. A measure of the ability of a satellite to see objects of different sizes. Higher resolution means that the satellite is able to discern smaller features. (27)
Ridge. A portion of a jet stream meander in which the wind direction changes from southwest to northwest (in the Northern Hemisphere) or northwest to southwest (in the Southern Hemisphere). (110) Also: a pile of broken sea ice caused by movement and collisions within a floe. (220)
Ring cloud. A circular cloud feature that is found near the center of a mid-latitude oceanic cyclone, especially when the storm's central pressure is lower than 970 mb. It often has strong winds associated with it. (141)
Rope cloud. A thin line of convective clouds that forms along the leading edge of a weakening cold front. (128)
Salinity. The amount of dissolved salts present in ocean water. (207)
Salt rejection. The forcing out of salt by sea ice as it ages. (217)
Sargasso Sea. A warm-water, high-salinity pool of water at the surface of the mid-Atlantic Ocean in the Northern Hemisphere. (209)
Satellite subpoint. The point on Earth that is directly below a satellite. (27)
Saturation. The point at which air can hold no more water vapor. Addition of more water vapor will cause condensation to occur. (57)
Scattering. The spreading of light into its component colors by particles in the atmosphere. Scattered blue light gives the sky its blue color. (7)
Sea breeze. A wind that blows from the sea to the shore. Sea breezes are caused by differences in heating between land and water. (89)
Sea breeze front. The boundary between cooler air from the ocean and warmer air from the land. Sea breeze fronts are often a favorable location for thunderstorm development. (89)
Sea ice. Ice formed from sea water. (217)
Seasonal ice zone. The region in the ocean that is covered with sea ice for part of the year. It is the transition zone between the summer and winter extent of sea ice. (218)
Sea surface temperature (SST). Surface temperature data collected over oceans using IR sensors on environmental satellites. These data are used in monitoring global climate change and improving ocean transportation, among other uses. (204)
Second year ice. Sea ice that has survived one melt season. (218)
Severe thunderstorm. A thunderstorm that contains dangerous weather such as tornadoes, strong

winds, and large hail. Severe thunderstorms often contain heavy rainfall, as well. (147)

Shelf water. Cool, low-salinity water that lies over the continental shelf. (209)

Shuga. An accumulation of spongy sea ice pieces a few centimeters across. Shuga often forms from grease ice or slush when the ocean is agitated, e.g., by waves or currents. (218)

Slope water. Cool, low-salinity water that lies over the continental slope, between the shelf water and the Gulf Stream. (209)

Slush. A viscous floating mass of sea ice formed from a mixture of snow and water. (218)

Smooth texture. A term describing the appearance of flat layers of clouds that have a uniform tone in a satellite image. (70)

Snow line. The boundary that marks the horizontal extent of snow cover over land surfaces. (50)

Solar subpoint. The point where the sun's most direct rays strike the Earth. (32)

Sounding. A vertical series of measurements taken at different heights above the ground. (63)

Source region. The region in which an air mass originates. This influences its temperature and moisture characteristics. (126)

Spectrum. The range of wavelengths of electromagnetic energy emitted by an object. (3)

Split flow. Divergence (spreading apart) of a jet stream. This creates a region where thunderstorms are most likely to form. (153)

Squall line. A nearly continuous line of thunderstorms that generally occurs along fronts and dry lines. (152)

Stable atmosphere. An atmosphere in which the lapse rate allows for very little vertical air motion. (64)

Standard atmosphere. An average profile of atmospheric temperatures used as a standard for comparison. (63)

Stationary front. A front that has stalled. (126)

Storm surge. An increase in water level in coastal regions that occurs during the approach of a hurricane. Storm surge and the associated flooding are the deadliest features of a hurricane. (185)

Stratiform clouds. Clouds characterized by a flat, sheetlike appearance. (70)

Stratocumulus clouds. Low-level clouds that are "puffy" like cumulus clouds but flattened and layered like stratus clouds. (73)

Stratopause. The boundary between the stratosphere and the mesosphere, located at approximately 50 km. (64)

Stratosphere. The second-highest layer of the atmosphere, located at an altitude of between about 12 and 50 km (7 and 30 miles). Some high ice crystal clouds exist in this layer; the stratosphere is also characterized by an increase of temperatures with altitude as a result of ozone heating. (64)

Stratus clouds. Low-level clouds that have a very flat, blanketlike appearance. (70)

Sublimation. The process of a solid changing into a gas. This process consumes energy. (57)

Subpolar low. An area of persistent low pressure located near 60° of latitude on both sides of the equator. (105)

Subsidence. The sinking of air. Associated with warming air and little cloud formation. (103)

Subtropical high (STH). A region of high pressure located approximately 30° north and south of the equator, caused by the sinking of cooled air from the equatorial region. (103)

Subtropical jet stream. The branch of the jet stream that is found at lower latitudes. (109)

Summer solstice. The position in earth's orbit in which the sun's direct rays are at their northernmost Earth position, on the Tropic of Cancer (23.5 ° C). Marks the first day of summer and the longest day of the year in the Northern Hemisphere. (59)

Sunglint. A bright area seen on VIS satellite imagery where the satellite is sensing sunlight reflected from water surfaces on Earth. (30)

Sun synchronous orbit. An orbit in which a satellite crosses a point on Earth at approximately the same local time each day. Polar orbiting weather satellites are in sun synchronous orbits. (10)

Surge region. A portion of a comma cloud in which the jet stream has caused the clouds to surge downwind. (132)

Synchronous Meteorological Satellite 1 (SMS 1). A prototype geostationary satellite program (mid-1970s) that led to the establishment of the Geostationary Operational Environmental Satellite (GOES) system. (17)

Synoptic scale. A term describing events that cover large areas of the atmosphere. Synoptic-scale events include large-scale movement of air masses, fronts, and high- and low-pressure systems. (125)

Telecommunications. A type of communication that includes computers linked together by telephone lines and modems, radio, telephone, and television. (22)

Television and Infrared Observational Satellite (TIROS). A series of polar orbiting weather satellites operated by the United States. (10)

Temperature. A measure of the amount of heat energy present in an object or a substance. (3)

Terminator. The boundary line between sunlight and darkness on the Earth. (28)

Texture. A characteristic of clouds as seen in satellite imagery. Cloud texture is a function of the amount of shadows on the cloud tops and can be smooth, lumpy, or fibrous. (70)

Thermal blanket. Atmospheric gases that act like a blanket to hold heat in the lower atmosphere. (49)
Thermal front. A boundary between ocean waters of different temperatures. (207)
Thermal infrared. The portion of the electromagnetic spectrum that includes emitted thermal, or heat, energy. (4)
Thermal instability. An atmospheric condition in which temperatures decrease with altitude. This allows air to rise and convection to occur. (130)
Thermosphere. The outermost layer of the earth's atmosphere, extending from an altitude of about 90 km (54 miles) to about 500 km (300 miles). This layer of the atmosphere gradually thins into outer space. (64)
TIROS Operational System (TOS). The program that operates the United States polar orbiting weather satellite system. (11)
Topographically induced upwelling. Upwelling that occurs when an ocean current flows over a bottom projection that forces the current to rise and transport subsurface water to the surface. (214)
Tornado. A funnel-shaped cloud that extends from a cumulonimbus cloud and reaches the ground. Tornadoes contain rapidly rotating winds that can reach speeds up to 500 km/hr (300 mph). (172)
Towering cumulus clouds. Cumulus clouds that exhibit extensive vertical development. (74)
Trade winds. A band of winds in each hemisphere that flow from the subtropical high (around 30° latitude) toward the equator. They were named by seagoing merchants who used them to sail from Europe to the New World. (104)
Translucent. Partially penetrable by light. A translucent object transmits a portion of the light that strikes it and absorbs or reflects the remaining portion. (7)
Transmissivity. A measure of the degree to which the atmosphere will allow energy at various wavelengths to pass through. (7)
Transmit. To allow radiation to pass through. (7)
Transpolar Drift Stream (TDS). An oceanic current in which water flows from the western Arctic Ocean across the North Pole toward the eastern Arctic. (220)
Transverse bands. Small-scale cirrus lines that appear along the main axis of a jet stream. They lie perpendicular to upper-level wind direction and are often associated with turbulence and high upper-level wind speeds. (114)
Trellis arrangement. A common geologic pattern formed by parallel river valleys in ridge and valley regions such as the Appalachian Mountains. In a trellis arrangement, rivers tend to run parallel to each other. (72)
Triple point. A point at which a cold, a warm, and an occluded front come together in a storm system. (134)
Tropical (T) air mass. A warm air mass that originates in lower latitudes. (126)
Tropical cyclone. A cyclonically rotating atmospheric vortex that ranges in diameter from a few hundred miles up to one or two thousand miles. It is associated with a central core of low pressure and with convective clouds organized into spiral bands, with a sustained convective cloud mass at or near the center. *See also* hurricane, cyclone, and typhoon. (184)
Tropical depression. An early stage in tropical cyclone development that occurs when a closed low-pressure system forms and wind speeds remain below 60 km/hr (39 mph). (187)
Tropical disturbance. The first stage of hurricane development, which starts out as a widespread area of cloudiness centered around a weak low-pressure area, with light winds and some precipitation. (186)
Tropical Number (T-number). A number that is assigned to a tropical cyclone to indicate the intensity of the storm. (187)
Tropical storm. A later stage in tropical cyclone development, which is characterized by an organized band of clouds around an area of low pressure, with winds between 60 and 120 km/hr (39 and 74 mph). (187)
Tropopause. The upper boundary of the troposphere, located at an altitude of approximately 12–15 km (7–9 miles). It is found at lower altitudes at the poles and at higher altitudes near the equator. (64)
Troposphere. The lowest layer of the atmosphere, extending from the ground to about 12–15 km (7–9 miles). All of the earth's significant weather takes place in this layer. (63)
Trough. The portion of a jet stream meander or wave in which the wind direction changes from northwest to southwest (in the Northern Hemisphere) or from southwest to northwest (in the Southern Hemisphere). (110) Also: the maximum downward displacement of a wave. (2)
Typhoon. A type of tropical cyclone that forms in the western Pacific Ocean and generally affects Asia. *See also* hurricane and tropical cyclone. (184)
Ultraviolet. A high-energy type of electromagnetic radiation. Ultraviolet radiation from the sun causes sunburn and is thought to cause skin cancer in humans. (3)
Universal Time Constant (UTC). A universally accepted unit of time that is based on the current time at the Prime Meridian (0° longitude) and is measured in a 24-hour format. UTC provides a unit of time that can be used during global studies of the Earth when local times can lead to confusion.
Unstable atmosphere. An atmosphere in which temperatures decrease rapidly with altitude. Heated air

is permitted to rise; therefore, a great deal of vertical mixing and convective cloud development occurs. (64)

Updraft. A strong upward current of warm air. (68) Also: rising air currents within a thunderstorm. (146)

Upwelling. A flow of colder, deeper water to the ocean's surface. It can be caused by winds, bottom topography, or diverging surface currents. Upwelling results in an upward flow of nutrient-rich waters, which makes it important to the fishing industry. (211)

Upwind. A term describing the direction from which the wind is coming. (92)

Urban heat island. A differential heating between a city and the surrounding countryside, resulting from a city's ability to retain heat more efficiently and continually generate heat. (49)

Vernal equinox. Position in the earth's orbit when the sun's most direct rays are on the equator during their northern migration. On the vernal equinox day and night are equal. Marks the first day of spring in the Northern Hemisphere. (61)

Vertical profile. A graph, plot, or chart showing temperature, pressure, wind direction, or other variables at different altitudes. (62)

Vertical wind shear. An atmospheric condition in which wind speeds increase with altitude. (130)

Visible and Infrared Spin Scan Radiometer (VISSR). The main sensor on the GOES satellite. It senses visible and infrared wavelengths while spinning at 100 rpm. (17)

Visible (VIS) light. The form of electromagnetic energy that human eyes detect. Visible light is composed of seven component colors: red, orange, yellow, green, blue, indigo, and violet. VIS sensors on meteorological satellites are used to locate and observe features in the Earth's atmosphere and on its land surfaces. (3)

Vorticity. Circulation, or spin, of air in the atmosphere. (132)

Warm-core eddy. *See* anticyclonic eddy. (210)

Warm front. The front created as a warm air mass forces itself over a colder air mass. Characterized by stratiform cloud types and steady light precipitation. Temperatures increase after the passage of the front. (126)

Warm-front type occlusion. An occluded front in which the coldest air lies ahead of the warm front, resulting in the cold front being lifted. (130)

Water cycle. The continuous cycling of water in the atmosphere. Water evaporates from plants, land, and water surfaces, condenses into clouds, and falls back to Earth as precipitation. (57)

Water vapor. Water in a gaseous state. (57)

Water vapor (WV) imagery. Imagery taken in the 6.7–7.3 micron range of infrared wavelengths that shows the distribution of water vapor in the high and middle portions of the atmosphere. (27)

Water vapor plumes (WVP). Large surges of moisture into the mid-latitudes from the tropics. They have been linked to heavy rainfall events in the United States. (167)

Wavelength. The distance between two successive crests or troughs of a wave. (2)

Weather Facsimile (WEFAX). Re-transmission of low-resolution imagery that is transmitted at a frequency of 1,691.0 MHz. (21)

Wind shear. An atmospheric condition in which wind direction and speed changes in the lower 15 km of the atmosphere. (187)

Wind stress. The force of the wind acting on a water surface, resulting in surface transport of water. (212)

Windward. A term referring to the upwind section of a feature. (92)

Winter solstice. Position in the earth's orbit when the sun's most direct rays are at their southernmost point, at the Tropic of Capricorn (23.5° N). Marks the first day of winter and the shortest day of the year in the Northern Hemisphere. (61)

X-rays. A very high energy form of electromagnetic radiation that is used in medicine to image internal portions of the body. (3)

Young ice. Newly formed sea ice that has consolidated into a layer 10–30 cm (4–12 in.) in thickness. (218)

ZA enhancement curve. A general-purpose IR enhancement curve that provides improved contrast between temperatures of meteorological significance. (36)

Zonal flow. Jet stream flow that is mostly west to east with very little meandering. Zonal flow leads to little mixing of warm and cold air and generally results in weaker storm development. (111)

References

Alder, R.F. 1979. Thunderstorm Intensity as Determined from Satellite Data. *Journal of Applied Meteorology*, 18:502–517.

Barnes, J.C., and Smallwood, M.D. 1982. *TIROS N Series Direct Readout Services Users' Guide.* U.S. Department of Commerce, Washington, D.C., 85 pp.

Beckman, S.K. 1981. Wave Clouds and Severe Turbulence. *National Weather Digest*, 6:30–37.

Berman, E.A., and Fletcher, J. 1991. *Guide for Using Satellite Imagery to Teach Science and Math.* Tri-Space, McLean, Va.

Brandli, H. 1992. Weathermen Try to Incorporate Real-Time Analysis and Forecasting. *Weather Satellite Report*, March, 1–10.

Carlson, T.N. 1980. Airflow through Midlatitude Cyclones and the Comma Cloud Pattern. *Monthly Weather Review*, 108:1498–1509.

Clark, J.D., ed. 1983. *The GOES User's Guide.* U.S. Department of Commerce, Washington, D.C., 155 pp.

Clark, R.M., and Feigel, E.W. 1981. *The WEFAX User's Guide.* U.S. Department of Commerce, Washington, D.C., 47 pp.

Dvorak, V.F. 1984. *Tropical Cyclone Intensity Analysis Using Satellite Data.* NOAA Technical Report NESDIS 11. U.S. Department of Commerce, Washington, D.C., 47 pp.

Dvorak, V.F., and Smigielski, F. 1994. *A Workbook on Tropical Clouds and Cloud Systems Observed in Satellite Imagery.* U.S. Department of Commerce, Washington, D.C.

Ellrod, G. 1985. Dramatic Examples of Thunderstorm Top Warming Related to Downbursts. *National Weather Digest*, 10, no. 2, 7–13.

Ellrod, G. 1992. Potential Applications of GOES I 3.9 µm Infrared Imagery. In *Proceedings of the Sixth Conference on Satellite Meteorology and Oceanography*, January 5–10, 1992, Atlanta, Ga., American Meteorological Society, Boston, Mass., 184–187.

Ellrod, G. and Field, G. 1984. The Characteristics and Prediction of Gulf Stream Thunderstorms. In *Tenth Conference on Weather Forecasting and Analysis*, June 25–29, 1984, Clearwater Beach, Fla. American Meteorological Society, Boston, Mass, 15–21.

Fett, R.W., et al. 1979. *Navy Tactical Applications Guide.* Vol. 2: *Environmental Phenomena and Effects.* NEPRF Technical Report 77–04.

Gallo, K.P., et al. 1993. Use of NOAA AVHRR Data for Assessment of the Urban Heat Island Effect. *Journal of Applied Meteorology*, 32:899–908.

Kidwell, K.B., ed. 1991. *NOAA Polar Orbiter Data Users' Guide.* U.S. Department of Commerce, Washington, D.C.

Lynch, J.S. 1987. *Satellite Interpretation Messages: A Users' Guide.* NOAA Technical Memorandum NWS NHC 39. U.S. Department of Commerce, Washington, D.C., 76 pp.

MacCallum, D.H., and Nestlebush, M.J. 1983. *The Geostationary Operational Environmental Satellite Data Collection System.* NOAA Technical Memorandum NESDIS 2. U.S. Department of Commerce, Washington, D.C., 21 pp.

Maddox, R.A. 1980. Mesoscale Convective Complexes. *Bulletin of the American Meteorological Society*, 61:1374–1387.

Matson, M., Scheider, S.R., Aldridge, B., and Satchwell, B. 1984. *Fire Detection Using the NOAA-Series Satellites.* NOAA Technical Report NESDIS 7. U.S. Department of Commerce, Washington, D.C., 34 pp.

Mayfield, M. 1976. Sahara Dust over the Atlantic. *Mariners Weather Log*, 19:346–347.

McGuirk, J., and Ulsh, D. 1990. Evolution of Tropical Plumes in VAS Water Vapor Imagery. *Monthly Weather Review*, 118:1758–1766.

Morse, B.J., and Ropelewski, C.F. 1983. *Spatial and Temporal Distribution of Northern Hemisphere Snow Cover.* NOAA Technical Report NESDIS 6. U.S. Department of Commerce, Washington, D.C., 32 pp.

Parke, P.S., ed. 1986. *Satellite Imagery Interpretation for Forecasters.* Weather Service Forecasting Handbook No. 6. U.S. Department of Commerce, Washington, D.C.

Rao, P.K., ed. 1990. *Weather Satellites: Systems, Data, and Environmental Applications.* American Meteorological Society, Boston, Mass.

Schwalb, A. 1982. *Modified Version of the TIROS N/NOAA A–G Satellite Series (NOAA E–J) —Advanced TIROS N (ATN).* NOAA Technical Memorandum NESS 116. U.S. Department of Commerce, Washington, D.C., 23 pp.

Schwalb, A. 1982. *The TIROS N/NOAA A–G Satellite Series.* NOAA Technical

Memorandum NESS 95. U.S. Department of Commerce, Washington, D.C., 75 pp.

Scofield, R.A. 1984. Satellite-Based Estimates of Heavy Precipitation. *Recent Advances in Civil Space Remote Sensing*, 481:84–91.

Scofield, R.A., and Robinson, J. 1992. The "Water Vapor Plume/Potential Energy Axis Connection" with Heavy Convective Rainfall. *In Symposium on Weather Forecasting/Sixth Conference on Satellite Meteorology and Oceanography*, January 5–10, 1992, Atlanta, Ga. American Meteorological Society, Boston, Mass., J36–J43.

Smigielski, F.J., and Ellrod, G.P. 1985. *Surface Cyclogenesis as Indicated by Satellite Imagery*. NOAA Technical Memorandum NESDIS 9. U.S. Department of Commerce, Washington, D.C., 30 pp.

Smigielski, F.J., and Mogil, H.M. 1991. Use of Satellite Information for Improved Ocean Surface Analysis. *In First International Symposium on Winter Storms*, January 14–18, 1991, New Orleans, La. American Meteorological Society, Boston, Mass., 137–144.

Smigielski, F.J., and Mogil, H.M. 1992. *A Systematic Approach for Estimating Central Pressures of Mid-Latitude Oceanic Storms*. NOAA Technical Report NESDIS 63. U.S. Department of Commerce, Washington, D.C., 64 pp.

Smith, W.L., et al. 1986. The Meteorological Satellite: Overview of 25 Years of Operation. *Science*, 231:455–462.

Summers, R.J. 1989. *Educator's Guide for Building and Operating Environmental Satellite Receiving Stations*. NOAA Technical Report NESDIS 44. U.S. Dept. of Commerce, Washington, D.C., 45 pp.

Weber, E.M., and Wilderotter, S. 1981. *Satellite Interpretation*. 3WW/TN-81/001. U.S. Air Force, Offutt A.F.B., Neb., 95 pp.

Weldon, R., and Holmes, S.J. 1991. *Water Vapor Imagery: Interpretation and Applications to Weather Analysis and Forecasting*. NOAA Technical Report NESDIS 57. U.S. Department of Commerce, Washington, D.C., 213 pp.

Wilkerson, W.D. 1991. *Dust and Sand Forecasting in Iraq and Adjoining Countries*. AWS/TN-91/001. Scott Air Force Base, Ill., 65 pp.

Williams, J. 1992. *The USA Today Weather Book*. Vintage Books, New York.

Yates, H., Strong, A., McGinnis, D., and Tarpley, D. 1986. Terrestrial Observations from NOAA Operational Satellites. *Science*, 231:463–470.

Young, S. 1993. *Summary of the Great Lakes Ice Conditions*. NOAA Technical Memorandum NESDIS 37. Satellite Services Division, Washington, D.C.

Zehr, R.M. 1992. *Tropical Cyclogenesis in the Western North Pacific*. NOAA Technical Report NESDIS 61. U.S. Department of Commerce, Washington, D.C., 181 pp.

Library of Congress Cataloging-in-Publication Data

Conway, Eric D.
An introduction to satellite image interpretation / by Eric D. Conway
p. cm.
Includes index.
ISBN 0-8018-5576-4 (hc : alk. paper). — ISBN 0-8018-5577-2 (pbk. : alk. paper)
1. Remote sensing. I. Title.
G70.4.C655—1997
621.36´78—dc21 96-53574
CIP